Psychophysics, Physiology and Models of Hearing

Cover designed by Katrin Tödter and Torsten Dau.

Psychophysics, Physiology and Models of Hearing

Oldenbury, Germany   31 August – 4 September 1998

editors

Torsten Dau

Volker Hohmann

Birger Kollmeier

Universität Oldenburg

World Scientific

Singapore • New Jersey • London • Hong Kong

Published by

World Scientific Publishing Co. Pte. Ltd.

P O Box 128, Farrer Road, Singapore 912805

USA office: Suite 1B, 1060 Main Street, River Edge, NJ 07661

UK office: 57 Shelton Street, Covent Garden, London WC2H 9HE

British Library Cataloguing-in-Publication Data
A catalogue record for this book is available from the British Library.

PSYCHOPHYSICS, PHYSIOLOGY AND MODELS OF HEARING

ISBN 981-02-3741-3

Printed in Singapore by Regal Press (S) Pte. Ltd.

PREFACE

Almost 40 years have passed since von Békésy was awarded the Nobel Prize for his pioneering work in hearing research - and we are still far away from a complete understanding of how our ear works. Among the many researchers who have contributed to the remarkable advances in hearing research since then, there is a clear-cut separation between those working on the physiology of hearing (mostly with animals) and those involved in the psychology of hearing, or psychoacoustics (mostly dealing with humans as subjects). However, a fairly recent development (mainly based on the availability of powerful computers) has shown its potential to link both fields closer together and to generate a more realistic and general view of the functioning of the auditory system: The development of comprehensive auditory models capable of quantitatively describing several effects with the same model (rather than to develop a special model exclusively for each single effect). Hence, the situation in hearing research resembles now the "classical" interaction and mutual reinforcement between experimental and theoretical physics. The topic of this book therefore is intentionally chosen as "psychophysics, physiology and models of hearing" in order to emphasize the close interaction between these fields. In the same way as applied physics is closely connected with experimental and theoretical physics, some important applications originating from this interaction are also covered, such as, e.g., speech perception and speech processing for the compensation of a hearing loss.

The current book summarizes the proceedings of a summerschool and symposium which was held from 31/08/1998 to 04/09/1998 in Bad Zwischenahn (close to the Universität Oldenburg). The aim of the conference was to bring together leading international scientists in the area with young, active researchers. It was organized by the editors of this book and the "Arbeitsgruppe Medizinische Physik" in order to clarify the field with invited lectures, to highlight new developments by contributed papers and posters, and to intensify the discussions between the various researchers with a large variety of academic backgrounds in a discussion-friendly environment. While the Summerschool "Psychophysics, physiology and models of hearing" from 31/08/ to 02/09/ was primarily financed by the Graduiertenkolleg (graduate school) Psychoakustik and the Sonderforschungsbereich 517 "Neurokognition" established at the Universität Oldenburg, the subsequent Symposium "Temporal processing in the auditory system: Psychophysics, physiology and models of hearing" was connected to the special research focus "Zeitverarbeitung im zentralen auditorischen System" (Temporal processing in the central auditory system) recently established by the Deutsche Forschungsgemeinschaft (DFG). The whole conference continues a series of conferences organized by Birger Kollmeier that started in 1992 in Seeburg ("Moderne Verfahren der Sprachaudiometrie", i.e., Modern methods in speech audiometry) and continued in 1995 in Bad Zwischenahn with the topic

"Psychoacoustics, speech and hearing aids". Although the history of these meetings is quite short and each meeting is a new adventure, certain elements have proven to be indispensable: The attractive setting of the meeting place close to the "Bad Zwischenahner Meer" with a rich potential of leisure activities; the limitation of the number of participants to not more than 60; the presence of leading international experts in the respective areas as well as some of the most talented and open-minded Ph.D. students and postdocs from various laboratories; the occurrence of some "hot topics" (that - among others - kept the discussions going throughout the whole days and nights) as well as the good food made the conference a very enjoyable event.

The current proceedings includes the written papers from nearly all authors who participated in the summerschool or the symposium. The ordering of the papers does not reflect the original ordering during the conference: Instead, a separation has been made between papers dealing with a physiological approach and with a psychological approach followed by papers with modelling issues and, last but not least, applications of the knowledge derived from experiments and theory. As far as possible, the papers are ordered from the periphery to more central elements of the auditory pathway or processing level. The editors hope that this ordering of the papers may help the reader to find his or her way through the book.

Many people and institutions have contributed to this conference and to this proceedings. First of all, the editors want to thank Ralph-Peter Derleth for helping to organize the conference and to prepare this book. Without him this all would not have been possible. Furthermore, we want to thank all other members of the Arbeitsgruppe Medizinische Physik and the Graduiertenkolleg Psychoakustik, especially Dirk Junius and Oliver Wegner, for their help as well as the excellent crew of the "Tagungsstätte der Angestelltenkammer Bremen" in Bad Zwischenahn for their hospitality. Financial support was provided by the Deutsche Forschungsgemeinschaft (DFG) via the Graduiertenkolleg Psychoakustik, Sonderforschungsbereich 517 "Neurokognition", and Schwerpunktprogramm 1046 "Zeitgebundene Informationsverarbeitung im zentralen auditorischen System". The hearing instrument manufacturers Phonak, Widex, GN Danavox and Starkey also supported the conference with a donation. On behalf of all participants, the organizers would like to acknowledge this support. Finally, we wish the reader as much pleasure and inspiration going through this book as we had during the course of this conference.

Oldenburg, February 1999

Torsten Dau, Volker Hohmann, Birger Kollmeier

CONTENTS

I. Psychological and Psychophysiological Acoustics

THE REPRESENTATION OF THE PITCH OF HARMONIC COMPLEXES AND ITERATED RIPPLED NOISE IN THE MAMMALIAN VENTRAL COCHLEAR NUCLEUS

IAN M. WINTER, LUTZ WIEGREBE AND ROY D. PATTERSON

CNBH, The Physiological Laboratory, Downing Street, Cambridge CB2 3EG, U.K.
E-mail: imw1001@cam.ac.uk

In this paper we present the results of a study comparing the encoding of the pitch of complex stimuli that differ greatly in their temporal structure. We used cosine-phase (CPH) and random-phase (RPH) harmonic complexes, and iterated rippled noise (IRN) with various fundamental frequencies or delays, respectively. Most neurones in the ventral cochlear nucleus were able to represent the pitch of harmonic complexes in their temporal response properties when all the components were in cosine-phase. However, some units failed to provide a robust representation of the pitch when the components were in random phase (as it is often the case for harmonic sounds in natural, echoic environments) or when the stimuli were IRNs. Based on first-order inter-spike interval statistics, sustained chopper units showed a bandpass periodicity tuning in response to IRN and RPHs. Surprisingly, bandpass periodicity tuning was also observed for onset-chopper units. Whereas 'best periods' for sustained chopper cells ranged from 4 to 10 ms, onset chopper units revealed best periods up to 16 ms thus covering also the lower pitch range. We hypothesise that sustained chopper *and* onset-chopper units play a role in the extraction of pitch of complex signals.

1 Introduction

Pitch is an important feature of many complex sounds, including speech. A variety of sounds with different frequency and phase spectra can elicit the same pitch in humans. Therefore it is believed that a common neural mechanism must be able to extract the pitch from these different sounds. In this study we used IRN and harmonic complexes with different phase spectra to investigate neural mechanisms of pitch coding in the cochlear nucleus.

Rippled noise is produced from Gaussian noise by delaying a copy of the noise and adding it back to the original. IRN is produced by iterating the delay-and-add process. The generation of IRN from Gaussian noise is illustrated in Figure 1. The delay-and-add process introduces temporal regularity into the noise structure. This temporal regularity of IRN is reflected in the stimulus spectrum as well as in its autocorrelation function. Figure 2 shows the spectrum (left panel) and the autocorrelation function (right panel) of the IRN stimulus plotted in the right panel of Figure 1. The position and height of the first peak in the autocorrelation function (h1) have been used by Yost [1,2] to explain the IRN pitch and pitch strength, respectively. The height of h1 increases with increasing number of iterations but is always smaller than unity, indicating that the delay-and-add process introduces a *quasi*-periodicity to the stimulus. Perfect periodicity would be reached only with an infinite number of iterations. A very similar, perfectly periodic stimulus can be obtained by randomising the phases of a harmonic complex according to a uniform distribution, which is a random-phase harmonic complex (RPH). A waveform sample of an RPH is plotted in the left panel of Figure 3, its magnitude spectrum is shown in the middle panel. A cosine-phase harmonic complex (CPH) with the same magnitude spectrum is produced by setting the phase of each harmonic to zero. A waveform sample of a CPH is plotted in the right panel of Figure 3. IRN as well as RPH and CPH were shown to produce a clear pitch sensation in human listeners. This is not only the case for broadband stimuli but also when the sounds are high-pass filtered above the frequency region within which the spectral ripple of the stimuli is resolved by the human peripheral auditory system. In this unresolved frequency region, the excitation patterns of the IRN stimuli are essentially indistinguishable from those of Gaussian noise.

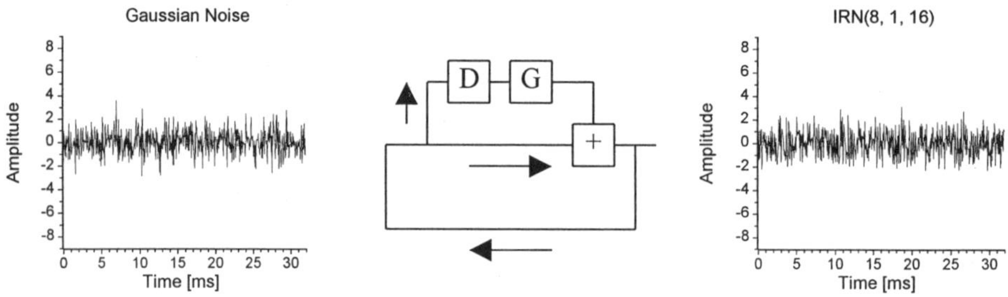

Figure 1: Generation of IRN from Gaussian noise. Delay (D) was 8 ms, gain (G) was 1 and the number of iterations was 16.

Thus, the pitch information must be extracted from temporal features, i.e., the temporal regularity in the spike timing. Cariani and Delgutte [3,4] investigated the representation of complex harmonic stimuli in the cat auditory nerve. They showed that the envelope modulation of the stimulus waveform is well preserved and that the pitch of the stimuli could be accurately predicted from an autocorrelation applied to the nerve-fibre response. Cariani and Delgutte mainly used amplitude modulated tones and sine-phase complexes. Both these classes of stimuli show a pronounced envelope modulation of the waveform. Inspection of an IRN waveform (Figure 1, right) and an RPH waveform (Figure 3,left) shows that this is not the case for two of the stimulus types used here. Thus, IRN and RPHs represent a more difficult challenge for temporal pitch-extraction mechanisms. Our results show that the autocorrelation of the response of many unit types in the ventral cochlear nucleus is able to encode the pitch of CPHs, RPHs and IRNs. However, sustained chopper units (Chop-S) and also onset-chopper units (On-C) showed a bandpass periodicity tuning based on their first-order inter-spike interval statistics as opposed to the autocorrelation which is an all-order statistic. This periodicity tuning was not only observed for CPHs but more interestingly for RPHs and IRNs despite the fact that the latter stimuli do not show pronounced envelope modulation.

2 Methods

2.1 Preparation

The data reported in this paper were recorded from pigmented guinea pigs. Animals were anaesthetised with urethane (1.5g/kg ip) and supplementary analgesia provided by either operidine (im 1mg/kg) or Hypnorm (im 1 mg/kg). All animals were given Atropine sulphate (0.06 mg/kg, sc) as a premedication. Additional doses of urethane and the analgesic were given when required.

The surgical preparation and stimulus presentation took place in a sound-attenuated chamber (IAC). All animals were tracheotomised and core temperature maintained at 38°C with a heating blanket. Following placement in the stereotaxic apparatus a midline incision of the scalp was made and the skin retracted laterally. The temporalis muscle on the left hand side of the skull was removed and the bulla exposed. The method of stereotaxic positioning follows that previously reported [5].

The compound action potential (CAP) was monitored with the use of a silver-coated wire placed on the round window of the cochlea. The signal was filtered and amplified (x10000). The CAP threshold was determined visually at selected frequencies (0.5, 1, 2, 3, 4, 5, 7, 10 and 15 kHz) at intervals during the experiment. If CAP thresholds had deteriorated by more than 10 dB and were not recoverable (for example, by removal of fluid from the bulla) the experiment was terminated.

2.2 Analyses

Recordings were made using tungsten-in-glass microelectrodes. Electrodes were advanced by an electronic microdrive (Kopf 650W) through the intact cerebellum in the sagittal plane at an angle of 45°. A wideband noise stimulus was used to locate the surface of the cochlear nucleus and to search for single units.

Upon isolation of a single unit, estimates of best frequency (BF) and threshold were obtained using audio-visual criteria. The spontaneous discharge was measured over a ten-second period. Single units were classified by their post-stimulus time histogram shape in response to suprathreshold BF tone–bursts.

PSTHs were generated in response to 250 short tone bursts (50 ms) at the unit's BF. Rise-fall time was 1 ms ($\cos^2$ gate) and the repetition rate was 4/s. Spikes were timed with 1-μs resolution (TDT ET1) and typically sound levels of 20 and 40 or 50 dB above threshold were used. All PSTHs shown in this paper include a 20-ms delay before the onset of the stimulus.

Neural responses to complex stimuli were analysed in terms of mean rate and first- and all-order inter-spike interval histograms (ISIH). The ISIH for white-noise stimulation with equal overall level to the pitch stimuli was subtracted from that of the pitch stimulus to determine to which extent the temporal regularity in the stimulus leads to a rearrangement of intervals. Periodicity sensitivity is expressed as the height of the peak at the interval corresponding to the stimulus period (or the IRN delay) in the first-order ISIH after noise subtraction.

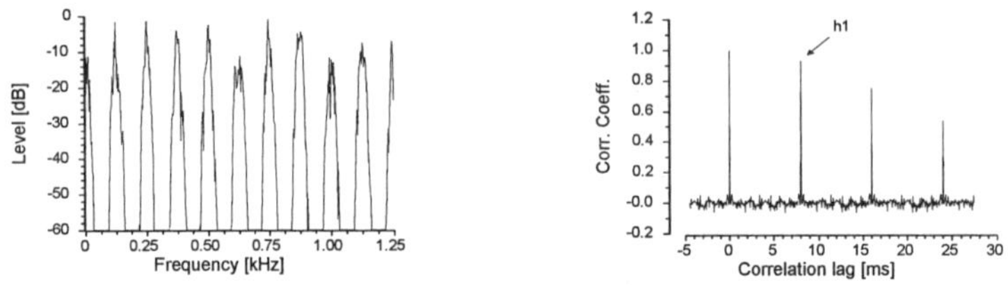

Figure 2: Spectrum (left panel) and autocorrelation function (right panel) of the IRN wave form shown in the lowest panel of Figure 1. The spectrum shows harmonic peaks at integer multiples of 125 Hz which is the reciprocal of the 8 ms delay used to generate the IRN. The normalised autocorrelation function shows peaks at integer multiples of 8 ms correlation lag. The first peak (apart from the peak at the zero lag) in the autocorrelation function, h1, has been used to model the pitch and pitch strength of IRN.

2.3 Stimuli

Stimuli consisted of IRN, RPH and CPH. The fundamental frequencies ranged from 31.25 Hz to 1 kHz in half-octave steps. IRN was generated as described in the introduction with delays ranging from 1 to 32 ms in half-octave steps (resulting in pitches equal to those of the harmonic complexes) a gain of unity and 16 iterations.

CPH and RPH were generated in the frequency domain by producing a harmonic magnitude spectrum and a uniformly random phase spectrum (for RPH) or a phase spectrum set to zero (for CPH). Applying an inverse fast Fourier transform to the magnitude- and phase spectrum results in the stimulus waveform. Stimuli were generated digitally using the Tucker Davis System II DSP Board and a sampling rate of 20 kHz. The stimulus duration was 409.6 ms including 10 ms cosine-squared ramps. The stimuli were refreshed for every presentation. Due to the non-deterministic nature of RPH and IRN, this results in different stimulus waveforms sharing the same regularity features. Irrespective of stimulus type and fundamental frequency, the digital stimulus energy was kept constant. After digital to analogue conversion, the stimuli were low-pass filtered at the Nyquist frequency

(TDT FT6), optionally filtered further by a TDT PF1 programmable filter, and attenuated (TDT PA4).

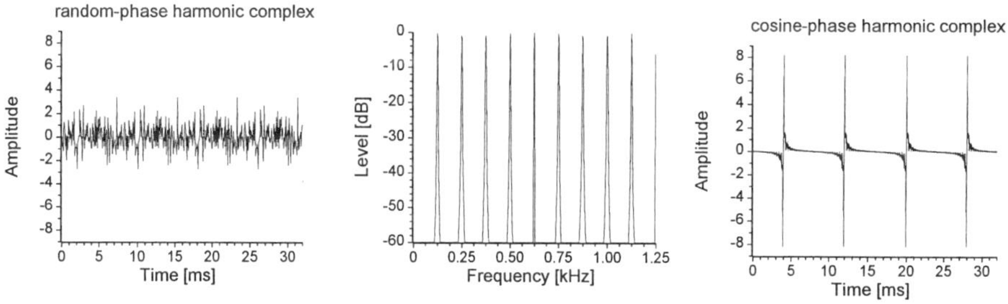

Figure 3: Harmonic complexes with a fundamental frequency of 125 Hz. The left panel shows a random-phase complex; the right panel shows a cosine-phase complex. The magnitude spectrum, common to both the waveforms plotted, is shown in the middle panel. The two waveforms have the same energy.

The stimuli were equalised (Phonics PEQ3600) to compensate for the speaker- and coupler frequency response before being fed into a Rotel RB871 power amplifier and a programmable passive end attenuation (0-75 dB). The signal was presented over a speaker (Mike Ravicz, MIT design) mounted in a coupler designed for the guinea-pig ear. The stimuli were acoustically monitored with a B&K 4134 microphone attached to a calibrated 1 mm probe-tube.

3 Results

3.1 The Basic physiological response types found in the mammalian ventral cochlear nucleus

There are three basic temporal adaptation patterns found in single units recorded in the ventral cochlear nucleus: Primarylike, Chopper and Onset (see Figure 4 for examples of each unit type). Each major group has been sub-divided and we now have a reasonably good view of the correspondence between unit type and its underlying morphology [6].

Each unit type in the cochlear nucleus can also be identified by its 1st order inter-spike interval histogram and discharge regularity. Single units that discharge in a very regular fashion (choppers) are characterised by a very low coefficient of variation (CV – standard deviation/mean). In contrast, Primarylike units discharge in an irregular manner and are characterised by CVs greater than 0.5. It is often not possible to measure a CV for onset units because of the low level of steady-state discharge but Figure 4 shows an onset unit with sufficient steady-state discharge rate to calculate the discharge regularity. The CV was 0.35, which puts it in the same category as transient choppers [7].

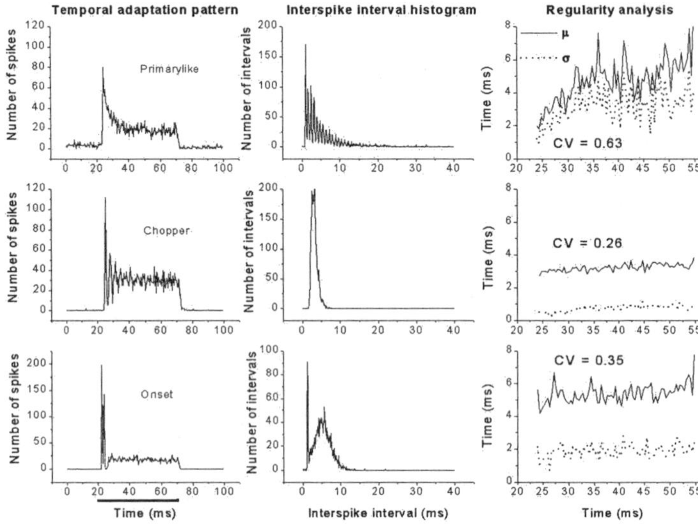

Figure 4: The three basic categories of response type in the mammalian ventral cochlear nucleus. The first column shows conventional peri-stimulus time histograms in response to suprathreshold best frequency tone bursts. The sound level was 20 dB above threshold for the Primarylike and Chopper unit but 50 dB above threshold for the Onset unit. The best frequencies (BF) were: Primarylike, 1.2 kHz; Chopper, 1 kHz and Onset 5 kHz. The solid bar indicates the time of the stimulus. The middle column shows the 1st order interspike interval analysis and the third column shows the regularity analysis which is a measure of a unit's temporal regularity. A CV less than 0.3 indicates a high degree of temporal regularity; a CV greater than 0.5 indicates a more random discharge pattern.

3.2 Responses to complex harmonic series

Units characterised by a sustained chopping pattern were able to represent the pitch of random phase harmonic complexes and iterated rippled noise when the f0 or 1/delay was close to the intrinsic chopping frequency of the unit. That is they showed a periodicity sensitivity with a bandpass characteristic which can be interpreted as periodicity tuning. This result is not entirely unexpected as chopper units may show a bandpass temporal modulation transfer function in response to SAM tones. Figure 5 shows an example of a chop-S unit. The top panel shows the mean chopping frequency (calculated from the regularity analysis). As can be seen from near threshold to approximately 15 dB above threshold the chopping period decreases from around 9 ms to a stable level at 6 ms. The periodicity sensitivity to the IRN stimulus shows a broad peak around a IRN delay of 8 ms which is close to the intrinsic chopping period (6ms). The middle panel of Figure 5 shows the bandpass characteristic of the periodicity sensitivity as a function of IRN delay. The bottom panel of Figure 5 shows the almost constant mean discharge rate as a function of the IRN delay. This means that the spectral ripple of IRN cannot be resolved in terms of a rate-place code in this 5.3 kHz unit. The preference for some delays in the IRN stimulus has also been observed in the first-order inter-spike interval analysis for units classified as On-C. Figure 6 shows the rate response of an On-C unit as a function of delay (equals fundamental period). The rate response to the RPH and IRN stimuli is independent of IRN delay or 1/f0 and equal to the mean rate response to white noise.

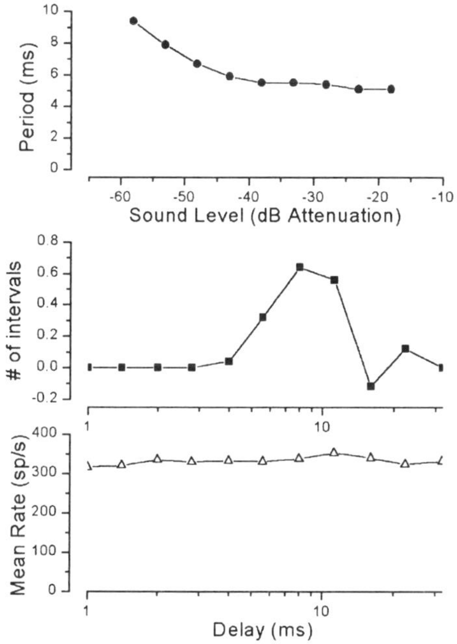

Figure 5. Upper panel: Chopping period as a function of sound level for a chop-S unit. BF = 5.3 kHz. The chopping period stabilises around 5 ms at levels 20 dB or greater above threshold. Middle panel: Periodicity sensitivity as a function of the delay of iterated rippled noise. Note the bandpass characteristic with a peak at 8 ms. Lower panel: Mean rate as a function of Delay for the same Chop-S unit.

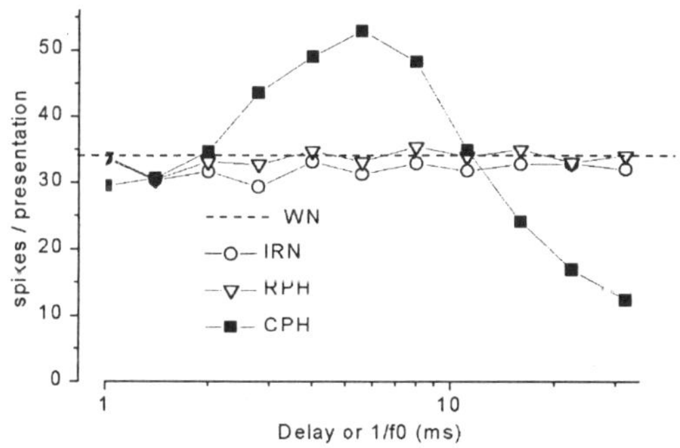

Figure 6. Mean-rate response of an On-C unit as a function of delay or 1/f0 for the three types of stimuli. The dotted line indicates the mean rate in response to white noise with the same energy as the periodic stimuli.

However, the rate response to the CPH stimulus appears to show some preference to intermediate f0s. The small rate response at short periods (<= 2 ms) can be explained by the inability of this unit to follow such quickly modulated waveforms (see Figure 2). The decrease in mean rate for the long delays is caused by the smaller number of peaks available in the same time span to which the cell could discharge.

Figure 7 shows both the first order and all-order periodicity sensitivity of the same unit to the three stimulus types. In response to the IRN and RPH stimuli the On-C unit clearly shows a preference for certain delays, and peaks at 8 ms. Such a preference is not evident in the all-order sensitivity, however, it should be noted that there is an excellent representation of the pitch for delays as long as 32 ms for the all-order analysis.

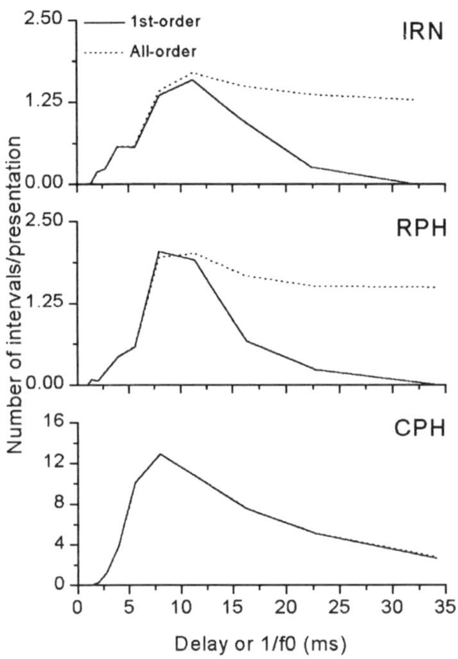

Figure 7. Periodicity tuning for an On-C unit (BF = 12.45 kHz). Based on first-order statistics this unit has a clear preference for delays around 8 – 11.2 ms.

If the response to IRN and RPH stimuli reflects temporal processing and not the response to resolved harmonics then high pass filtering the stimuli, above the region in which resolved harmonics may occur, should have little effect on their response.

The effect of high-pass filtering at 1 kHz on the response to delays of 8, 11.2 and 16 ms is shown in Figure 8. There is little difference between the filtered and non-filtered conditions indicating that the response is not dependent on the resolvability of harmonics.

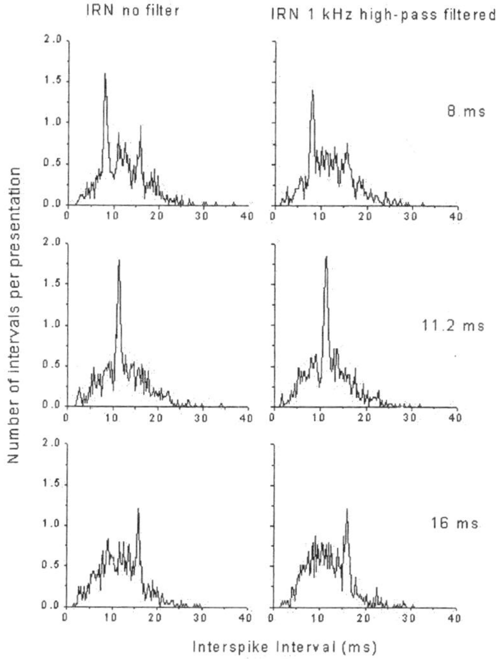

Figure 8: Effect of high-pass filtering on the temporal response to IRN stimuli with delays of 8, 11.2 and 16 ms for the same unit as in Figure 6 and 7.

4 Discussion

Our results provide further evidence for the role of chopper and onset unit responses in the encoding of the pitch of complex sounds. It is, of course, possible that the pitch may be carried in the all-order information in the neural responses, in which case Primarylike units may also be important [8,9]. Chopper units in the posteroventral cochlear nucleus and the dorsal cochlear nucleus have previously been shown to exhibit intrinsic oscillations [10]. The oscillation frequency is related to the best modulation frequency (as measured by the units response to sinusoidally amplitude modulated tones). In this study we have extended this observation for a sub-group of chopper units by demonstrating that they represent the periodicity of RPH and IRN in their first order interval statistics; both RPH and IRN reveal much less envelope modulation in their waveforms than SAM tones or CPHs. However, the range of best delays or fundamental periods to which the chopper units respond is limited in our data set to 4-10ms. It must also be pointed out that the chopper units that respond best are classified as sustained choppers (see Row 2, Figure 4). These chopper units show a near constant chopping rate and a very low CV value (< 0.3). Transient chopper units do not appear to encode the pitch of complex sounds as well.

A novel finding has been the delay preference of units classified as On-C. These units have been implicated in the encoding of pitch [11-15] but to date such intrinsic preferences have not been reported. The range of preferred delays was between 4 and 16 ms.

5 Acknowledgements

Supported by a grant from the MRC (G9703469) and the "Deutsche Forschungsgemeinschaft".. We thank Mark Radon for rewriting all our DOS based, data acquisition, programmes into visual C++.

References

1. Yost, W.A. (1996a) The pitch of iterated rippled noise. J. Acoust. Soc. Am. 100(1), 511-518.
2. Yost, W.A. (1996b) The pitch strength of iterated rippled noise. J. Acoust. Soc. Am. 100(5), 3329-3335.
3. Cariani, P.A., and Delgutte, B. (1996a) Neural correlates of the pitch of complex tones. I. Pitch and pitch salience. J. Neurophysiol. 76, 1698-1716.
4. Cariani, P.A., and Delgutte, B. (1996b) Neural correlates of the pitch of complex tones. II. Pitch shift, pitch ambiguity, phase-invariance, pitch circularity, rate pitch, and the dominance region of pitch. J. Neurophysiol. 76, 1717-1734.
5. Winter, I. M. and Palmer, A. R. (1990) Responses of single units in the anteroventral cochlear of the guinea pig. Hear. Res., 44, 161-178.
6. Rhode, W. S. and Greenberg, S. (1992) Physiology of the cochlear nuclei. In: The mammalian auditory pathway: Neurophysiology. Springer Verlag, New York.
7. Young, E. D., Robert, J.-M. and Shofner, W. P. (1988) Regularity and latency of units in ventral cochlear nucleus: Implications for unit classification and generation of response properties. J. Neurophysiol. 60, 1-29.
8. Shofner, W. P. (1991) Temporal representation of rippled noise in the anteroventral cochlear nucleus of the chinchilla. J. Acoust. Soc. Am. 90, 2450-2466.

9. Shofner, W. P. (1998) Evidence that primarylike units are the major CN subsystem which encodes pitch related information in their temporal discharge. Abstracts of the twenty-first midwinter research meeting of the ARO. p96.
10. Kim, D. O., Sirianni, J. G. and Chang, S. O. (1990) Responses of DCN-PVCN neurons and auditory-nerve fibres in unanaesthetised decerebrate cats to AM and pure tones: Analysis with autocorrelation/power spectrum. Hear. Res. 45, 95-113.
11. Kim, D. O., Rhode, W. S., and Greenberg, S. R. (1986) Responses of cochlear nucleus units to speech signals: Neural encoding of pitch, intensity and other parameters. In: Auditory frequency selectivity. Eds: B. C. J. Moore and R.D. Patterson, New York, Plenum, pp. 281-288.
12. Palmer, A. R. and Winter, I. M. (1992) Cochlear nerve and cochlear nucleus response to the fundamental frequency of voiced speech sounds and harmonic complex tones. In: Auditory physiology and perception, edited by Y. Cazals, L. Demany and K. Horner. Oxford, Pergammon, pp231-239.
13. Palmer, A. R. and Winter, I. M. (1993) Coding of the fundamental frequency of voiced speech sounds and harmonic complex tones in the ventral cochlear nucleus. In: Mammalian cochlear nuclei: Organization and function. Edited by M. A. Merchan, J. Juiz, D. A. Godfrey and E. Mugnaini. New York: Plenum. Pp373-384.
14. Rhode, W. S. (1995) Interspike intervals as a correlate of periodicity in cat cochlear nucleus. J. Acoust. Soc. Am. 97, 2414-2429.
15. Rhode, W. S. and Greenberg, S. R. (1994) Encoding of amplitude modulation in the cochlear nucleus of the cat. J. Neurophysiol. 71, 1797-1825.

DYNAMICS OF NEURONAL EXCITATION AND INHIBITION IN THE AUDITORY BRAINSTEM

RUDOLF RÜBSAMEN AND CORNELIA KOPP

Neurobiology Lab., Zoological Department of the University of Leipzig, Germany
E-mail: rueb@rz.uni-leipzig.de

GERD-JOACHIM DÖRRSCHEIDT

Dept. of General Zoology and Neurobiology, Ruhr-Universität Bochum, Germany
E-mail: doe@ruhr-uni-bochum.de

Simultaneous recording and differential analysis of pre- and postsynaptic spike activity in spherical bushy cells of the anteroventral cochlear nucleus in the Mongolian gerbil was used to evaluate the interaction of acoustically driven excitation and inhibition in auditory signal processing. The data show that in 2/3rd of the units' acoustically evoked neuronal activation, even when pure tones are used as stimuli, is due to an integration of excitatory and inhibitory afferents. This inhibition can effectively narrow the bandwidth of acoustically evoked responses, enlarge the dynamic range of responsiveness, and increase the acuity of the coupling of the discharges to transients of the acoustic stimuli..

1 Introduction

Neurons in the anteroventral cochlear nucleus (AVCN), representing second order neurons in the central auditory system, receive their excitatory input directly from auditory nerve fibers. Two-to-five fibers converge on a single AVCN neuron and form large axosomatic terminals named 'Calyces of Held'. It is commonly accepted that the specific function of these large terminals is to guarantee a reliable pre-to-postsynaptic transmission, and that in this regard the AVCN resembles a nuclear relay transmitting information from the cochlea to the superior olivary complex and the inferior colliculus [1, 2]. This notion has been questioned after terminals on the same AVCN neurons were shown by immuncytochemical labeling to be containing the inhibitory transmitters GABA and glycin [3], and when the pharmacological effectiveness of these transmitters had been demonstrated using slice electrophysiology [4]. However, we yet don't know to what extent the inhibitory input originates from outside the central auditory system, and to what extent it conveys an auditory input. With regard to the latter possibility, we lack detailed information in which way such an interaction between excitation and inhibition affects neuronal responsiveness. In our present investigation on Mongolian gerbils we used *in vivo* electrophysiology to quantify inhibitory influences on acoustically driven neuronal responsiveness.

2 Methods

Extracellular recordings were acquired from neurons at the rostral pole of the AVCN, the representation site of frequencies between 0.3 to 4 kHz. After characterizing the excitatory response area of each recorded unit, its corresponding suppression/inhibition frequency-intensity domains were evaluated by employing the two-tone stimulus paradigm. These were characterized by the range of frequency-intensity settings of acoustic stimuli which caused a significant reduction of discharge activity elicited by stimulation at the units' characteristic frequencies. From these measurements it was possible to evaluate within a single record the unit's spontaneous discharge activity, its excitatory response, and the response reduction due to inhibition (Fig.1).

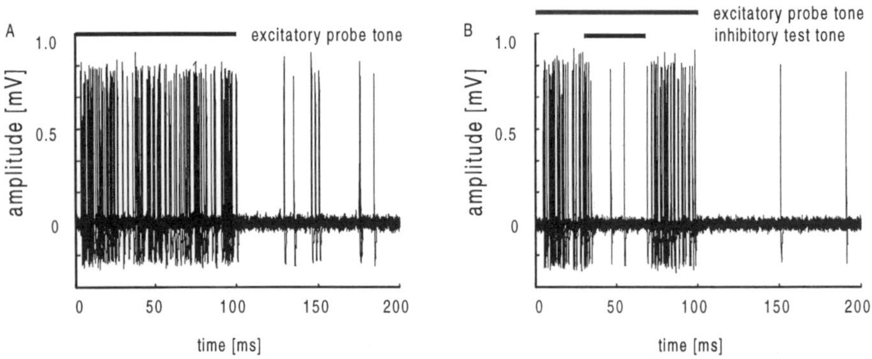

Figure 1. Two-tone stimulus paradigm. **(A)** PSTH of a primary-like discharge activity of an AVCN unit recorded during 20 repetitions of a 100 ms pure tone (excitatory probe tone) presented at the unit's characteristic frequency 20 dB above threshold. **(B)** Same excitatory stimulus as in A, but, in addition, a second tone (inhibitory test tone) was presented to the same ear with a delay of 30 ms. The test tone was set to a frequency 1 octave above CF and was presented at 50 dB above the units threshold.

In these records, the waveforms of the extracellularly recorded discharges of AVCN neurons show the typical biphasic courses indicating action potentials of the nuclear's principal cells. In a large number of these units spikes are preceded by somewhat smaller 'prepotentials', which indicate action potentials of a presynaptic afferent fiber (i.e. auditory nerve fiber) innervating the neuron (Fig. 2). While in these records each spike, without exception, is preceded by a prepotential, additionally a number of prepotential-like signals (resembling prepotentials by amplitude and duration) are picked up, which are not followed by large spikes. These prepotential-like signals are considered action potentials of the presynaptic fiber, which do not trigger postsynaptic spikes (Fig. 2). Thus, this specific recording condition makes it possible to simultaneously acquire the discharge activity of a presynaptic afferent nerve fiber and the postsynaptic spike of the AVCN neuron innervated by that fiber. By a differential analysis of these activities with special regards to the elicitation of both acoustically evoked excitatory and inhibitory inputs the effect of neuronal inhibition on auditory stimulus processing can be quantified.

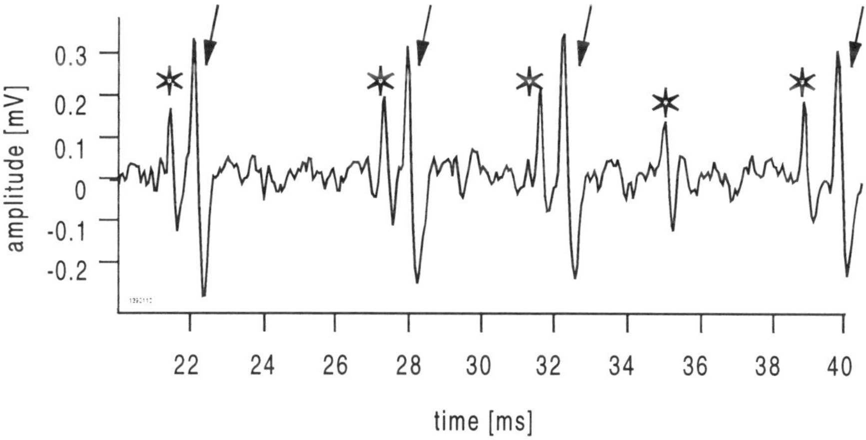

Figure 2. Extracellular single unit activity recorded at a rostral low-frequency site of the anteroventral cochlear nucleus (AVCN). Each biphasic action potential (arrows) is preceded by a prepotential (asterisks). The large-amplitude signals reflect the discharge activity recorded from the cell soma (postsynaptic side), and the prepotential the discharge activity of an afferent Calyx of Held (presynaptic side). Here, also a 'prepotential-like' signal was recorded, a presynaptic spike not followed by a postsynaptic one.

3 Results and Discussion

About 2/3rd of the recorded AVCN units showed marked differences in the response areas acquired pre- and postsynaptically (Fig. 3). Typically, the bandwidth of activation was narrower and the increase in responsiveness at higher stimulus levels was reduced on the postsynaptic side. The cause for both phenomena is the elicitation of acoustically evoked inhibition (Fig. 4B) which clearly can be separated from sideband suppression, a cochlear bound effect already recognizable in the presynaptic activity (Fig. 4A).

For a quantitative analysis of the relation between excitation and inhibition, we applied the principal component analysis to the neuronal signals (Fig. 5). Thereby it is possible to identify and distinguish the different basic components of signals with complex waveforms [5]. One advantage of the principal component analysis is that it does not depend on any assumptions regarding characteristic features of the signals. In most AVCN units the waveforms of the discharges can be accounted for by the first two principal components (Fig. 6A). The analysis of the degree to which PC1 and PC2 contribute to the actual recorded signal waveforms yields bimodal distributions under all three stimulus conditions (Fig. 6 B-D). Postsynaptic action potentials are characterized by higher PC1-weights, solitary prepotential-like signals by lower PC1-weights. The number of elements in these two clusters vary with the stimulus conditions.

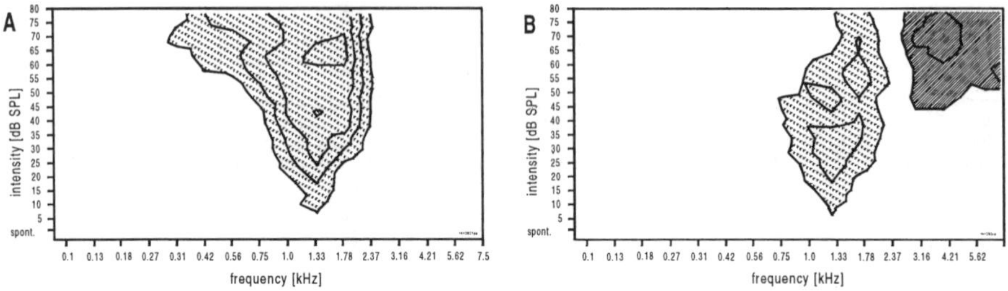

Figure 3. Response areas of an AVCN unit recorded presynaptically **(A)** and postsynaptically **(B).** The response area is broader on the presynaptic side, as is the stimulus induced response strength (indicated by the number of contour lines). In the postsynaptic record an inhibitory area is shown flanking the high-frequency border of the excitatory response area.

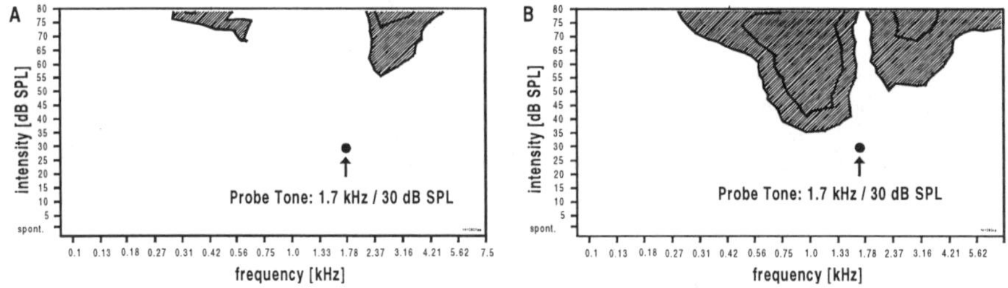

Figure 4. Sideband suppression **(A)** and inhibitory areas **(B)** of the same unit shown in Fig. 3. measured by the use of two-tone stimulation. In both cases the excitatory probe tone was set at 1.7 kHz and 30 dB SPL. The suppression sidebands disappear while the inhibitory sidebands persist when the probe tone level is further increased (not shown). Note that the inhibitory sidebands cover most of the excitatory response area characterized by the single tone stimulation in Fig. 3B.

During single tone stimulation at CF, both the absolute number and the percentage of solitary prepotential-like signals are higher than during spontaneous activity (Fig. 6B&C). This is an indicaton for an increase in the presynaptic excitation, but also shows that concurrently an inhibitory circuitry is activated which effectively limits the number of postsynaptic spikes. When the inhibitory influence is further increased by the use of two-tone stimulation, the share of the postsynaptic action potentials is even more reduced (Fig. 6D). The fact that under these stimulus conditions the percentage of presynaptic discharges remains high quote for an unaltered afferent excitatory input and also shows that the inhibition is effective at the AVCN neuron itself. This inhibition can effectively narrow the bandwidth of acoustically evoked neuronal responses, enlarge the dynamic range of responsiveness and increase the acuity of the coupling of the discharges to transients of the acoustic stimuli.

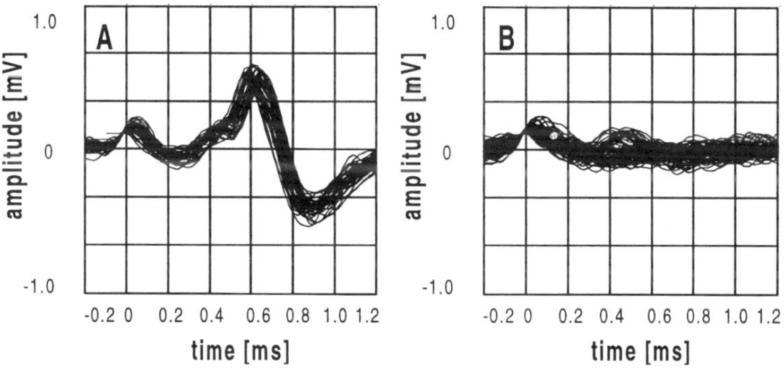

Figure 5. Superimposed waveforms of 180 signals. All records are triggered by the onset of the prepotential (resp. prepotential-like) signals. For clarity those signals which consist of prepotentials and instantaneously succeeding postsynaptic spikes **(A)** are graphically separated from the prepotential-like signals which were not followed by postsynaptic spikes **(B)**.

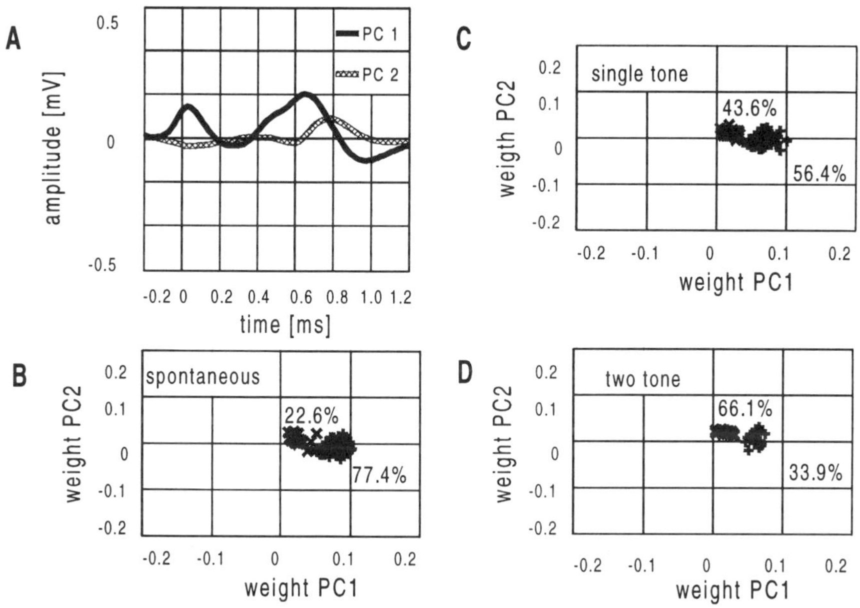

Figure 6. (A) Principal component analysis of the extracellularly recorded discharges of the unit shown in Fig 5. Already the first two eigenvalues (PC1 & PC2) give a description of the waveforms of the entire ensemble with an efficiency of 98%. **(B-D)** The signals are easily separated into two clusters by a cluster algorithm using the PC1- and PC2 weightening factors as variables. The cluster characterized by the higher PC1-weight is composed of signals which show postsynaptic action potentials plus preceding prepotentials, and the cluster with the lower PC1-weight consist of the isolated prepotentials of

presynaptic origin. The relation between the numbers of the respective signals change depending on the stimulus conditions, **(B)** spontaneous discharge activity, **(C)** pure tone stimulation at CF, and **(D)** additional activation of inhibition using two-tone stimulation. Note that --in comparison with the spontaneous activity-- during excitatory single tone stimulation both the absolute number and the relative percentage of isolated prepotential-like signals increase. This shows that the presynaptic excitation is increased but concurrently also an inhibitory influence is activated which reduces the number of postsynaptic spikes (see also Fig. 3B).

Acknowledgements:

This work was supported by Deutsche Forschungsgemeinschaft,
DFG GRK 250/1-96 and DFG Ru 390/15-1

References

1. Pfeiffer, R. R., Classification of response patterns of discharges for units in the cochlear nucleus: tone-burst stimulation. *Exp. Brain Res.* 1 (1966) pp. 220-235.
2. Rhode, W. S. and Smith, P. H., Encoding time and intensity in the ventral cochlear nucleus of the cat. J. Neurophysiol. 56 (1986) pp. 262-286.
3. Osen, K. K., Anatomy of the mammalian cochlear nuclei: A review. In J. SYKA, R. B. MASTERTON (eds.) Auditory pathway –structure and function. Plenum Press, New York, pp. 65-76.
4. Wickesberg, R. E., Oertel D., Delayed, frequency-specific inhibition in the cochlear nuclei of mice: A mechanism for monaural echo suppression. J. Neuroscience 10 (1990) 1762-168.
5. Glaser, E. M., Ruchkin, D. S., Principles of neurobiological signal analysis. Academic Press, New York, San Francisco, London (1976).

PERIODICITY CODING IN THE AUDITORY CORTEX: WHAT CAN WE LEARN FROM LEARNING EXPERIMENTS?

HOLGER SCHULZE AND HENNING SCHEICH

Leibniz-Institute for Neurobiology, Brenneckestr. 6, 39118 Magdeburg, Germany
E-mail: Holger.Schulze@IfN-Magdeburg.de

GERALD LANGNER

Institute for Zoology, Technical University Darmstadt, 64287 Darmstadt, Germany
E-mail: gl@neuro.bio.tu-darmstadt.de

Many natural sounds are characterized by periodic envelope or amplitude modulation (AM). The perceptual qualities corresponding to this signal characteristic are rhythm for low and pitch for high modulation frequencies (f_m). We demonstrate that f_m ranges corresponding to these two perceptual qualities seem to be represented by different codes in the primary auditory cortex (AI) of the Mongolian gerbil: Low f_m are represented by a synchrony code whereas high f_m are represented by a rate-place code. For the rate-place code we show that in response to AM neurons integrate over spectral ranges much wider than expected on the basis of their classical frequency receptive fields. Furthermore, we present preliminary data from learning experiments where gerbils were trained to discriminate between AM with different f_m. It seems that the different cortical codes used to represent AM with different perceptual qualities are also reflected in the learning curves: learning performance is better for low f_m ranges, where phase-locking is still present, than for high f_m ranges that are encoded by a rate-place code.

1 Introduction

Acoustic communication signals, like human speech and animal vocalizations, are characterized by periodic envelope or amplitude modulation (AM). The perceptual quality associated with this signal periodicity varies as a function of modulation frequency (f_m): AM signals of low f_m, up to about 100 Hz, evoke percepts with a temporal quality (rhythm and roughness) [4], at higher f_m percepts have a spatial quality (periodicity pitch) [2]. A recent study has provided evidence that these two different perceptual qualities might be based on different cortical codes for stimulus periodicity, one that is temporal and codes low f_m (synchrony code) and one that is spatial and codes high f_m (rate-place code) [3]. Here we characterize these two cortical codes and propose the hypothesis that there may also be two different learning mechanisms that operate on these different codes. To obtain a first estimate for the validity of this hypothesis we trained gerbils to discriminate between AM of different f_m. Our assumption was that if indeed two different learning mechanisms were realized in the auditory cortex, this might be reflected in the learning performance of the animals.

2 Methods

Neuronal responses from single and multi-units to pure tones and 100% sinusoidally amplitude modulated tones were recorded extracellularly from the left primary auditory cortex (AI) of anaesthetized (n=6) and unanaesthetized (n=7) adult male Mongolian gerbils. Pure tones were used to determine frequency response functions (FRF = a plot of spike over tone frequency). From the FRF the frequency response range (FRR = range of tone frequencies that increased spike rate) and best frequency (BF = maximum of FRF) were determined. AM-experiments were carried out where (1) the carrier frequency of the AM (f_c) was set to the BF of the unit, and (2) where high f_c were used so that the AM spectrum was completely above the unit's FRR. F_m ranged from 5 Hz to 5kHz. Responses to AM were quantified using rate modulation transfer functions (rate-MTF = a plot of spike rate over f_m) or synchronization MTFs (sync-MTF = a plot of vector strength (VS) over f_m). MTF-maxima are referred to as best modulation frequency (BMF). Q_{3dB} factors (=BF or BMF devided by the bandwidth of the corresponding FRF or MTF 3dB below the maximum, respectively) were used to measure sharpness of tuning.

AM discrimination was investigated using a footshock (UCS = unconditioned stimulus) motivated shuttle box avoidance go/no go procedure. Animals were trained to discriminate between two AM of different f_m and a f_c of 2 kHz. Six groups of gerbils, with 3 specimens each, were trained (A to F in Fig. 5). F_m to be discriminated were 10 vs. 20, 20 vs. 40, 40 vs. 80, 80 vs. 160, 160 vs. 320, and 320 vs. 640 Hz. In each group, the AM with the lower f_m served as the CS+ (= conditioned stimulus followed by the UCS), the other AM was the CS- (not followed by the UCS). 15 training sessions with 60 trials each were carried out in each animal. Crossings of the hurdle during a 4 s presentation of the CS+ or CS- (= conditioned responses; CR+ and CR-, respectively) were counted in each session and plotted as a function of training session to quantify learning performance.

3 Results

3.1 Synchrony code for stimulus periodicity

For unanaesthetized gerbils 17 % of the units tested (n= 310) responded to AM with a f_c equal to the unit's BF with discharges phase-locked to the f_m. Significant phase locking was observed up to 100 Hz f_m. The highest sync-BMF found was 45 Hz (cf. Fig. 4A); VS ranged from 0.15 to 0.81. An example is shown in Fig. 1. The tone response of the unit is given as response plane [1] (A) and FRF (C). Its BF of 1 kHz was used as the f_c in the AM experiment shown in B and D (sync-

MTF). The unit showed significant phase locking up to 80 Hz f_m ($P < 0.01$, Raleigh test) with a sync-BMF of 10 Hz. Phase locking in anaesthetized animals was generally weaker (VS ranging from 0.15 to 0.53) and confined to even lower f_m, up to 65 Hz only. A systematic topographic representation of sync-BMFs within AI was not found, neither in anaesthetized nor in unanaesthetized animals.

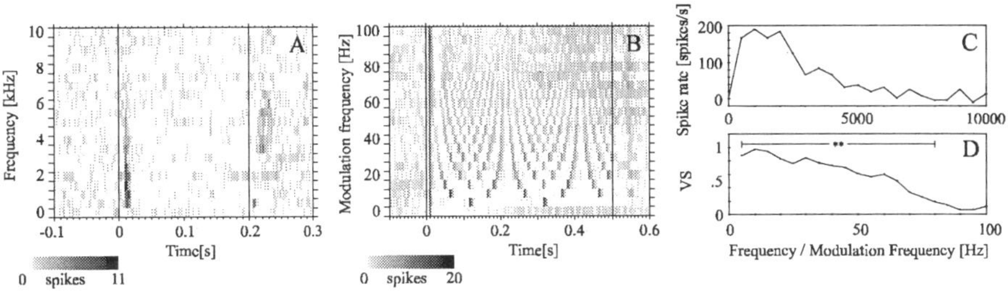

Figure 1. Tone response (A,C) and AM response (B,D) of a unit showing phase locking to f_m of AM. Vertical lines in response planes (A,B) indicate begin and end of stimulation.

3.2 Rate-place code for stimulus periodicity

However, a systematic representation of rate-BMFs (periodotopy) within the low frequency region of AI (up to about 3 kHz BF) was found using AM with a spectrum completely above a unit's FRR. Within this AI region over 70% of the units (n=232) responded to such AM in both anaesthetized and unanaesthetized animals with discharges tuned to a certain range of f_m. Fig. 2 shows responses of two such units (top and bottom row). The responses to tones (left panels) and to AM (middle panels) and the corresponding FRFs and rate-MTFs (right panels) are shown. The unit shown in the upper row responded to pure tones between 0.5 and 2 kHz. Fig. 2B shows the AM response of that unit ($f_m = 0$ to 1.5 kHz; $f_c = 7$ kHz). Although none of the AM was spectrally within its FRR, the unit responded with discharges tuned to f_m ranging from 800 to 1050 Hz. The AM response showed a longer latency (25.4 ms) and was more sharply tuned (Q_{3dB}=6.9) than the tone response (latency=18.6 ms; Q_{3dB}=1.6). The bottom row shows a unit that was tuned to low tone frequencies (0.1 to 0.4 kHz) but to high f_m (0.5 to 1.4 kHz, rate-BMF = 1.25 kHz; f_c=7 kHz). The AM response again showed longer latency (37.1 ms) and sharper tuning (Q_{3dB}=7.4) than the tone response (latency=20.6 ms; Q_{3dB}=2.1). For the population of units the AM response had generally longer latency (Fig. 3B) and sharper tuning (Fig. 3C) than the tone response. BF and rate-BMF of a given unit could differ by more than 2 octaves (Fig. 3A). A comparison of this rate-place code with the synchrony code showed that they covered almost distinct f_m ranges: The distributions of sync-BMFs (up to 45 Hz) and rate-BMFs (50 Hz to about 3 kHz) showed no overlap (Fig. 4A,B). Furthermore, whereas the synchrony code showed

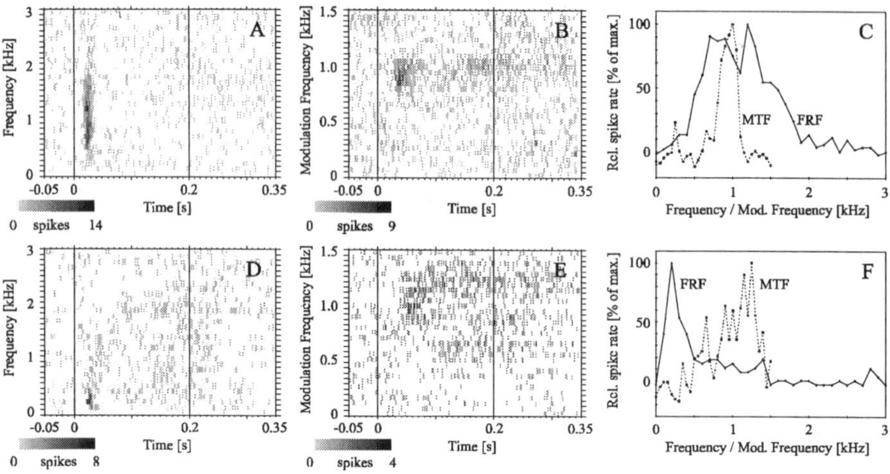

Figure 2. Responses of two units to tones (A,D) and AM with a spectrum completely above the units' FRRs (B,E; fc=7kHz in both cases). C and F show the corresponding FRFs and rate-MTFs. Vertical lines in response planes (A,B,D,E) indicate begin and end of stimulation.

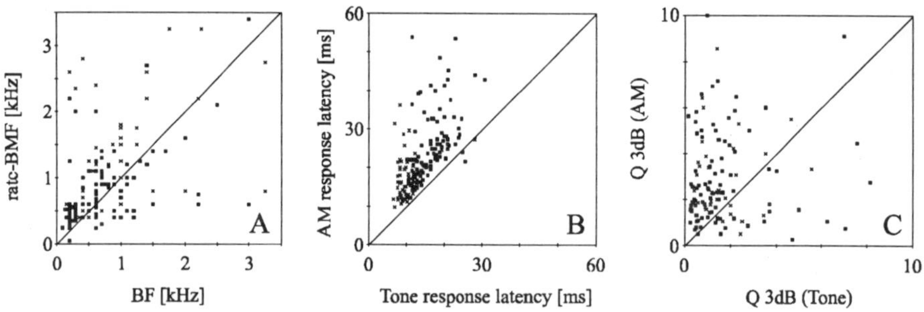

Figure 3. Scatterplots of rate-BMF vs. BF (A), AM response latency vs. tone response latency (B), and Q_{3dB} of AM responses vs. Q_{3dB} of tone responses (C) for the population of units tested. Squares show data from anaesthetized, dots from unanaesthetized animals.

no topographic organization, the rate-place code showed a clear periodotopic gradient oriented roughly orthogonal to the tonotopic gradient (Fig. 4C) [3].

3.3 Behavioral discrimination of stimulus periodicity

To quantify learning performance for AM discrimination we used two measures: A gerbil was considered to discriminate between CS+ and CS- if the CR+ for a given session was at least 3 times as high as the CR-. As a measure of how fast gerbils

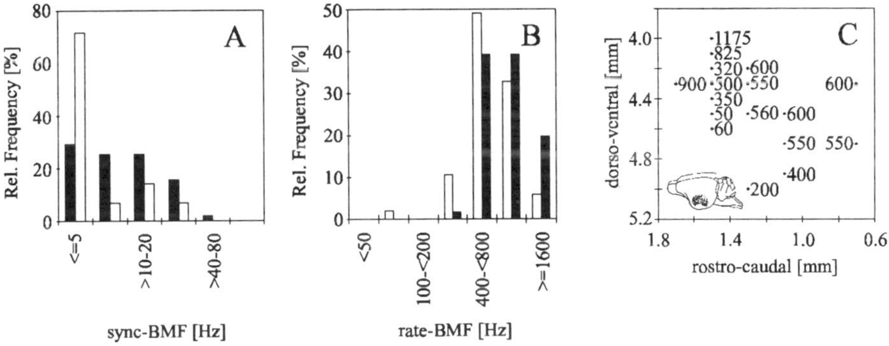

Figure 4. Frequency distribution of sync-BMFs (A) and rate-BMFs (B). Filled and open bars show data for unanaesthetized and anaesthetized animals, respectively. C shows the topographic distribution of rate-BMFs (periodotopy) in one animal. The inset shows the location of the left auditory cortex in the temporal lobe of the forebrain of the gerbil.

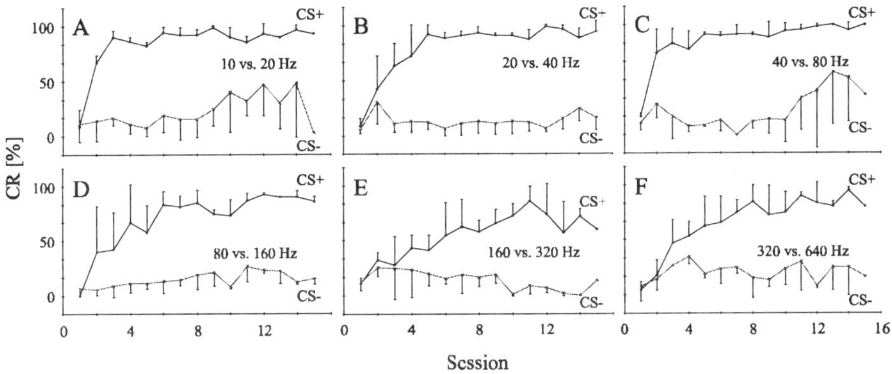

Figure 5. Discrimination of AM tones. Shown are group mean values and standard deviations (n=3 per group) of conditioned responses as a function of training session.

learn to discriminate between CS+ and CS- the first session where this criterion was reached was determined. Furthermore, we determined the maximal difference between CR+ and CR- that was reached for each group to obtain a measure of discrimination performance. Groups A to D reached the criterion for discrimination at session 2 or 3, E and F only around session 6 (Fig. 5). Discrimination performance in A to C was 91 ±1%, in D to F 76 ±2%. So for A to C, when AM is represented by the temporal synchrony code in AI, discrimination learning is fast and discrimination performance is good, whereas when AM is represented by the topographic rate-place code (E and F), learning takes longer and discrimination performance is worse. For the intermediate group D, where an AM that is encoded temporally has to be discriminated from an AM that is encoded spatially, learning is still as fast as in A to C, but discrimination performance is as poor as in E and F.

4 Conclusions

We demonstrated that in AI two codes for stimulus periodicity are realized, one that is temporal and codes for periodicities that create a temporal percept (rhythm) and one that is spatial and codes for periodicities that create a percept with a spatial quality (pitch). Because units can be tuned to high f_m but low tone frequencies (cf. Fig. 2, bottom row) and because the periodotopic gradient is roughly orthogonal to the tonotopic gradient (cf. Fig. 4C) it is unlikely that distortion products in the cochlea can account for the responses shown. Therefore, and because units responded to AM with spectra far above their FRR, it is more likely that AI units integrate information over wide spectral ranges. Finally, on the basis of our behavioral data it seems that the two codes for stimulus periodicity are also reflected in the animals' learning performance: Learning is faster and learning performance is better for periodicities that are encoded temporally than for those that are encoded spatially.

5 Acknowledgements

This study was supported by the Human Frontier Science Program, and by the Schwerpunktprogramm „Zeitliche Informationsverarbeitung im zentralen auditorischen System“ of the Deutsche Forschungsgemeinschaft (DFG). We wish to thank K. Buckisch for animal training.

References

1. Gerstein G.L., Butler R.A. and Erulkar S.D. Excitation and inhibition in the cochlear nucleus. I. Tone-burst stimulatuion. *J. Neurophysiol.* **31** (1968) pp. 613-636.
2. Ritsma R. J., Existence region of the tonal residue. *J. Acoust. Soc. Am.* **34** (1962) pp. 1224-1229.
3. Schulze H. and Langner G., Periodicity coding in the primary auditory cortex of the Mongolian gerbil (Meriones unguiculatus): two different coding strategies for pitch and rhythm?, *J. Comp. Physiol. A* **181** (1997) pp. 651-663.
4. Terhardt E., Über die durch amplitudenmodulierte Sinustöne hervorgerufene Hörempfindung. *Acustica* **20** (1968) pp. 210-214.

EFFECTS OF SOUND PRESSURE LEVEL, STIMULUS DURATION, AND BINAURAL CORRELATION ON VIRTUAL SOUND SOURCE CODING IN THE INFERIOR COLLICULUS OF THE GUINEA PIG

S. J. STERBING

Dept. Zoology and Neurobiology, Ruhr University Bochum, 44780 Bochum, Germany
E-mail:sterbing@neurobiologie.ruhr-uni-bochum.de

K. HARTUNG

Institute of Communication Acoustics, Ruhr University Bochum, 44780 Bochum, Germany
E-mail:hartung@ika.ruhr-uni-bochum.de

A great majority of neurons in the central nucleus of the inferior colliculus is spatially tuned for broadband sounds as revealed by free-field or virtual sound source (VSS) stimulation. The shape of the receptive fields can be explained by the shape of the monaural spectra and the interaural differences (time difference, level difference) of the individual head-related transfer functions (HRTF). Variation of overall sound pressure level had little effect on the preferred directions of the neurons. The size of the receptive fields, however, changed in a variety of ways. Constancy, expansion, contraction or combinations of both effects were found. Stimulation with binaural non-correlated VSS led to a loss of spatial tuning in many low-frequency neurons. Most high-frequency neurons showed no differences of the receptive fields under both conditions. In order to get information about the integration time of the inferior colliculus neurons, different stimulus duration was also tested. More that half of the neurons lost spatial selectivity when the stimulus duration was shorter than 5ms.

1 Introduction

Single neuron electrophysiology in the mammalian as well as in the avian auditory midbrain (mammal: central nucleus of the inferior colliculus (ICc), barn owl: nucleus mesencephalicus lateralis dorsalis (MLD)) revealed that a high percentage of the neurons prefer certain sound source directions [e.g. 3, 4]. While the amount of tested loudspeaker positions is restricted to the frontal hemisphere in most free-field studies due to the experimental setup (head holder, electrode drive), the technique of "virtual auditory environment" offers the opportunity to present sounds from any possible space position. The ICc receives many binaural and monaural projections. Among others there is input from the lateral superior olive (LSO), which is involved in processing interaural level differences and from the medial superior olive (MSO), whose neurons act as coincidence detectors and process interaural time differences [e.g. 6]. The aim of the present study was to "knock out" the coincidence detector by presenting non-correlated virtual sound sources, and to get information of the stability of the neuronal spatial receptive fields in dependence of overall sound pressure level. Furthermore, we presented virtual sound source stimuli of different duration.

2 Methods

The individual head-related transfer functions of 19 pigmented guinea pigs were measured for 122 directions of the upper hemisphere in an anechoic room (details see [3]). White noise (200 Hz – 16 kHz) virtual spatial sounds, sounds of different duration were generated off-line. Additionally, a stimulus set of binaural non-correlated VSS was calculated. During the electrophysiological experiment these stimuli were presented via individually calibrated insert earphones. Each of the 122 positions was presented pseudorandomly five times. The guinea pigs were anesthetized with ketamine/thiazine mix. The single neuron activity in the ICc was recorded with glass microelectrodes. The number of spikes for each direction was tested for statistically significant differences against the neighbored directions (Kruskal-Wallis-Test). The receptive fields were defined by half-maximum response. If no single position accumulated the maximum number of spikes, the center of the receptive field was taken as best position.

3 Results

3.1 Sound pressure level

A subset of 30 out of 90 neurons was stimulated with VSS of different overall sound pressure level (SPL). For 43% of the neurons, the receptive fields did not change with decreasing SPL, 33% showed a decrease and 6% an increase of RF size. A slight shift of the best position was observed in four neurons. The size and position of the receptive fields remained stable. Another three neurons showed both a change of the RF size and a change of best position.

3.2 Binaural correlation

It is assumed that a mechanism of coincidence detection is involved in sound localization. Especially neurons with low characteristic frequency are sensitive to the fine structure on the signal envelope. It was tested, whether the spatial receptive fields of the neurons become altered when a non-correlated virtual spatial stimulus is presented: 89% percent of the high-frequency neurons (characteristic frequency (CF > 2.5 kHz, n=18) showed no response changes under the non-correlated stimulus condition. Only 11% lost spatial tuning. The tuning of low-frequency neurons (CF $\leq$ 2.5 kHz, n=18) was affected by non-correlated signals to a higher degree. Only 28% of these were spatially tuned under both types of stimulation, 61% lost spatial tuning, and 11% showed a shift of the receptive field.

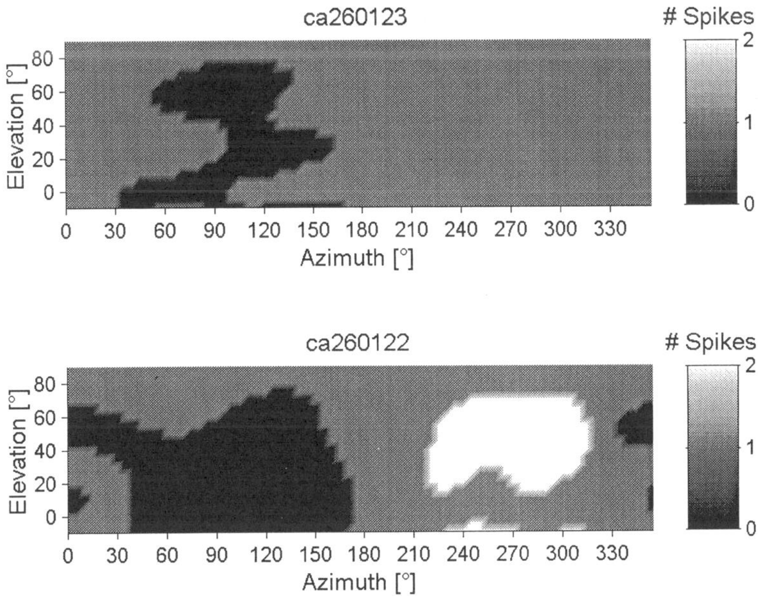

Figure 1. Directional preference of an example neuron stimulated with VSS of different sound pressure level. Neuronal activity of the neuron is plotted as a function of direction of incidence. The gray scale bar on the right side of each diagram indicates the median spike count. White areas mark directions with maximum activity. Upper panel: 80 dB SPL, lower panel: 60 dB SPL.

3.3 Stimulus duration

A subset of 36 neurons was stimulated with virtual sound source signals of different duration. Bursts of 50ms, 25ms, 10ms, 5ms, 2ms and 1ms were presented. Each stimulus had 10% rise/fall time. 59% of the neurons, which were spatially tuned to the longer signals, lost spatial tuning when the duration was 5ms or shorter. One neuron did show spatial selectivity only when the stimulus was longer than 25ms.

4 Discussion

Only few data are available for discussing the influence of sound pressure level on neuronal spatial tuning [e.g. 2,5]. Brugge and coworkers used non-individual virtual auditory stimuli for single unit recordings in the cat primary auditory cortex (AI). They found that the receptive fields of most AI neurons expand with increasing SPL. Surprisingly, the best position of most neurons was located near the "acoustical axis" of the cat. In the present study, best positions were distributed over the entire hemisphere, and were also located at e.g. extreme rear positions. In several other studies based on free-field stimulation, the majority of neurons showed receptive fields (only azimuth) of constant size, when stimulus intensity was varied [5]. The same holds for our data. The use of individual HRTF for creating virtual auditory space leads obviously to results, which are more comparable to those obtained in free-field. However, from human psychocacoustics it is known that localization performance decreases from intermediate to very low sound pressure levels [1], while only few changes occur in a wide suprathreshold range. In previous studies spatial sounds of different duration have never been tested. In the present study, many neurons did not show spatial tuning when the stimulus was shorter than 5ms. This suggests that clicks are inappropriate for studying spatial tuning for a high percentage of neurons.

5 Acknowledgements

This work was supported by DFG grant Ho 450/23-1-347.

References

1. Blauert J., Sound localization in the median plane, *Acustica* **22** (1969/70) pp. 205–213.
2. Brugge J. F., Reale R. A. and Hind J. E., The structure of spatial receptive fields of neurons in primary auditory cortex of the cat, *J. Neurosci.* **16(14)** (1996) pp. 4420–4437.
3. Hartung K. and Sterbing S. J., Generation of virtual sound sources for the electrophysiological characterization of auditory spatial tuning in the guinea pig, In: J. SYKA (ed.) Acoustical signal processing in the central auditory system , Plenum Press, New York, 1997 pp. 408–412.
4. Knudsen E. I. and Konishi M., A neural map of auditory space in the owl, *Science* **200** (1978) pp. 795–797.
5. Rajan R., Aitkin L.M., Irvine D. R. F. and McKay J., Azimuthal sensitivity of neurons in primary auditory cortex of cats. I.types of sensitivity and the effects of variations in stimulus parameters, *J. Neurophysiol.* **64**(3) (1990), pp. 872-887.
6. Yin T. C. T. and Chan J. C. K., Interaural time sensitivity in medial superior olive of cat, *J. Neurophysiol.* **64** (1990) pp. 465–488.

MONAURAL BRAINSTEM PROCESSING IS CONTROLLED BY INHIBITION

C. KOPP[1], S. DEHMEL[1], G.J. DÖRRSCHEIDT[2], R. RÜBSAMEN[1]

[1] *Zoologisches Institut der Universität Leipzig, Talstrasse 33, 04103 Leipzig, Germany*
[2]*Lehrstuhl für Allgemeine Zoologie und Neurobiologie der Ruhr-Universität Bochum*
E-mail: ckopp@rz.uni-leipzig.de

Monaural auditory brainstem processing within the superior olivary complex (SOC) and cochlear nucleus, specifically the neuronal coding of temporal, spectral and intensity cues are a prerequisite for both sound localization and differentiation of acoustic objects. Neurons which are innervated strictly monaurally are located in the anteroventral cochlear nucleus (AVCN), the medial nucleus of the trapezoid body (MNTB) and in the superior paraolivary nucleus (SPN) (66% of the neurons). In all three nuclei, 58% of the investigated neurons receive a prominent inhibitory input. Different sets of acoustic stimuli were used to characterize the interaction between excitation and inhibition. In 19% of the AVCN neurons, inhibition causes an upper threshold in the response areas, while in a number of MNTB neurons thresholds are elevated due to inhibition. In another population of AVCN neurons (33%) the inhibitory influence increases the dynamic range. Many SPN neurons show a very broad excitatory or inhibitory frequency response. Cells which are dominated by inhibition are further characterized by a strong off-response. Both, excitatory and inhibitory SPN neurons show strong synchronized responses to AM-stimuli. This indicates that transients in sound amplitude are the main features processed by these neurons.

Introduction

Neurons in the AVCN and the MNTB, the major monaural nuclei of the auditory brainstem, get their input through large synapses of Held, which give rise to presynaptic potentials preceding each postsynaptic action potential [1]. These prepotential units are therefore thought to provide a fast and secure 1:1 signal transmission. However, recent immunocytochemical and pharmacological studies have shown that neurons in the AVCN and the MNTB are also the target of extensive inhibitory inputs. In the present study we investigate the effectivity of these inputs in the coding of sound features.

The SPN is part of the SOC, but to date only very little is known about it´s physiology [2,3]. Studies with neurotracers show that the CN on both sides (excitatory) and the ipsilateral MNTB (inhibitory) give rise to projections into the SPN, which itself is a major source of ipsilateral projections to the inferior colliculus (IC) [4]. The goal of this study was to compare the signal processing by SPN neurons with that in AVCN and MNTB neurons.

1 Methods

Extracellular single cell recordings with glass micropipettes (3M KCl, 15-25 MΩ) were performed in an *in vivo* preparation of adult Mongolian gerbils. Acoustic stimuli were presented through custom made earphones. Pre- and postsynaptic potentials were separated by different threshold criterions. The response areas of the units were measured using random presentations of tone pulses (100ms duration, 5ms rise-fall time: "single tone") within predefined frequency and intensity ranges. Spikes were counted during stimulus time. From these records characteristic frequency (CF), threshold, bandwidth (BW) and rate-intensity function (RIF) were determined. The influence of inhibition was characterized using the two-tone stimulus paradigm (2TS). The detail of the 2TS were as follows: An excitatory probe tone was set at the unit´s CF and presented above threshold. A second stimulus, the inhibitory test tone, varied in a frequency-intensity range from two octaves above to two octaves below CF.

2 Results

2.1 Coding of sound intensity

In 19% of low frequency AVCN neurons, inhibition causes an upper threshold in the response areas (Fig.1B), not seen in response areas of the excitatory afferent fiber (Fig.1A). The RIF of the afferent auditory fiber (ANF) is not influenced by inhibition and thereby shows a monotonic increase with SPL before reaching a plateau. However, the RIF of AVCN neurons with upper thresholds are characterized by a strong non-linearity (Fig.1C). In a number of MNTB neurons the threshold of the excitatory response areas are elevated due to inhibitory inputs (Fig.2). In a further population of AVCN neurons (33%) the inhibitory influence increases the dynamic range of responsiveness by leveling the RIF (Fig.3).

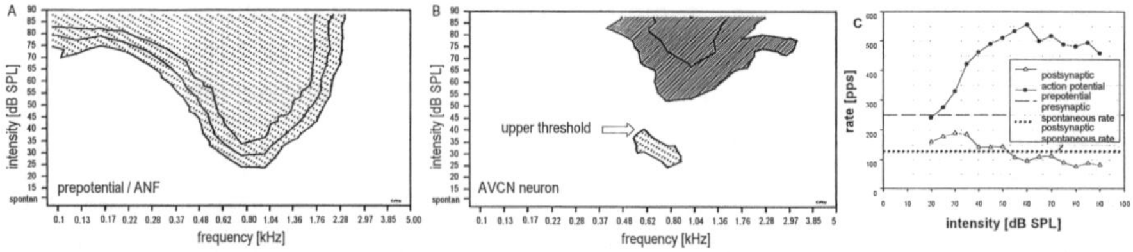

Figure 1. Response areas and rate-intensity functions of an AVCN neuron and it´s presynaptic input (ANF). **[A]** The presynaptic response area only shows excitation. **[B]** The postsynaptic response area shows a prominent inhibitory area at higher SPLs (dark shade). The RIFs are taken at the frequency marked by dotted lines in A and B. **[C]** The RIF of the presynaptic ANF (circles) shows a monotonic increase. Due to inhibition, postsynaptically a subtle increase in spike rate at low SPLs is followed by a decrease even below the spontaneous rate (dotted line).

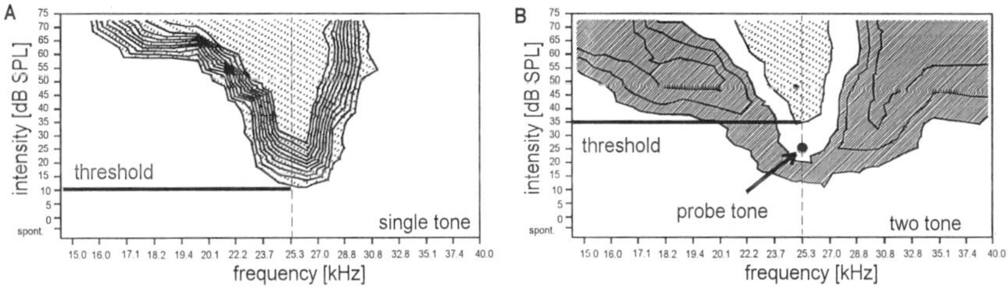

Figure 2. Response areas of an MNTB neuron. **[A]** Excitatory response area acquired during single tone presentation. **[B]** During 2TS (test tone as in [A] & probe tone at CF and 15dB hearing level) the influence of inhibition can be seen which covers large portions of the excitatory response area and additionally extends towards high and low frequency ranges. This inhibitory influence causes a threshold shift of 25 dB.

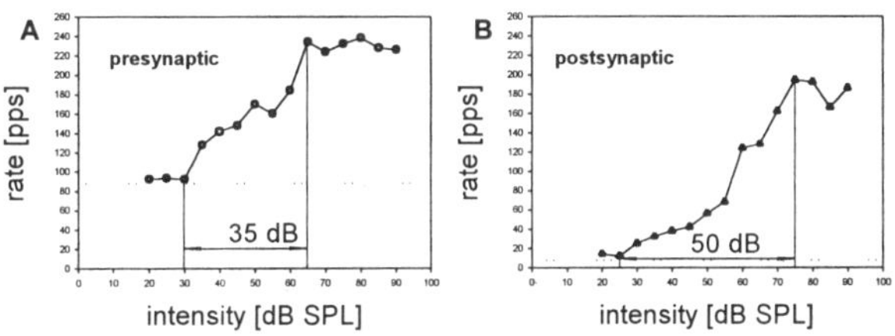

Figure 3. The comparison of RIFs of a single AVCN neuron **[B]** and of its afferent fiber **[A]** shows the pre-to-postsynaptic increase in dynamic range of about 15 dB.

2.2 Coding of transients in sound amplitude and duration coding

Sixty-six percent of all investigated cells in the SPN showed responses to tone pulses presented to the contralateral ear only. Twenty-four percent of these monaural units showed inhibitory response areas or mixed inhibitory/excitatory response areas. All monaural neurons with exclusively inhibitory response areas are characterized by broad frequency tuning (Q_{30}: median = 0,9) and elevated absolute threshold levels (43.3 ± 13.1 dB; Fig.4). Most of these cells exhibit a strong, short train of rhythmic activity after the end of the stimulus (off-chopper PSTH; Fig.4A).

Off-responses can also be seen in response to stimuli with AM (Fig.4B,C). AM coding was shown by half of the SPN neurons tested with AM stimuli (11/22); among these are monaural, contralateral inhibited neurons with a broad tuning and other neuron types (not shown here: binaural and monaural, excitatory, inhibitory and mixed excitatory and inhibitory).

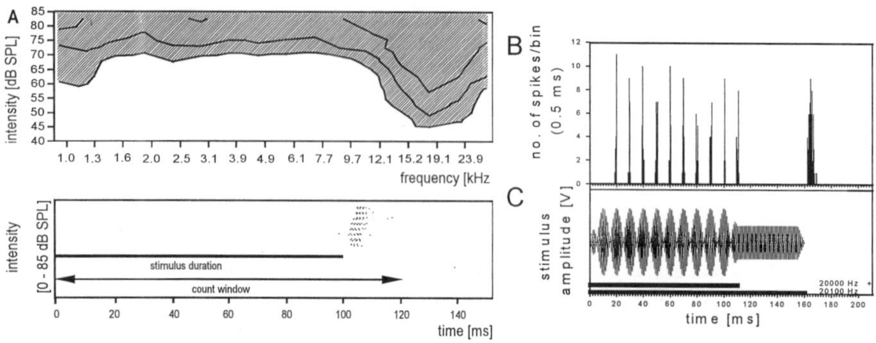

Figure 4. A SPN neuron that shows typical response properties of a 0.I cell. **[A]** Inhibitory response area (top) and dot display (bottom). **[B]** PSTH of the response to the beat stimulus (AM) shown below at 66 dB SPL peak level. **[C]** Oscillogram of the stimulus: beats produced by simultaneously presenting tones with equal intensity and different frequency (20,0 +20,1 kHz).

3 Discussion

The results described above deal with inhibition in the monaural auditory brainstem. Acoustically evoked inhibition influences the tuning, the RIF or the temporal response pattern of the neurons investigated in this study.

The AVCN neurons described above are characterized by presynaptic potentials (prepotentials) preceding each spike [1]. Based on anatomical studies, which show auditory nerve fibers with giant synaptic endings aiming at spherical bushy cells (SBC), the prepotential units can be identified as SBCs [5]. In addition to the excitatory input the SBCs receive a prominent inhibitory input [6]. In one third of SBCs inhibition influences the monaural coding of sound intensities by increasing the dynamic range compared to that of the presynaptic afferent fiber (fig.3). The mean dynamic range of low frequency auditory nerve fibers of the gerbil is about 36 dB [7]. The increased dynamic ranges of a number of SBCs are always correlated with a flattening of the RIF due to the reduction of discharge rate at higher sound intensities. Thus , the reduction of postsynaptic activity caused by inhibition could be one mechanism to increase the dynamic range of a unit. A further increase of the amount of inhibition, leads not only to non-monotonic RIFs, but in a number of units to upper thresholds, and by that to a restriction of the dynamic range. In this way, inhibition may establish intensity filters for either low (AVCN) or high (MNTB) sound intensities.

The elevated thresholds found in the monaurally broadly tuned, contralaterally inhibited SPN neurons could be caused by a convergence of the inputs from high threshold neurons found in the MNTB. Hints for this assumption also come from the large extension of the dendritic trees of the SPN cells reaching into the ventral acoustic stria, as seen in GOLGI impregnations [Kretzschmar,R., not published]. The broad tuning of the SPN neurons is important for a frequency independent and therefore timely precise coding of amplitude fluctuations across stimulus frequencies. Up to now we can only speculate what behavioral relevance or contribution to auditory signal processing the AM coding of SPN cells could have. Certainly, the coding of sound duration, and a special case of this - the coding of AM stimuli, should be important for the encoding of inter- and intraspecies communication signals in general.

Acknowledgements

The study was founded by DFG GRK 250/1-96.

References

1. Ryugo, D.K. and Fekete, D.M. (1982) Morphology of primary axosomatic endings in the anteroventral cochlear nucleus of the cat: A study of the endbulbs of Held. J. Com. Neurol. 210, pp. 239-257
2. Brückner, S. (1995) Kodierung interauraler Zeitdifferenzen durch Neurone des medialen Kernes des oberen Olivenkomplexes bei der mongolischen Wüstenrennmaus. Unpublished doctoral dissertation, Ludwig-Maximilians-Universität, München
3. Spitzer, M.W. and Semple, M.N. (1995) Neurons sensitive to Interaural Phase Disparity in Gerbil Superior Olive: Diverse Monaural and Temporal Response Properties. J Neurophysiol 73:1668-1690
4. Nordeen, K.W., Killackey, H.P. and Kitzes, L.M. (1983) Ascending Auditory Projections to the Inferior Colliculus in the Adult Gerbil, *Meriones unguiculatus.* J Comp Neurol 214:131-143
5. Pfeiffer, R.R. (1966) Anteroventral cochlear nucleus: Waveforms of extracellularly recorded spikes. Science 154:667-668
6. Wenthold, R.J., Parakal, M.H., Oberdorfer, M.D. and Altschuler, R.A. (1988) Glycine receptor immunoreactivity in the ventral cochlear nucleus of the guinea pig. J. Comp. Neurol. 276, pp. 423-435
7. Ohlemiller, K.K., Echteler, S.M. and Siegel, J.H. (1991) Factors that influence rate-versus-intensity relations in single cochlear nerve fibers of the gerbil. J. Acoust. Soc. Am. 90 (1), pp. 274-287

TEMPORAL PROPERTIES OF DPOAE AND SFOAE

M. BARTSCH, M. MAUERMANN*, S.UPPENKAMP° AND B. KOLLMEIER

Universität Oldenburg, AG-Medizinische Physik,
26111 Oldenburg, Germany

**E-mail: mani@medi.physik.uni-oldenburg.de*

°CNBH, Physiology Department, University of Cambridge,
Downing Street, Cambridge, CB2 3EG, England

1 Introduction

Psychophysical experiments give evidence for a fast acting compression in the auditory system [2], but it is still questioned whether this is mostly a result of the cochlear preprocessing or of additional neuronal compression. Distortion product otoacoustic emissions (DPOAE) are a direct result of the nonlinear preprocessing of sound in the cochlea. The onset duration of DPOAE might therefore serve as a direct quantitative indicator for the time constant of the compression caused by the cochlear nonlinearity. An experimental paradigm described by Brass and Kemp [1] was used during this study to record both, DPOAE and the stimulus frequency emission (SFOAE) for the f_2 primary. The onset and temporal development of DPOAE and SFOAE have been analysed. The temporal growth and decay of DPOAE and SFOAE are compared to determine whether SFOAE should be interpreted as "zero-order" distortion products or rather as a reflection emissions, similar to TEOAE at low levels. To get further information about possible differences in the temporal properties of DPOAE and SFOAE, responses to modulations of the f_1 primary have been analysed.

2 Methods

An insert ear probe ER-10C is used for the recording of OAE. The digital signals are D/A converted using 16 bit converters on a signal processing board (Ariel DSP-32C) in a PC. The microphone output is connected to a low-noise amplifier SR560 and then converted to digital using the 16 bit A/D converters on the Ariel board. The DSP is used also for online analysis and signal-conditioning of the residual signal.

Experiment 1 - DPOAE build up time: DPOAE at $2f_1$-f_2 were recorded in two subjects. The frequency f_2 served as continuous stimulus tone, tone pulses at different frequencies f_1 with 32 ms plateaux and varying ramps (2, 4, 8, 16 ms length) as probe tones. In addition, responses to tone pulses without additional presentation of the stimulus tone were recorded.

The temporal envelopes of the $2f_1$-f_2 frequency in the residual signal and of the f_1 tone pulse were calculated using a least squares fit method [4]. The amplitudes of these frequency components were computed for overlapping frames of 1.5 ms length. The resulting envelopes were normalised to their first maximum within 30 ms (compare Figure 1). This allows a qualitative comparison of the 'relative' onset slopes of the f_1 tone pulse and the $2f_1$-f_2 component.

Experiment 2 - Influence of f1 pulse length: DPOAE were measured and analysed as in Experiment 1, but f_1 tone pulses with fixed ramps of 2 ms duration and different length of the plateaux (0, 2, 4, 8, 16, 32, 64 ms) were used. The maximum levels within the first 20 ms were compared (Figure 2).

Experiment 3 - Temporal properties of DPOAE vs SFOAE: In addition to the results from Experiment 2, SFOAE were recorded at the f_2 frequency (stimulus and probe tone at the same frequency). The latencies and the onset/offset slopes of the DPOAE and SFOAE envelopes were compared (Figure 3).

3 Results and Discussion

Experiment 1: For an onset duration of 1 ms for f_1 the relative onset slope of the DPOAE is shallower than for the f_1 pulse measured in the ear canal, i.e. that the DPOAE need more time to build up than 1 ms. The relative slopes of f_1 and DPOAE onset become more and more similar with increasing duration of the f_1 onset ramps up to 4 ms ramps, for which the relative slopes of f_1 tone pulse and DPOAE are almost identical. For onset ramps longer than 4 ms the DPOAE increases even faster than the primary tonepulse. Therefore 4 ms might be regarded as the minimum interval required for the nonlinear compression to be built up.

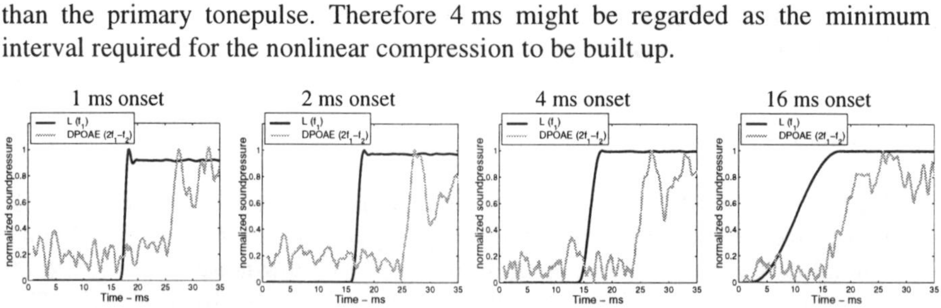

Figure 1. The influence on DPOAE onset for different durations of the hanning shaped onset of the f_1 tone pulse (f_1 = 1282.1 Hz, f_2 = 1558.8 Hz; results for 1, 2, 4 and 16 ms onset duration.). Black line: Normalised envelope of the f_1 tone pulse in the ear canal. Grey line: Normalised envelope of the $2f_1$-f_2 DPOAE. The relative slopes of DPOAE and f_1 pulse envelopes can directly be compared. Subject MM right ear.

Experiment 2: The signal amplitude estimated by the LSF-method is influenced by the length of the analysis frame, especially if parts of the analysis frame cover the onset part of the pulse. If the DP envelope would directly follow the f_1 pulse envelope, the maximum DP level should be reached when the whole analysis frame lays on the plateau of the envelope. This is possible for the 2 ms plateau pulse, but for f_1 plateau length up to 4 ms the maximum level is still inreasing followed by a decrease in level for a 8 ms plateau for both subjects observed in this study.
In good agreement with the results from Experiment 1 a time constant of 4 ms for the DPOAE built up time is found in this experiment as well.

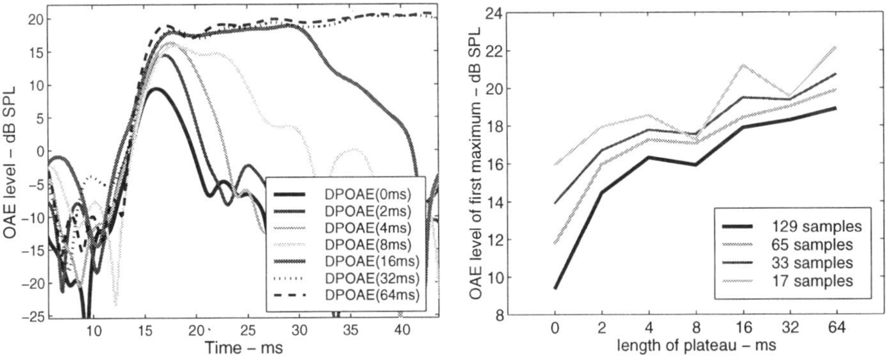

Figure 2. Maximum DPOAE level in dependence on the f_1 probe tone duration. Left: DPOAE envelopes for different plateaux lengths of probe tone f_1 (see legend), computed with the LSF algorithm for a frame length of 129 samples. Right: Maximums of the DPOAE envelopes within the first 20 ms in dependence of the plateaux duration, computed with the LSF algorithm for different frame lengths (see legend). The frequencies and subject are the same as in Figure 1.

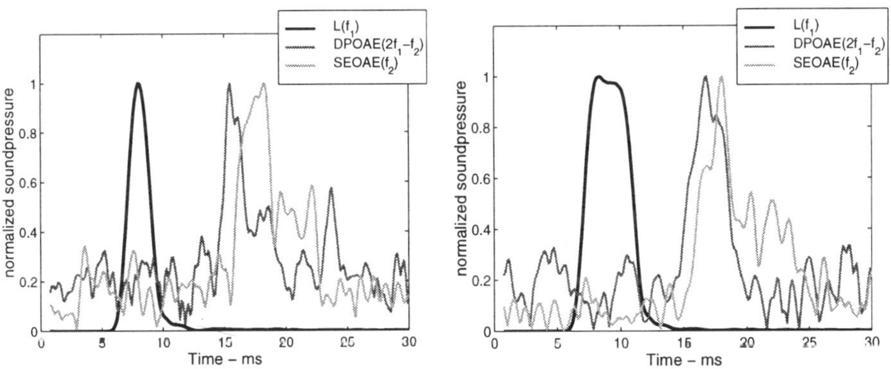

Figure 3. Comparison of the envelopes of the f_1 probe tone in the ear canal, DPOAE and the SFOAE (at the f_2 frequency). Envelopes of the probe tone pulses at f_2 and at f_2 look almost identically in the ear canal. Thus only the envelope of the f_1 probe tone is shown. Subject MM right ear.

Experiment 3: The latencies of the SFOAE are longer than the latencies of the DPOAE, although for this signal configuration the roundtrip travel time in the cochlea is slightly shorter for the SFOAE than for the DPOAE. In addition the onset of the SFOAE is less steep than that one for the DPOAE (Figure 3). SFOAE need more time to build up. This suggests to regard SFOAE more as a reflection emission similar to TEOAE at low levels than as a zero order distortion product.

4 Conclusions

The data presented so far allow the following conlusions:

- The compressive nonlinearity in the cochlea acts faster than 4 ms.
- The generation mechanisms for SFOAE and DPOAE are different.

Acknowledgements

Supported by Deutsche Forschungsgemeinschaft, DFG Ko 942/11-2.
Thanks to G. Long and C. Talmadge for helpful hints and many useful discussions.

References

1. Brass, D. and D. Kemp (1991), *J Acoust Soc Am* **90** pp. 2415–2427.
2. Carlyon, R. P. and A. J. Datta (1997), *J. Acoust. Soc. Am.* **101** (1997) pp. 3636–3647.
3. Heitmann, J., Waldmann, B., Schnitzler, H. U., et al. (1998), *J. Acoust. Soc. Am.* **103**: 1527-1531.
4. Long, G. and C. Talmadge (1997), *J Acoust Soc Am* **102** (1997) pp. 2831–2848.

PRE-ATTENTIVE AUDITORY PROCESSING IS REVEALED BY ERPS

E. SCHRÖGER

Institut für Allgemeine Psychologie, Univ. Leipzig, Seeburgstr. 14, 04103 Leipzig, Germany
E-mail: schroger@rz.uni-leipzig.de

The recording of event-related brain potentials (ERPs) is a useful tool for studying central auditory processing in humans since it can be utilized in basic and clinical research to study the brain´s pre-attentive auditory functioning without contamination by high-level, task-related processing- For example, the mismatch negativity (MMN) component of the ERP can tap mechanisms of involuntary attention switching underlying the conscious registration of sound change and features of passive auditory sensory memory creating the informational basis of active memory. This pre-attentive brain response is believed to be elicited by the outcome of a comparison process that registers a difference between the neural representation of the actual input and the memory trace of the invariances inherent to the recent stimulation.

1 Introduction: Benefits of ERPs

During the last decade the recording of scalp-recorded event-related potentials (ERPs) has became an important tool in cognitive neuroscience to illuminate the processing underlying human cognition with its manifold of functions serving perception, action, memory, and emotion [9]. ERPs are characterized by a unique bundle of features combining the advantages from modern brain image techniques with those from traditional behavioral measurement techniques:

- ERPs directly reflect neural activity (most likely postsynaptic potentials), that is, this kind of activity that is believed to be the main constituent of our information processing system. Many other brain imaging methods such as positron emission tomography (PET), single photon emission computed tomography (SPECT), functional magnetic resonance imaging (fMRI), or event-related optical signaling (EROS) do not directly tap neural activity but changes in metabolic activity following neural activity.
- The recording of ERPs provides an on-line measure of neural processing that allows to track the time course of a variety of transient, short-lived processes. Since ERPs can be measured at submillisecond level, its temporal resolution is superior to most modern brain imaging techniques. Its temporal resolution is, of course, also superior to behavioral measures (such as response time or accuracy of the response) which usually give information after the process of interest (e.g. tone-pitch discrimination) has been terminated.
- Although the spatial resolution of ERPs is markedly lower than that of many other brain imaging techniques, source analysis methods based on multi-channel EEG recordings (such as BESA, LORETA, or SCD) may yield estimates for the locations of the electrical activity generating particular ERP components being

sufficient for many purposes. Especially the ability to reveal particular components as being generated in different regions of the brain may be helpful in many circumstances. It should be noted that differences of the location of generators of ERP components cannot only be shown via dipole analysis (assuming models with a variety of constraints) but also via the comparison of the distribution of potential fields or via a test for additivity of ERP components recorded in different experimental conditions (cf. Fig. 2). It should also be noted (a) that in the case of transient neural processes methods revealing a higher spatial resolution than ERPs cannot easily be taken to localize neural activity and (b) that the inference from the area of metabolic activity (as, e.g., determined with fMRI) to the area of the neural process of interest is not trivial (e.g. in the case of neurons with distant projections [5]).

- The recording of ERPs is in contrast to PET and SPECT noninvasive and its costs are compared with PET, fMRI, or MEG lower by a factor of about 50.
- The recording of ERPs is (in contrast to most other brain imaging methods) not confined to blocked designs. The flexibility in the experimental design that can be achieved is superior even to those that can be achieved by "event-related" fMRI [4] and almost approaches that of designs being utilized in behavioral studies. However, since the signal-to-noise ratio of ERPs mostly is worse than with behavioral studies, experimental sessions have to be longer.
- Most importantly, the recording of ERPs does (similar to most other brain imaging techniques and opposed to behavioral measurement techniques) not require a behavioral response by the participant. Thus it is possible to record brain responses even when the participant is not able to behaviorally respond (e.g. with infants or comatose patients) or when the stimulation is task-irrelevant. Thus the recording of ERPs represents a unique tool to study processing of task-irrelevant information (i.e. pre-attentive processing).

2 ERP Components

In most circumstances ERPs are not recorded as a means per se but a particular component of the ERP is measured as indicator for its underlying neural process. Thus, particular ERP components may serve as a tool to study particular functions of our auditory system. However, these tools have to be developed carefully, including the optimization for recording, measuring, analyzing, and interpreting the component in question. Before giving an example of such an attempt, it should be clarified what an ERP component is: It is not necessarily identical to a particular positive or negative deflection of the ERP but vice versa the ERP might be thought to be constituted by a manifold of components that may follow each other serially in time and that may also overlap in time resulting in an superimposition of the electrical activity of its neural generators (thus increasing or decreasing the ERP depending on the orientation of the electrical source). An ERP component may be defined via (a) its morphology, (b) its distribution, and, most important, (c) its specificity to experimental manipulations [3]. There exists a variety of ERP components that may be classified according to different criteria. One possible classification is according to whether the elicitation of the component requires allocation of attention to the stimulation or not. A component requiring that the subject attends to the stimulation may be termed attention-dependent or attentive and a component that is elicited even when the subject does not attend to the stimulation attention-independent or pre-attentive (which does, of course, not rule out that such a component may be modulated by attention). Within the realm of auditory processing, there exist several ERP components that fall within the latter category (e.g. P1, N1, P2, and mismatch negativity (MMN)). All those components may be recorded even when the subject ignores the auditory stimulation and may therefore utilized to study pre-attentive processing. Among those pre-attentive components, the MMN is the one reflecting the most "cognitive" processing since it involves representations of sensory memory (see below).

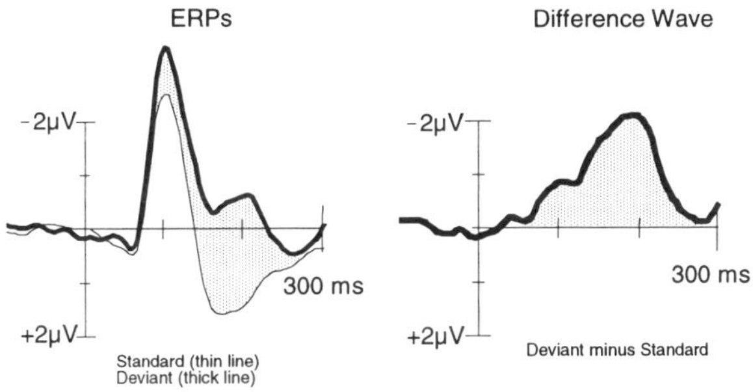

Figure 1. Left: ERPs elicited by an infrequent deviant (thick) and a frequent standard (thin) sound at frontocentral lead. Right: Deviant minus standard difference wave. The dotted area indicates the MMN.

3 Mismatch Negativity (MMN)

The MMN is usually obtained in passive oddball paradigms where a frequent standard sound and an infrequent deviant sound are randomly delivered to subjects not paying attention to the stimulation [6, 11]. It can best be identified in the deviant minus standard difference waves in which the ERP to the standard is subtracted from the ERP to the deviant (Fig. 1). MMN reveals a frontocentral, often right-hemispheric preponderant distribution and occurs between 100 and 300 ms relative to the onset of the deviation. When nose is used as reference, it usually inverts polarity at leads positioned below the Sylvian fissure. Scalp-recorded electroencephalographic studies and magnetoencephalographic studies in humans, intracranial electrical recordings in animals and humans [1] and a recent fMRI study [7] revealed that MMN is generated within or in the vicinity of the primary auditory cortex. Source-current density analyses gave rise to an additional generator located in the right-frontal cortex [2].

This pre-attentive brain response most likely is generated by the outcome of a comparison process that registers a difference between the neural representation of the actual input and the sensory memory trace of the invariances inherent to the recent standard stimulation. More specifically, the elicitation of MMN involves a series of processing stages: It assumes (a) that features of the acoustic input are extracted and encoded into a short-lived representation R, (b) that invariants inherent in the recent stimulation are encoded into a representation R' of sensory memory, and (c) that a set of processes is active which may metaphorically be described as a comparison between the actually generated representation R and the memory trace R'. If a discrepancy is detected between R and R' by this comparison, a mismatch signal is generated that is proportional to the amount of the difference between the two representations. There exists evidence supporting the hypothesis that R is part of the so-called short auditory store lasting about 100-200 ms, whereas R' is part of the long auditory store maintaining information for about 10 s. It may be speculated that R is a read-out of information already computed at brainstem level by more central levels [8] and that R' is computed in cortical areas since it includes integration of information over a period of several seconds. Provided that the MMN exceeds a variable threshold the irregularity in stimulation may invoke subsequent processes: (A) If subjects intend to detect these irregularities and, therefore, listen attentively to the sounds, the threshold is set low enabling weak MMN signals to surpass the threshold and the subsequent N2b and P3b components (indicating the consciousness registration of the irregularity) are elicited; (B) If the subject does not pay attention to the sounds, the threshold is set high and only strong MMN signals may surpass the threshold giving rise to an involuntary attention switch towards the perturbating sound (indicated by the P3a component of the ERP). Moreover, a series of studies delivered evidence suggesting that the sensory memory traces underlying the elicitation of MMN form the basis of the traces underlying the conscious discrimination of sounds.

The MMN signal itself has been interpreted as either indicating a call for attention [10] or as the updating of R' when encountering a deviant sound [12] both assuming the involvement of a memory comparison mechanism. However, enhancements in the electric brain response to an infrequent deviant sound compared with the response to a frequent standard sound may be explained without recurrence to a memory comparison mechanism: It may simply be explained as the result of a differential state of refractoriness of afferent elements specifically responding to the features of the deviant and those specific to features of the standard. For particular cases like MMN elicited by infrequent intensity decrements or MMN elicited by irregular feature conjunctions this new-afferent-elements-hypothesis seems to be rather unlikely (except one believes in an ampliotopic organization of auditory cortex and in an almost infinite set of highly specialized neurons). However, by controlling for the state of refractoriness via appropriate control conditions it has been shown that also changes in simple features (such as location) are registered via memory comparison mechanism [11].

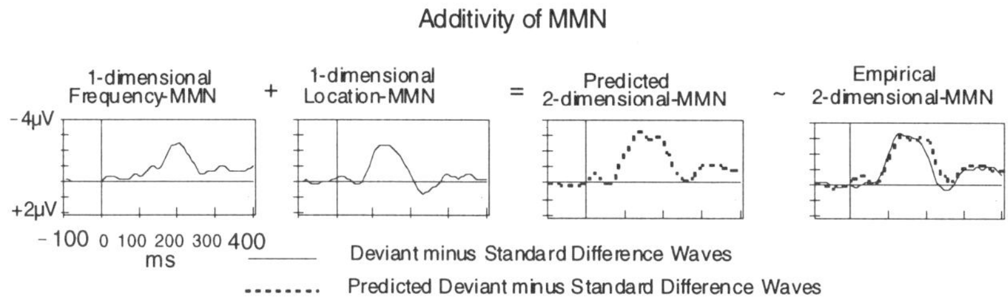

Figure 2. The MMN to a 2-dimensional change in frequency and location is highly similar to the sum of the corresponding 1-dimensional deviants (data measured at Cz). This additivity of MMN demonstrates that pre-attentive change detection may operate in parallel for different stimulus dimensions and that the MMN for changes in frequency and in location are generated in different areas of the brain.

4 Applications of MMN

Since the nature of MMN has been unraveled to a large degree, the MMN may be taken as a tool to study the brain´s pre-attentive processing of sounds. For example, the functioning of pre-attentive change detection and sensory memory have been illuminated in a large number of studies. With respect to pre-attentive change detection the additivity of the sum of the MMNs elicited by two 1-dimensional deviants (e.g. frequency and location) relative to the MMN elicited by the 2-dimensional deviant (e.g. a simultaneous change in frequency and location) suggests that several features of a sound can be mismatched in parallel (Fig. 2). With respect to MMN research on sensory memory, a variety of studies determining the temporal dynamics and the contents of auditory sensory memory have been performed. It has been shown that the contents of R' may be simple stimulus features such as the pitch or the loudness of a sinusoidal tone (1st- order feature) but also the combination of features such as, for example, the conjunction of a particular pitch and a particular intensity (2nd-order feature). The processes creating R' do even reveal some anticipatory power, so that extrapolated events which have not been present in the previous stimulation may be represented in R' (3rd-order feature). In a series of studies it has been shown that the memory traces underlying MMN may contain the spectrotemporal information of tonal patterns and even (to some degree) information about the time course of duration and intensity information in such patterns.

One promising future realm for utilizing MMN may be a kind of "inner" psychophysics relating the physical aspects of the acoustic input to the brain´s automatic central processing (in analogy to the traditional "outer" psychophysics relating physical aspects to behavior). It would be, for example, interesting to trace back classical psychophysical findings (such as loudness summation) to pre-attentively operating cortical processes. Another promising realm is research on the formation of auditory objects including the illumination of two basic mechanisms involved, namely, streaming and grouping.

5 Acknowledgements

Supported by the DFG (Schr 375/7-1).

References

1. Alho, K. (1995). Cerebral generators of electrical and magnetic mismatch responses to changes in sounds. *Ear Hear.*, *16*, 38-51.
2. Deouell, L. et al. (1998). Mismatch negativity in dichotic listening: Evidence for inter-hemispheric differences and multiple generators. *Psychophysiol.*, *35*, 355-365.
3. Donchin, E. et al. (1978). Cognitive psychophysiology: In E. Callaway, P. et al. (Eds.), *Event-related brain potentials in man* (pp. 349-441). New York: Academic Press.
4. Friston, K. J. (1997). Imaging cognitive anatomy. *Trends Cognit. Sci.*, *1*, 21-27.
5. Kutas, M. and Federmeier, K. D. (1998). Minding the body. *Psychophysiol.*, *35*, 135-150.
6. Näätänen, R. (1992). *Attention and brain function.* Hillsdale, NJ: Erlbaum.
7. Opitz, B. et al.. (in press). Combining electrophysiological and hemodynamic measures of auditory oddball. *Psychophysiol.*.
8. Rübsamen, R. Dynamic of neural excitation and inhibition in the auditory brainstem. Paper presented at the *Symposium "Temporal processing in the auditory system"*, Bad Zwischenahn, Germany, September 1998.
9. Rugg, M. and Coles, M. G. N. (Eds.). (1995). *Electrophysiology of mind.* Oxford: Oxford University Press.
10. Schröger, E. (1996). A neural mechanism for involuntary attention shifts to changes in auditory stimulation. *J. Cognit. Neurosci.*, *8*, 527-539.
11. Schröger, E. (1998). Measurement and interpretation of the Mismatch Negativity (MMN). *Beh. Res. Meth., Instr., & Computers*, *30*, 131-145.
12. Winkler, I. et al. (1996). Adaptive modeling of the unattended acoustic environment reflected in the mismatch negativity event-related potential. *Brain Res.*, *742*, 239-252.

For extensive reference lists on particular issues of this paper contact schroger@rz.uni-leipzig.de.

ERP INDICATIONS FOR SUSTAINED AND TRANSIENT AUDITORY SPATIAL ATTENTION WITH DIFFERENT LATERALIZATION CUES

ANDREAS WIDMANN AND ERICH SCHRÖGER

Universität Leipzig, Institut für Allgemeine Psychologie,
Seeburgstr. 14-20, 04103 Leipzig,
E-mail: widmann@rz.uni-leipzig.de

The presented study was designed to investigate ERP effects of auditory spatial attention in sustained attention condition (where the to-be-attended location is defined in a blockwise manner) and in a transient attention condition (where the to-be-attended location is defined in a trial-by-trial manner). Lateralization in the azimuth plane was manipulated (a) via monaural presentation of left- and right-ear sounds, (b) via interaural intensity differences, (c) via interaural time differences, (d) via an artificial-head recording, and (e) via free-field stimulation. Ten participants were delivered with frequent Nogo- and infrequent Go-Stimuli. In one half of the experiment participants were instructed to press a button if they detected a Go-stimulus at a predefined side (sustained attention), in the other half they were required to detect Go-stimuli following an arrow-cue at the cued side (transient attention). Results revealed negative differences (Nd) between ERPs elicited by to-be-attended and to-be-ignored sounds in all conditions. These Nd-effects were larger for the sustained than for the transient attention condition indicating that attentional selection according to spatial criteria is improved when subjects can focus to one and the same location for a series of stimuli.

1 Introduction

Spatial attention is a well explored phenomenon in the visual domain: Visual spatial attention can be manipulated towards a certain spatial location for example by symbolic arrow cues in combination with a probability bias towards the cued location like in Posner's trial-by-trial-cueing paradigm [1]. This leads to shorter reaction times (RTs) to targets at attended (validly cued) compared to unattended (invalidly cued) locations. This RT difference is due to differential processing consisting of costs for unattended as well as benefits for attended locations. In the event-related brain potential validly cued visual targets elicit an enhanced negativity [2]. This effect is called negative difference or Nd. Also in the auditory domain effects of spatial attention can be shown both on behavioral level [3,4] and on electrophysiological level in dichotic listening [5] and trial-by-trial cuing tasks [6].

The presented study was designed to investigate Nd effects of auditory spatial attention in a <u>sustained</u> attention condition (where the to-be-attended location is defined in a blockwise manner, comparable to a dichotic listening task) against a <u>transient</u> attention condition (where the to-be-attended location is defined in a trialwise manner, comparable to a trial-by-trial cuing task) for different <u>lateralization</u> cues. Lateralization in the azimuth plane was manipulated (a) via monaural presentation of left- and right-ear sounds, (b) via interaural intensity difference, (c) via interaural time difference, (d) via an artificial-head recording, and (e) via free-field stimulation.

2 Methods

16 paid volunteers participated in the experiment. Five of them had to be excluded due to poor eye fixation. Thus 11 participants (7 females) with a mean age of 29.5 years remained in the sample. They were delivered with bursts of white noise with a duration of 100 msec (5 msec rise-and-fall time). There were frequent nogo- (10%, 0.1-10 kHz bandpass filtered) and infrequent go-stimuli (90%, 1-10 kHz bandpass filtered). In one half of the experiment the stimuli followed an ambiguous arrow-cue (duration 100 msec, SOA 600 msec). Participants were instructed to press a button if they detected a go-stimulus at a predefined side (sustained attention). In the other half the arrow-cue pointed to the left or to the right and participants were required to detect go-stimuli at the cued side (transient attention). Lateralization was manipulated either by monaural presentation (Mono), by interaural intensity difference (6 dB, IID), by interaural time difference (0.3 msec, ITD), by the artificial-head-recording of a free-field presentation (Rec), or by the free-field presentation itself (Free). In the Mono, IID, ITD, and Rec conditions auditory stimuli were presented via headphones, in the freefield condition stimuli were presented via two loudspeakers placed at eye-level at a distance of about 2 m plus and minus 45 degrees to the left and right, respectively. The EEG was recorded from the standard electrode locations of the 10-20 system, and the left and the right mastoids. EOG was recorded from the outer canthus and above the left eye. All electrodes were referred to the tip of the nose.

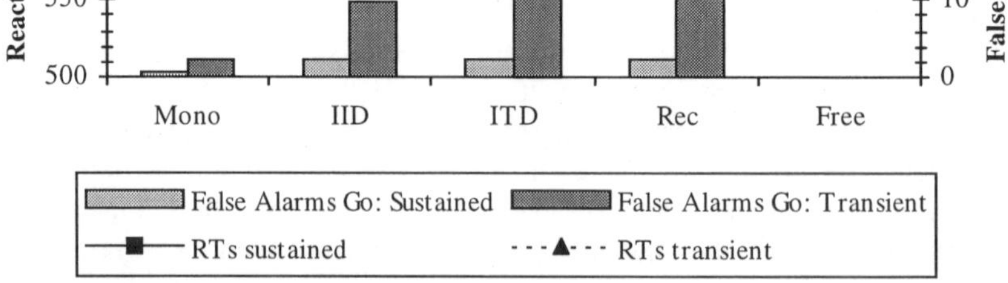

Figure 1. Reaction times (msec) and false alarm rates to go-stimuli (%) separate for sustained and transient attention condition

3 Results

The mean reaction time was 611 msec with a hit rate of 93.9%. The reaction times were longer (a) for the transient attention condition (590 vs. 631 msec) and (b) for the interaural time difference lateralization cue (see Figure 1), but both effects did not reach the level of statistical significance. The false-alarm rate to invalid go-stimuli (go-stimuli coming not from the predefined or cued side) is significantly higher in the transient than in the sustained attention condition (1.5% vs. 7.8%, F(1,10)=8.88, p=.014).

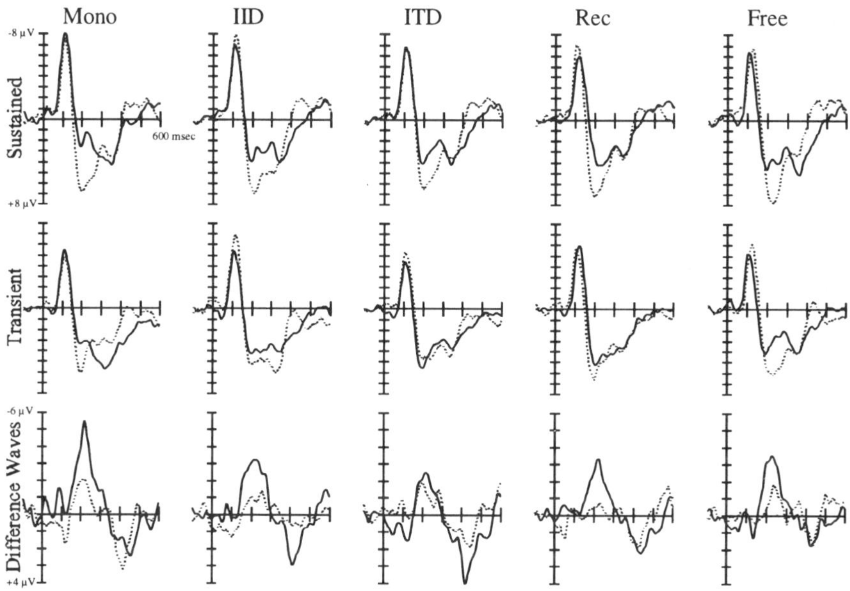

Figure 2. Event related potentials (µV) in sustained (row 1) and transient attention condition (row 2) to attended (solid lines) and unattended stimuli (dotted lines) and difference waves (row 3, sustained attention solid, transient attention dotted lines) for the different lateralization cue conditions (columns) at electrode location CZ

ERP-results revealed Nds between ERPs elicited by valid (to-be-attended) and invalid (to-be-ignored) sounds in all conditions peaking at central electrode locations (see Figure 2). The mean amplitudes of the Nds (time window 160 to 280 msec relative to stimulus onset) are significantly attenuated in the transient attention condition (Cz: -2.5 vs. -0.8 µV, F(1,10)=5.13, p=.047). No statistically significant difference between the Nds elicited by the different lateralization conditions was found, although visual inspection of the ERP data shows a slightly enlarged Nd in the monaural and the free-field conditions.

4 Discussion

Since Nds were elicited in all conditions the ERP-results show that all of the presented lateralization cues can be used as criteria for the attentional selection by the auditory spatial information processing system. However, the enhancement of the Nd-effects for the sustained against the transient attention condition is indicating that attentional selection according to spatial criteria is improved when subjects can focus to one and the same location for a series of stimuli. This is also suggested by the lower false alarm rate to go stimuli in sustained relative to transient attention.

Moreover, the results demonstrate that interaural time and intensity differences can be used either separately or in combination as a tool for the investigation of the auditory attentional system, although a stronger lateralization and a more natural setting (like in freefield presentations) may yield larger effects. In respect to the absence of a statistically significant difference between the lateralization cue conditions further research is needed because this null-result might be due to low statistical power.

References

1. Posner, M.I., Snyder, C.R.R., & Davidson, B.R., Attention and the detection of signals. *Journal of Experimental Psychology: General* **109** (1980) pp. 160-174.
2. Eimer, M., "Sensory gating" as a mechanism for visuospatial orienting: Electrophysiological evidence from trial-by-trial cuing experiments. *Perception & Psychophysics* **55** (1994) pp. 667-675.
3. Schröger, E., Event-related potentials to auditory stimuli following transient shifts of spatial attention in a Go/Nogo task. *Biological Psychology* **36** (1993) pp. 183-207.
4. Spence, C.J., & Driver, J., Covert Spatial Orienting in Audition: Exogenous and Endogenous Mechanisms. *Journal of Experimental Psychology: Human Perception and Performance* **20** (1994) pp. 555-574.
5. Näätänen, R., *Attention and brain function* (Erlbaum, Hillsdale NJ, 1992).
6. Schröger, E., & Eimer, M., Covert spatial orienting in audition. 'Cost-Benefit' analyses of reaction times and event-related potentials. *The Quarterly Journal of Experimental Psychology* **50A** (1997) pp. 457-474.

DIPOLE SOURCE ANALYSIS OF BINAURAL INTERACTION IN THE BRAIN-STEM

H. RIEDEL, M. GRANZOW, AND B. KOLLMEIER
AG Medizinische Physik, Carl von Ossietzky Universität, D-26111 Oldenburg, Germany

1 Introduction

The earliest point along the auditory pathway where binaural processing can take place is the superior olivary complex because it is the first nucleus to receive input from both ears [1]. In order to study binaural processing at this early stage in humans we measured the potential distribution at the surface of the head when subjects were hearing clicks with directional information. The time interval of interest here is 0–10 ms after stimulus onset; the curves obtained by such measurements are known as the auditory brain-stem response (ABR). It is not feasible, however, to use the EEG itself to describe the processing. Rather, one has to infer the charge (or current) distribution from the surface potential. This "inverse problem" is known to be ambiguous. Therefore, the charge distribution has to be parametrised. We make the *ansatz* of a dipole. The litmus test of a genuine binaural process is the analysis of the so-called binaural difference curve (BDC) [2]. This quantity is obtained by subtracting the sum of the monaural responses from the binaural response, symbolically: $B - (L + R)$.

2 Methods

The participating subjects were normal hearing and lay in a sound treated and electrically shielded room during ABR recording. 10000 responses to stimuli (clicks of 100 μs duration) were recorded from 32 channels, differentially amplified, AD-converted and stored to hard disk. FIR filtering [3] and averaging with an improved weighted averaging scheme [4] was done off-line. Interaural time differences (ITDs) of -0.4, zero and +0.4 ms and interaural intensity differences (ILDs) of -10, zero and +10 dB were used. Here, negative ITDs and ILDs correspond to a lateralisation of the stimulus to the left hand side, whereas stimuli which are perceived as coming from the right have positive ITDs and ILDs. ABRs were measured at 60 dB nHL. All combinations of 3 ILDs and 3 ITDs yielded 15 stimulus conditions recorded for all subjects, 9 binaural and 6 monaural, the latter ones are needed to calculate the BDC.

For the source analysis of the data a simple spherical volume conductor was used as head model [5,6]. First a distributed source model was applied to the data. It shows that it is reasonable to assume a single source in the centre of the head model. Therefore a rotating dipole (3 degrees of freedom) resulting in a time dependent moment vector was fitted to the data. A right handed coordinate system with the origin in the centre of the head, $+x$ towards the right ear, $+y$ towards the nose and $+z$ towards the vertex was used.

The trajectory in the frontal plane (x-z-plane) in the time interval between 5 and 7 ms after stimulus onset was was used for further analysis.

3 Results

Fig. 1 visualises the quality of the rotating dipole fit by comparing measured and model data. In the time interval [0,10] ms the mean deviation in rms-value is 17.9 nV, significantly higher than the mean noise level of 9.5 nV.

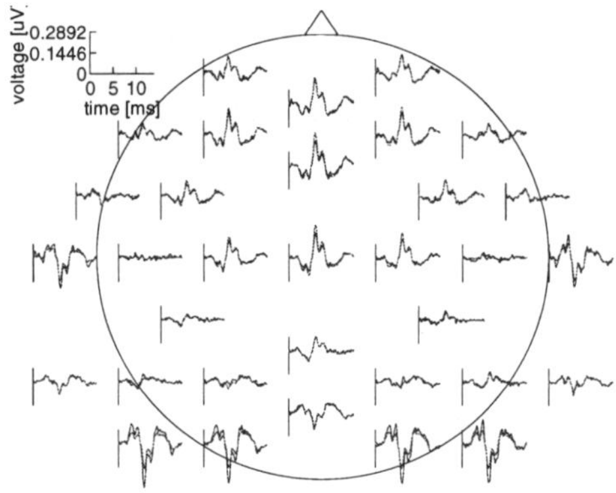

Figure 1. Responses to the diotic stimulus (ILD = 0 dB, ITD = 0 ms) for subject hr. Data are plotted average referenced. Measured data are compared with model data.

When the fitting is done with a moving dipole, i.e. location and moment of the dipole are fitted for every time instance (6 degrees of freedom), the optimal location is found to move in close vicinity of the centre of the head, and the rms-deviation is lowered to 11.8 nV. Another frequently used error measure, the residual variance, amounts to 13.6 % in this example.

To compare the fits of the equivalent dipoles corresponding to different stimulus conditions trajectories of the moments in the frontal (x-z) plane are plotted in Fig. 2. The time interval from 5 to 7 ms after stimulus onset was chosen because the largest components of the binaural interaction are found in this interval.

In the symmetric condition (sub-plot in the middle) the trajectory exhibits almost no horizontal component. In the left and right plot of the middle row the result for an ITD of -0.4 and +0.4 ms, respectively, is plotted. The trajectories are mirror images of each other, both exhibit x-components. The same holds for the trajectories below and above in the middle column, which correspond to ILDs of -10 and +10 dB, respectively.

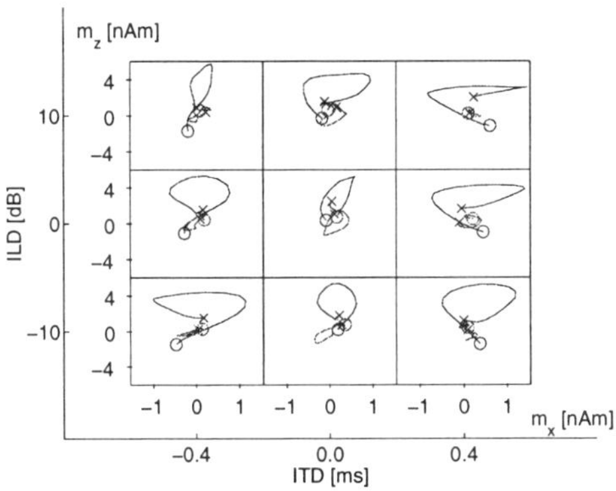

Figure 2. Trajectories of the equivalent dipole in the frontal plane in the time interval from 5 ms (circle) to 7 ms (cross) for the nine different binaural stimuli. In each sub-plot the moment m_z is plotted against m_x with time as the parameter. Because the dwell time is 0.1 ms 21 independent fits were carried out for each sub-plot and connected by lines. The large trajectories correspond to the binaural responses those with smaller amplitude to the BDC.

Moreover, stimuli which psychoacoustically elicit a similar lateralisation also show similar trajectories. An ILD of +10 dB and an ITD of +0.4 ms both gave an impression of about 50 degrees to the right for all subjects. If ITD and ILD are combined synergistically, the trajectories exhibit a more pronounced eccentricity in the horizontal direction, reflecting a larger x-component of the equivalent dipole. In the case of an antagonistic combination of ITD and ILD (so-called time-intensity-trading) this component nearly vanishes (sub-plots on the diagonal of the matrix). Again, this is in agreement with the psychoacoustic finding that this type of stimulus produces a central but somewhat blurred image.

The fits of the BDC are shown as red curves. The small moments of the equivalent dipole corresponds to the low energy of the BDC (typically an order of magnitude smaller than the binaural response). In spite of their minute amplitude, they show the same tendency as the trajectories of the binaural stimulation, at least in so far as the turn is clockwise for stimuli which are perceived as coming from the right side and counter-clockwise for left stimuli.

4 Discussion

As mentioned above, the results of a distributed-source-fit proved the legitimacy of the dipole *ansatz*. We further investigated the properties of the dipole fit by computing the deviation scan.

Fig. 3 shows the residual variance if the location of the dipole is varied in the $z = 0$-plane. The presence of a single global minimum lends further credence to the dipole model.

While differences in the processing of stimuli coming from different directions could clearly be established and the analysis of the binaural component at least qualitatively reveals similar systematics, it remains to be tested whether the differences between the equivalent dipoles for the BDC-data are significant compared to the residual noise.

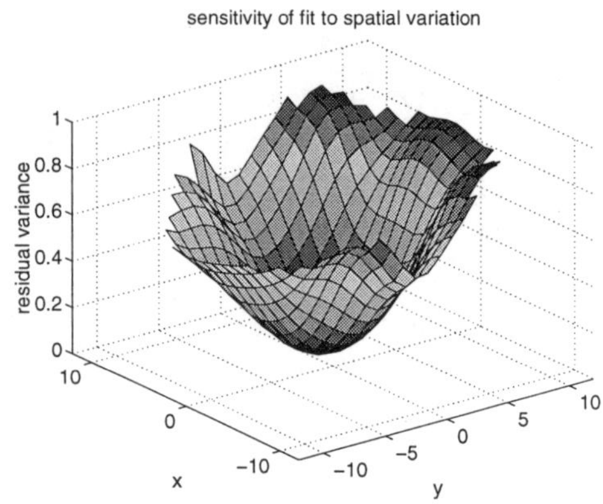

Figure 3. Residual variance of the dipole fit for locations in the horizontal plane corresponding to $z = 0$. x and y are sampled with a spacing of 0.5 cm. Fit to data at a single time point of a binaural response.

Using the three channel Lissajous' trajectory, Polyakov and Pratt [7] found significant differences in the BDC between symmetric and asymmetric stimuli. Differences between ITDs and ILDs, however, were not significant.

References

1. R Nieuwenhuys, J Voogd, and C van Huijzen. *The human central nervous system: a synopsis and atlas.* Springer Verlag, Heidelberg, third edition, 1988.
2. R. A. Dobie. Binaural interaction in auditory brainstem responses: Theoretical and methodological considerations. *J. Acoust. Soc. Am.*, 71(4):1031–1033, April 1982. Letters to the Editors.
3. M Granzow, H Riedel, and B Kollmeier. Single-sweep-based methods to improve the quality of auditory brain stem responses. part I: optimized linear filtering. submitted to J. Acoust. Soc. Am.
4. H Riedel, M Granzow, and B Kollmeier. Single-sweep-based methods to improve the quality of auditory brain stem responses. part II: averaging methods. submitted to J. Acoust. Soc. Am.
5. B Lütkenhöner. *Möglichkeiten und Grenzen der neuromagnetischen Quellenanalyse.* Lit Verlag, Münster, 1992.
6. D. A. Brody, F. H. Terry, and R. E. Ideker. Eccentric dipole in a spherical medium: generalized expression for surface potentials. *IEEE Trans Biomed Eng*, 20(2):141–3, March 1973.
7. A Polyakov and H Pratt. Evidence for spatio-topic organization of binaural processing in the human brainstem. *Hear. Res.*, 94:107–115, 1996.

AMPLITUDE MODULATION PROCESSING IN THE AUDITORY SYSTEM: INVESTIGATIONS USING ENVELOPE FOLLOWING RESPONSES

G. MAUER, W.H. DÖRING

Technical University RWTH Aachen, ENT-Department, Pauwelsstr. 30, D-52057 Aachen, Germany

E-mail: mauer@alpha.imib.rwth-aachen.de

Most of our knowledge about processing of amplitude modulated signals in the auditory system is based on single nerve recordings in animals and psychoacoustical experiments in humans. An objective investigation method in humans is the recording of amplitude modulation following responses (AMFRs). AMFRs are a subset of envelope following responses (EFRs) and can be recorded from the human scalp. In AMFRs continuous amplitude modulated (AM) stimuli with carrier frequencies f_c and modulation frequencies f_m are used. Responses arise at frequency f_m and are largest for modulation frequencies in the range below 100 Hz with maxima depending on the subject's age and state of arousal. Responses are assumed to be evoked by neural excitation originating mainly from a basilar membrane region corresponding to the carrier frequency f_c since stimulus bandwidth is very narrow. However, in masking experiments we could show that also regions of the basilar membrane related to frequencies up to several octaves above the carrier frequency contribute to the responses.

1 Introduction

In the last two decades several papers were published concerning AMFRs. Most of them deal with the application of AMFR recordings as a diagnostic tool. These contributions focus mainly on optimization of stimuli parameters, AMFR-thresholds in normal and hearing impaired subjects, the effects of aging and the influence of state of arousal. In summary, AMFRs seem to contain important information about frequency specific hearing loss even in the low frequency range. In particular, AMFRs have been proved to be recordable in sleeping children, too. On the other hand many questions about the origin of AMFRs are still unanswered. Characteristics of AMFRs depend on the modulation frequency of the stimulus. Generators are assumed to be located in the inferior colliculus and in the auditory cortex, but the published results are contradictory.

In this paper investigations of the frequency specifity of AMFRs are reported and results of masking experiments are presented. High-pass filtered noise with various cut-off frequencies was used to mask AM stimuli with carrier frequencies f_c=240 Hz and f_c=480 Hz and modulation frequency f_m =40 Hz. The magnitude of the AMFRs is evaluated in various masking conditions in order to determine effective stimulation bandwidth, as responses are assumed to originate mainly from a basilar membrane region corresponding to the stimulus carrier frequency f_c because of the very narrow stimulus bandwidth $f_c \pm f_m$.

2 Methods

Four normal hearing subjects (24 - 32 years, controlled by standard pure tone audiometry) participated in the study. They were seated in a reclining chair in a sound treated room. Subjects were instructed to read in order to maintain a constant state of arousal.

Scalp potentials were recorded differentially between vertex and ipsilateral mastoid using silver cup electrodes, contralateral mastoid served as ground. For each parameter set 2000 sweeps of 124 ms were continuously recorded, phase locked to the envelope of the sustaining acoustic stimuli.

Amplitude-modulated sinusoidal tones with carrier frequencies of 240 Hz and 480 Hz, modulation frequency of 40 Hz and modulation depth of 80 percent were presented monaurally via headphone (Sennheiser HDA 200). Stimulus level in this masking study was fixed to 60 dB HL.

Masking noise was compensated for the headphone transfer function by inverse filtering. The resulting filtered noise was weighted to yield uniform masking noise and filtered using ten different digital high-pass FIR filters (2000 taps, cut-off frequencies: 141 up to 1600 Hz in half-octave steps, 1600 Hz to 6400 Hz in one-octave steps). The slopes of the high-pass filtered noise were set to at least 100 dB/oct. for lower and much larger for higher cut-off frequencies. Masking noise was attenuated according to the desired SNR and added to the AM-stimuli.

AMFRs were recorded in three conditions: unmasked, masked with high-pass filtered noise, and masked with uniform masking noise. The level of the noise was adjusted by the following procedure: first in each subject for each carrier frequency the individual level of the uniform masking noise was determined that subjectively was just masking the stimulus. Then the corresponding spectral magnitude level was decreased by 20 dB and fixed for the following set of recordings. Intentionally the responses were not totally suppressed in order to remain observable for low cut-off frequencies, too.

Each set of 2000 recorded sweeps was divided in ten subensembles, averaged, and transformed to the frequency domain. The component synchrony measure (CSM) [2] was calculated to assess statistically the probability that a response exists above noise floor. The CSM value at the expected main response component of 40 Hz (corresponding to f_m) was evaluated. Only AMFRs with CSM values exceeding the 1‰ significance level were regarded for further evaluation.

3 Results

Figure 1 shows the results of the study for frequencies f_c = 240 and f_c = 480 Hz in four subjects. At the horizontal axis the cut-off frequencies of the high-pass filters are plotted that were used to generate the masking noise (0: unfiltered; no: without masking). At the vertical axis the AMFR magnitudes are indicated. Since interindividual response magnitudes can differ strongly they were normalized to the maximum individual value for each carrier frequency. Both, normalized individual values for up to four subjects (symbols) and mean values (lines) are shown. Note that individual results were only accepted, if their CSM-values were above 1‰ significance level.

The shapes of the two plots are similar for both carrier frequencies, except of a horizontal shift according to the carrier frequency used. Response magnitudes were widely reduced when the cut-off frequency of the high-pass filter was lower or in the range up to two octaves higher as the stimulus carrier frequency f_c. The response magnitude rapidly rose when the cut-off frequency was increased further. Magnitudes kept on rising until the unmasked condition was reached. There was a difference in response magnitude even in the case that the masker cut-off frequency was changed from 6400 Hz to the unmasked condition.

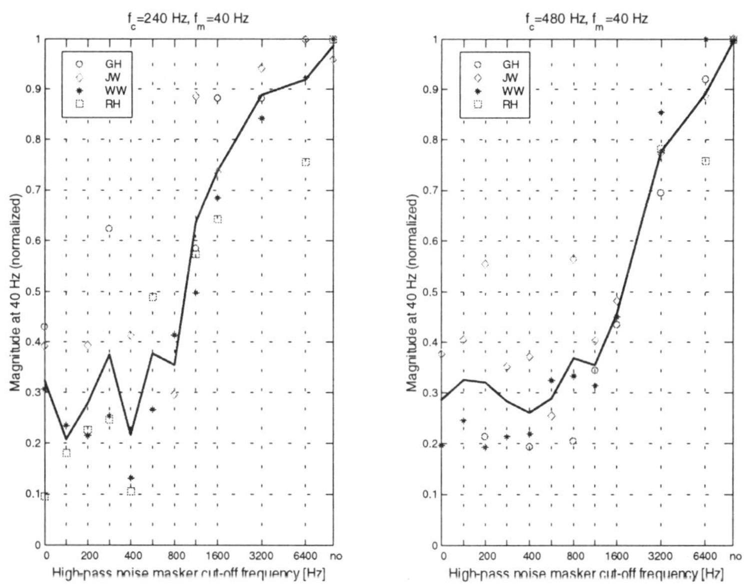

Figure 1. AMFR magnitudes in four subjects as a function of high-pass noise masker cut-off frequency. Carrier frequency of 240 Hz (left) and 480 Hz (right), modulation frequency 40 Hz.

4 Discussion

The spectral bandwidth of the AM-stimuli used in this study was in the range of 80 Hz. Since the bandwidth of transient signals generally used to evoke auditory potentials is much broader, AM-stimuli are assumed to be favorable in order to evoke frequency specific potentials. This idea is supported by studies comparing pure-tone audiograms with AMFR thresholds for different stimulus carrier frequencies in subjects whose hearing impairment is strongly frequency dependent [1]. In this studies there was a good correlation between subjectively and objectively determined thresholds. Another argument for frequency specifity of AMFRs is given in masking experiments published by Griffiths [3] that are quite similar to those in the present study. However, in Griffiths's experiments the cut-off frequencies were varied in a region up to one octave above carrier frequencies only. In 25 out of 36 recordings AMFR magnitudes increased further on for cut-off frequencies higher than carrier frequency. Griffiths explained this effect of non saturating augmentation by the insufficient slope of his high-pass filtered noise maskers. However, in the present study the slopes of the masking noise for the higher cut-off frequencies were much larger and the frequency range was expanded up to four octaves above carrier frequency. Nevertheless, the augmentation of AMFR magnitudes continued for the complete range of cut-off frequency variations.

A reasonable explanation for the results of our study is that regions of the basilar membrane related to frequencies up to several octaves above the AM carrier frequency may contribute to the AMFRs. In examinations concerning the amplitude modulation processing in the auditory system as well as in diagnostics this effect should be taken into consideration. In further studies models of cochlear mechanics shall assist to elucidate stimulus-response characteristics and effective stimulation bandwidth of AMFRs in different masking conditions.

References

1. Aoyagi M., Kiren T., Yoshinori K., Suzuki Y., Fuse T. and KoikeY., Frequency specifity of amplitude-modulation following response detected by phase spectral analysis. Audiology 32 (1993) pp. 293-301
2. Fridman J., Zappulla R., Bergelson M., Greenblatt E., Malis L., Morrell F. and Hoeppner T., Application of phase spectral analysis for brain stem auditory evoked potential detection in normal subjects and patients with posterior fossa tumors. Audiology 23 (1984) pp. 99-113
3. Griffiths S. K. and Chambers R. D., The amplitude modulation-following response as an audiometric tool. Ear Hear 12 (1991) pp. 235-241

ON THE RELATIONSHIP BETWEEN AUDITORY EVOKED POTENTIALS AND PSYCHOPHYSICAL LOUDNESS

OLIVER WEGNER, TORSTEN DAU AND BIRGER KOLLMEIER
Arbeitsgruppe Medizinische Physik, Carl von Ossietzky Universität Oldenburg, Germany

1 Introduction

Wegner et al. (1998) developed optimized chirp stimuli to examine the electrophysiological correlate of loudness summation at brainstem level. They found that loudness does not seem to be directly related to the auditory brainstem response (ABR) amplitude. Nousak & Stapells (1998) examined loudness-growth functions for tone pulses. They compared psychophysical data with amplitude data of ABR wave V and middle latency responses (N_a-P_a). Their results were similar to those obtained by Wegner et al. (1998) suggesting a higher compression as normally derived from psychophysical data and as reflected in common loudness models. Several authors stated that the cortical auditory evoked potentials (CAEP) – namely the N_1-P_2-component – reflects an integration process of neural activity over several milliseconds (e.g., Davis, 1974; Eddins, 1998).
The present study examines whether effects of loudness summation observed in psychoacoustical experiments are reflected in CAEP. In particular, the role of synchronisation and effects of temporal integration at this stage of processing are investigated.

2 Method

The experiments were carried out with a PC based ERA-system (ZLE-Systeme). The digitally generated stimuli (40 kHz, 12 bit) were presented through an insert earphone (Etymotic Research ER-2) to the subjects. Electroencephalic activity was recorded from the scalp via silver/silver chloride electrodes, attached to the vertex (positive) and the ispilateral mastoid (negative). The forehead served as site for the ground electrode. Interelectrode impedances were maintained at < 5 kΩ. Responses were amplified (88 dB) and filtered (analog: 1 - 100 Hz, 6 dB/oct; digitally: 16 Hz lowpass, 2nd order).

2.1 Experiment 1

Psychophysical loudness-matching experiments were performed using a 2-AFC 1-up

1-down procedure with an "interleaved" presentation of the different signal bandwidths. The signals were 100 ms long chirps with linearly changing instantaneous frequency (Neumann et al., 1994). The bandwidth of the chirps was in the range from 3 to 19 bark. The reference stimulus was a chirp with a bandwidth of 9 bark at a level of 58 dB SPL. On average the 3-bark wide chirp was adjusted at a 23 dB higher level than the 19-bark wide chirp to be equal loud.

CAEP elicited by these bandlimited chirps were recorded. They are shown in the left panel of Fig. 1. In addition, corresponding responses were recorded at a constant level of 58 dB SPL at each bandwidth condition, shown in the right panel of Fig. 1.

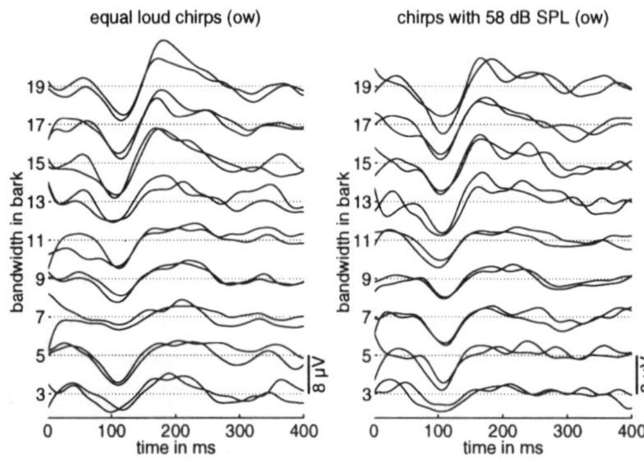

Figure 1. CAEP evoked by bandlimited chirps with constant loudness (left panel) and with constant sound pressure level (right panel), respectively. The chirp duration was 100 ms. The recording time started at stimulus onset. The waveforms are the average of 32 responses for each buffer. The interstimulus interval (ISI) was 5.504 s. Subject: ow.

The N_1-P_2 amplitude of the CAEP increases with increasing chirp bandwidth (left panel). This suggests that an integration of neural activity across frequencies occurs at this level of processing. The results also show that equal loudness does not seem to be reflected in an equally large N_1-P_2 amplitude of the CAEP (right panel). However, only small differences in potential amplitude occur for the different conditions tested.

2.2 *Experiment 2*

This experiment examines the role of synchronization at cortical level. CAEP evoked by single rising and falling chirps were recorded. In this case, exponentially rising chirp stimuli were used which were developed to compensate basilar-membrane dispersion. These chirps have been shown to maximize synchronization at brainstem level (ABR) (Dau et al., 1998). Fig. 2 shows ABR (left panel) and CAEP (middle panel) obtained with single rising chirps (top) and single temporally reversed (falling) chirps (bottom). In addition, CAEP were recorded with a 200 ms long series of these chirps (right panel of fig. 2). The repetition rate was 20 Hz.

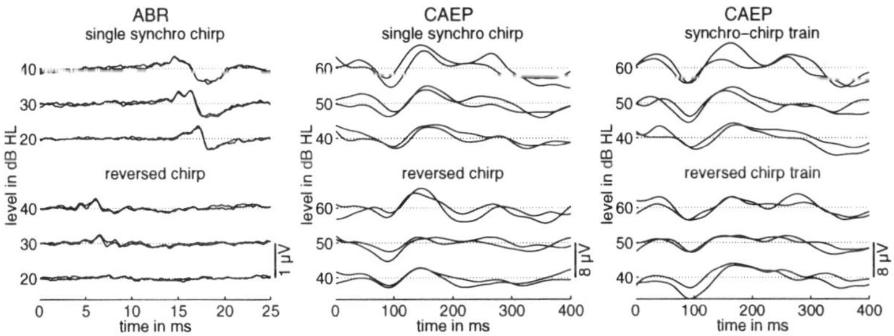

Figure 2. Potentials evoked by the "exponential" broadband chirp stimulus and the time-reversed chirp stimulus. *Left*: ABR for stimulation levels of 20 - 40 dB HL. Waveforms are the average of 2000 responses for each buffer. The ISI was 50 ms. *Middle*: CAEP for stimulation levels of 40 - 60 dB HL. Waveforms are averaged over 32 responses and the ISI was 5504 ms. *Right*: CAEP evoked by a 200-ms series of chirps at a repetition rate of 20 Hz. Stimulus level, averages and ISI as in the middle panel. Subjects: cr/os.

The data show that for the generation of CAEP neural synchronization is much less important than for the generation of ABR: In most conditions, the chirp stimuli evoke the same potential amplitude as the temporally reversed chirp stimuli. The results also show that neural activity is not temporally integrated when a series of multiple short-duration chirps is presented.

2.3 *Experiment 3*

As mentioned in the introduction, there seems to be some evidence that the N_1-P_2-component of the CAEP reflects an integration process of neural activity over several milliseconds (e.g., Davis, 1974; Eddins, 1998). The aim of this experiment was to derive the time constant of this integration process.

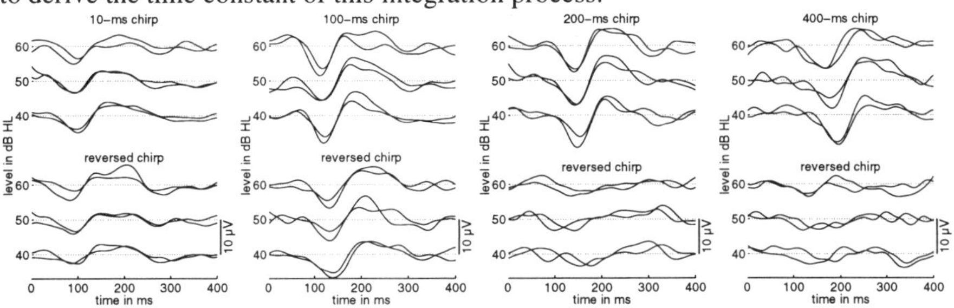

Figure 3. Results of experiment 3. CAEP evoked by rising and falling linear chirps with levels of 40 - 60 dB HL. Four different durations were tested: 10 ms, 100 ms, 200 ms, and 400 ms. Other parameters were the same as in Fig. 3 (right panel).

The stimuli used here were chirps with linearly changing instantaneous frequency as in experiment 1. The chirp bandwidth was 9 kHz. Four different durations were used: 10, 100, 200 and 400 ms. Each stimulus was presented with rising and with falling instantaneous frequency at levels of 40, 50 and 60 dB HL, respectively.

The data are presented in Fig. 3. They show that neural activity appears to be integrated for stimulus durations up to about 100 ms, whereas for longer durations, no further integration seems to occur.

3 Summary and discussion

- An integration of neural activity across frequencies occurs at the level of processing where N_1-P_1 amplitude of the CAEP is generated. However, loudness summation does not seem to be directly reflected in the potential amplitude.
- The fact that only small differences in potential amplitude occur for the different conditions tested in exp. 1 may be due to the relatively high stimulation level at which neural activity possibly is near saturation. Similar experiments should be performed at lower stimulation levels.
- For the generation of CAEP, neural synchronization is much less important than for the generation of ABR.
- Neural activity at cortical level as reflected in CAEP amplitude, appears to be integrated up to about 100 ms, whereas it remains constant for longer durations.
- Further experiments are currently investigated to clarify the role of onsets and offsets for the generation of CAEP and the influence of the spectral shape of the stimulus.

References

Dau, T., O. Wegner, V. Mellert und B. Kollmeier (1998). "Auditory brainstem response (ABR) with optimized chirp signals compensating basilar-membrane dispersion." submitted to J Acoust Soc Am.

Davis, H., Ed. (1974). “Relations of peripheral action potentials and cortical evoked potentials to the magnitude of sensation.” Sensation and measurement. Boston, Reidel Publishing.

Eddins, A. C. (1998). “Spectro-temporal influence on auditory cortical evoked potential thresholds.” 16th International Congress on Acoustics and 135th Meeting of the Acoustical Society of America, Seattle, Washington.

Neumann, J., S. Uppenkamp und B. Kollmeier (1994). "Chirp evoked otoacoustic emissions." Hear Res **79**: 17-25.

Nousak, J. und D. R. Stapells (1998). “Loudness and the ABR/MLR in noise-masked normal-hearing subjects.” Association for Research in Otolaryngology mid-winter meeting, St. Petersburg Beach.

Wegner, O., T. Dau, J. Verhey und B. Kollmeier (1998). “Untersuchungen zu Lautheits-Korrelaten in akustisch evozierten Potentialen.” Fortschritte der Akustik – DAGA 98, Zürich, Deutsche Gesellschaft für Akustik e.V.

ALTERNATIVE DEFINITIONS OF THE MISMATCH NEGATIVITY

J. DAMASCHKE, H. RIEDEL, AND M. GRANZOW
AG Medizinische Physik, Carl von Ossietzky Universität, D-26111 Oldenburg, Germany

The Mismatch Negativity (MMN) is an event related potential component generated by pre-attentive difference-detection/comparison processes in the auditory cortex of the human brain [1] if a deviant stimulus is presented embedded in a sequence of standard stimuli. Traditionally, the MMN is defined as the difference between the response to the deviant stimulus and the response to the standard stimulus (i.e. deviant minus standard) [2]. If the MMN is interpreted as an indication for and a measure of the detection process, this traditional definition may not be adequate because it not only measures the field caused by neurons detecting the difference between the stimuli but also by any difference in the processing of the two stimuli, be it related to an explicit comparison or not. Several alternative definitions of the MMN are proposed and the resulting MMNs are compared.

1 Definitions

The present study compares the three following definitions of the Mismatch Negativity (MMN):

- **Definition**(*i*): Responses to deviant stimuli minus responses to standard stimuli (traditional definition).

- **Definition**(*ii*): Two mismatch experiments are performed, with the role of the standard and the deviant interchanged. Then deviants from one and standards from the other measurement are subtracted. Possible advantage: No physical difference between the stimuli.

- **Definition** (*iii*): Two mismatch experiments are performed. One traditional measurement and one measurement in which the standards are ommitted. Then deviant responses from the second experiment are subtracted from the deviants of the traditional experiment. Possible advantages: All responses are measured in the same state of refractoriness; no physical difference between the stimuli.

2 Methods

Subjects: Three normal hearing subjects (24 to 35 years) took part in the experiment. All of them had previous experience in EEG recordings. They were seated on a reclining chair in a sound attenuated and electrically shielded cabin and asked to avoid body movements and to ignore the auditory stimuli while reading a self-selected text.

Stimuli: There were two different stimuli : A 1002 Hz tone and a 1027 Hz tone, both lasting 50 ms including 5 ms rise and 5 ms fall times. They were presented at a level of 60 dB with an inter-stimulus interval of 640 ms.

Procedure: In the first part of the experiment the 1002 Hz tone was the standard with a probability of about 85 per cent and the 1027 Hz tone was the deviant with a probability of about 15 per cent. Both tones were presented in a pseudo-randomised sequence of 1800 sweeps where at least three standards followed a deviant. In the second part the probabilities were inverted so that the 1027 Hz tone became the standard and the 1002 Hz tone became the deviant. Additionally, one sequence was presented where only the deviants where applied and the standards were omitted.

EEG recording: The EEG was recorded from 30 scalp electrodes positioned according to the international 10-20 system, all referenced to Cz. The recording filters were set to 0-200 Hz. The EEG was digitized by a NeuroScan data acquisition unit at a rate of 1000 Hz. Sweeps were 550 ms in duration. There was no online artifact rejection during the recording.

Analysis: All analyses were made off-line. FIR-Filter of order 100 and an artefact rejection of $\pm$ 25 μV were applied to the data. Each stimulus type was averaged seperately. The evoked responses immediately following a deviant were treated as an additional stimulus type and excluded from further analysis.

3 Results

The following figures show the results for subject A. Results of subject B differ significantly within one definition for subject C even extremly. Therefore it was not possible to compare different definitions. In all figures the MMN amplitude in μV is plotted versus time in ms.

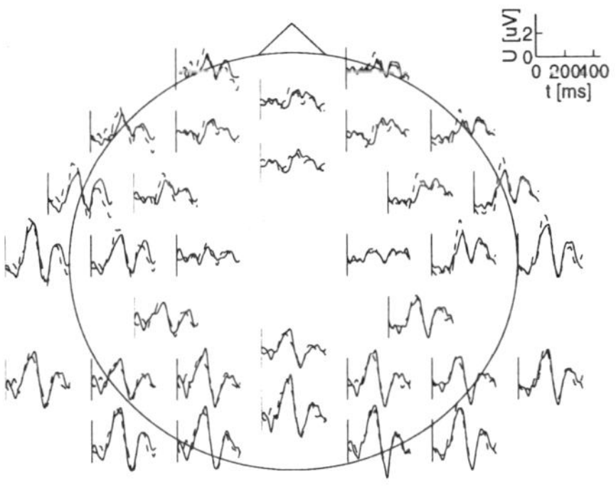

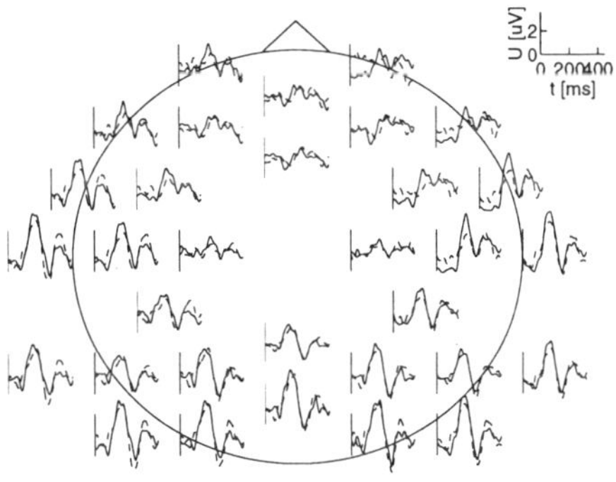

Fig.1 (left): MMNs calculated according to the traditional definition for subject A. The solid curve shows the MMN where the 1002 Hz tone was the standard stimulus and the 1027 Hz tone the deviant stimulus. The dotted curve shows the inverted condition, i.e. the 1002 Hz tone being the deviant stimulus and the 1027 Hz tone being the standard stimulus.
Fig.2 (right): MMNs calculated according to definition (ii) for subject A. The solid curve shows the MMN of the 1002 Hz tones, the dotted curve shows the MMN obtained with the 1027 Hz tones.

As shown in Fig. 1 and Fig. 2 subject A, there is no significant difference between the amplitudes of the MMN for the two different conditions within one definition. When Cz is used as the reference electrode the MMN amplitude is largest at occipital and temporal electrodes.

Fig. 3 compares definition (i) and (ii) for subject A. The solid curve shows the MMN of definition (i) where the standard stimulus was a 1002 Hz tone and a 1027 Hz tone was the deviant. The dotted curve shows the MMN of definition (ii). In this case both stimuli were 1027 Hz tones. As there is no significant difference between the two curves both definitions seem to yield similar results for subject A.
Fig. 4 compares definition (i) and (iii) of the MMN for subject A. The solid curve shows the MMN of definition (i) where the standard stimulus was a 1027 Hz tone and a 1002 Hz tone was the deviant. The dotted curve shows the MMN of definition (iii). In this case both stimuli were 1002 Hz tones.

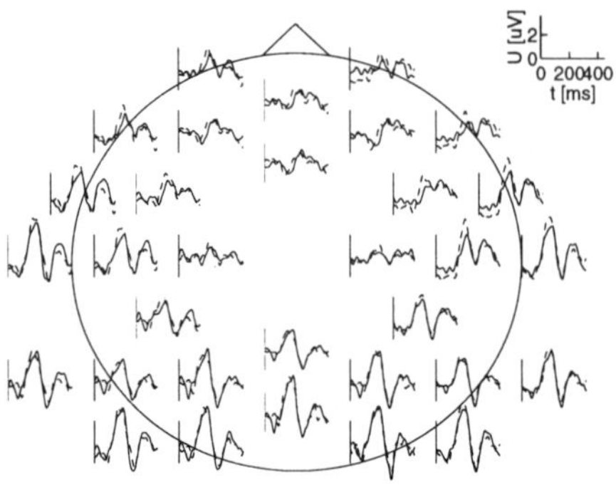

Fig.3: Comparison of definition (i), solid curve, and definition (ii), dashed curve, of the MMN for subject A.

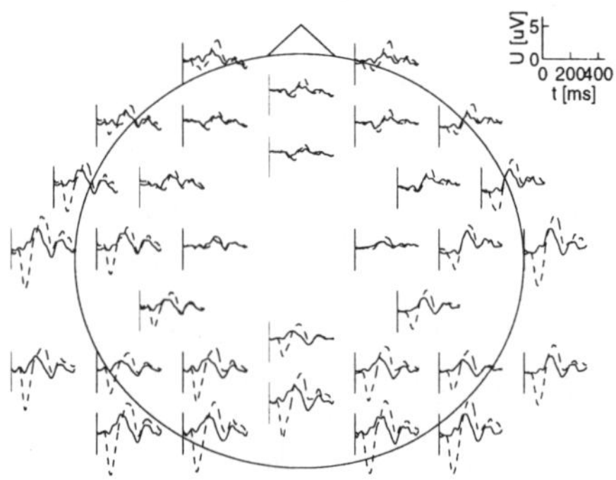

Fig.4: Comparison of definition (i), solid curve, and definition (iii), dashed curve, of the MMN for subject A.

4 Conclusions

- Only one subject showed a reproducible MMN in the traditional definition. For this subject definitions (*i*) and (*ii*) were in agreement.
- The other two subjects showed no reproducible MMN, so further comparison between definitions was not possible.
- Definition (*iii*) was not useful because the "standard" showed a large N100 "contamination".
- The alternative definitions were not found to be superior to the traditional one.

References

1. R Näätänen and K Alho. Generators of electrical and magnetic mismatch responses in humans. *Brain Topography*, 7(4):315–320, 1995.
2. R Näätänen. The mismatch negativity: A powerful tool for cognitive neuroscience. *Ear & Hearing*, 16:6–18, 1995.

II. Psychological Acoustics: Perceptual Processing of Complex Sounds

THE PERCEPTION OF PREPHONEMATIC ACOUSTIC SIGNALS ALTERS WITH AGE

P. BUNGERT

Max-Planck-Institut für Neuropsychologische Forschung, Inselstr. 22-26, 04103 Leipzig, Germany

Tagesklinik für kognitive Neurologie, Liebigstr. 22a, 04103 Leipzig, Germany

E-mail: bungert@tk251.neuro.uni-leipzig.de

G.J. DÖRRSCHEIDT

Lehrstuhl für Allgemeine Zoologie und Neurobiologie der Ruhr-Universität Bochum, PF 102148, 44780 Bochum, Germany

R. RÜBSAMEN

AG Neurobiologie, Zoologisches Institut der Universität Leipzig, Talstr.33, 04103 Leipzig, Germany

Presently we are establishing two sets of psychoacoustic tests (basic tests and integrational tests) which should allow the differentiation of peripheral and central auditory processing in normal-hearing and hearing-impaired subjects. In a baseline study performed on one hundred, 20-70 years old normal-hearing subjects we used these tests to evaluate age-dependent alterations in auditory processing. The basic tests measured (1) the monaural audiogram, (2) the frequnecy resolution and the limits of simultaneous (3) frequency and (4) intensity discrimination. The integrational tests reported here evaluate the limits of binaural (5) frequency, (6) intensity and (7) temporal discrimination. These measures, together with tests under successive stimulus conditions, show that temporal processing and processing of stimulus frequency worsens with increasing age, while the processing of stimulus intensity remains stable.

1 Introduction

The influence of age on human hearing capabilities has become a major topic of interdisciplinary research [2]. Most of the efforts however have concentrated on the auditory periphery, giving minor space to the fact that hearing problems of the elderly listeners, e.g. in speech comprehension, are to a great extend caused by central disabilities [1]. As part of a study aiming at the hearing capabilities of brain-damaged subjects, mainly stroke patients, we have tested some integrational psychoacoustical paradigms with normal-hearing subjects of different age.

2 Methods

One hundred normal-hearing, naive 20-70 years old subjects were tested in two sessions of two hours each. The subjects were divided into five age-groups of one decade and each group consisted of ten women and ten men.
The subjects sat in a sound-proof booth and were stimulated monaurally or binaurally through earphones. The test procedures were automatized using a psychoacoustic setup and software from Tucker-Davis-Technologies (TDT).
The audiogram was measured using a yes/no (heard/not-heard) criterion. For all other experiments a three-interval-forced-choice method was used. From three acoustic signals offered, one signal had to be detected as differing from the two others (the reference signal presented twice). The acoustic signals were 250 ms long and had rise/fall-times of 10 ms.

3 Results

The acquisition of the audiogram in quiet was the basis for all other psychoacoustic tests. On the on hand, the frequency specific threshold values were used as a reference for all discrimination tests, which typically were performed at 30 dB HL. On the other hand, the results were used to exclude subjects from the test, which had substantial hearing deficits.

1.1 Frequency discrimination

Two subtests were developed to evaluate the limits of frequency discrimination: binaural frequency discrimination and successive frequency discrimination.
It is well known, that frequency discrimination depends on many factors, like test frequency, hearing level and the kind of test presentation. Fig.1 shows the frequency difference limens (DLs), with the age of the subjects as a parameter. Under all conditions frequency discrimination worsens with increasing frequency ($p<0.001$). However the ANOVA results don´t show a significant effect of age (binaural $p=0.223$; successive $p=0.075$) on frequency discrimination. The test with binaural presentation of (interaural) frequency differences revealed significant lower threshold values for 0.25 kHz and 0.5 kHz stimuli.

1.2 Intensity discrimination

In the present study intensity discrimination was investigated using two test variations - a) intensity discrimination under binaural signal presentation and b) intensity discrimination with successively presented signals.

In Fig.2 the intensity discrimination limens for these conditions are plotted over the age of the subjects. Neither test frequency nor age has a significant influence on the limens. A pairwise comparison at the different frequencies reveals higher thresholds under the binaural test conditions than under successive presentation.

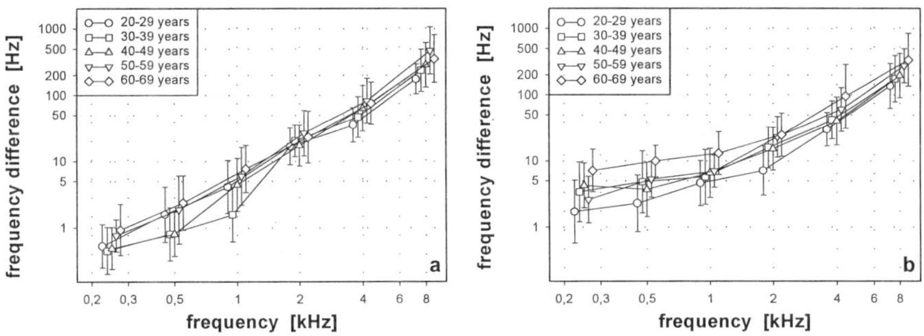

Figure 1a-b. Frequency discrimination limens determined with two different paradigms **a)** binaural signal presentation, **b)** successively presented signals.

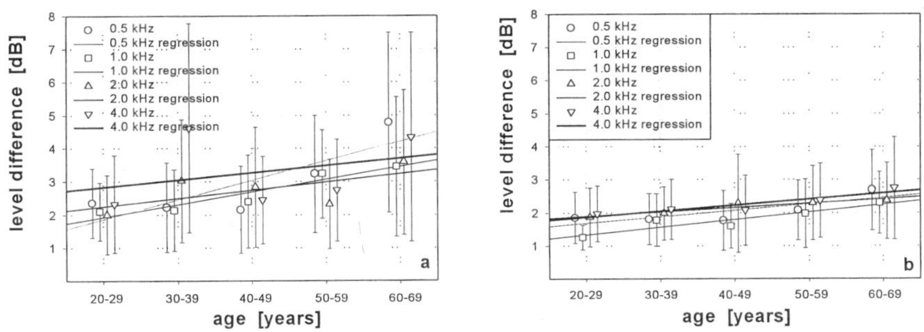

Figure 2a-b. Intensity discrimination limens determined with two different paradigms **a)** binaural signal presentation, **b)** successively presented signals.

1.3 Temporal Processing

Fig.3 shows measures of temporal processing as a function of the age of the subjects using two different test procedures: binaural and successive duration differences. An ANOVA analysis revealed that the duration difference limens of the elderly subjects were significantly larger than those of younger subjects (successive - dF=4, F=10.637; $p<0.001$; binaural - dF=4, F=17.729, $p<0.001$). The ANOVA results also revealed significant effects of stimulus condition: the duration difference limens measured under successive stimulus presentation don`t show any dependence on test-frequency (p=0.825), in contrast to the binaural condition where they strongly depend on it (dF=4, F=30.446, $p<0.001$). For 0.5 kHz and 1.0 kHz the temporal resolution tended to be better under the binaural test condition.

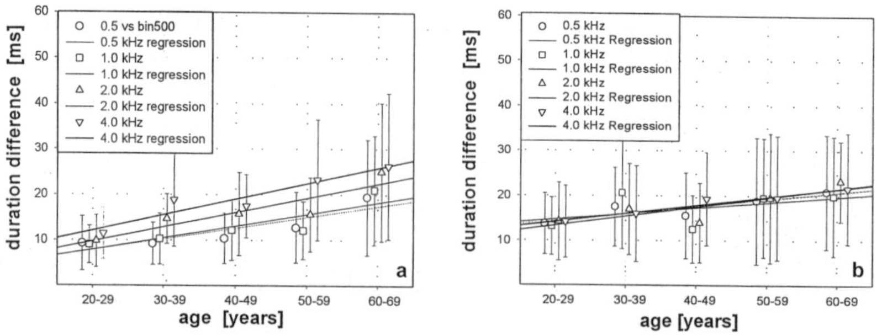

Figure 3a-c. The ability of temporal processing tested as duration difference limens **a)** binaural signal presentation, **b)** successive test condition.

4 Discussion

The paper reports discrimination limits (DLs) for the three parameters describing the pure-tone pulses used (frequency, intensity and duration), as a function of age and frequency. With one exception (duration), the two paradigms of stimulus presentation (binaural and successive) reveal comparable DLs. With the age-dependent peripheral hearing loss balanced by stimulating at constant hearing level, age has a significant impact only on the duration DLs. Frequency shows its known effect on the frequency DLs [3] easily traced back to the tonotopic mapping found in the cochlea and all investigated auditory nuclei. The improvement of the DLs observed for low frequencies under binaural stimulus presentation could indicate that decision is supported by the detection of *binaural beats*. Besides the effect on frequency DLs frequency only affects the DL for duration under binaural stimulation. We could only speculate about the underlying mechanism.

It has to be expected that with the two test paradigms the test of patients with (often one-sided) brain damage will not reveal this similar outcome observed with normal-hearing subjects. Under the two paradigms, discrimination is based upon the function of different parts of the auditory pathway. Therefore, we hope to learn from the *differences* in their outcome about the localization of the dysfunction.

References

1. Auerbach, S.H.; Allard, T.; Naeser, M.; Alexander, M.P.; Albert, M.L. (1982) Pure word deafness. Analyses of a case with bilateral lesions and a defect at the prephonemic level. *Brain* 105, 271-300
2. Marshall, L. (1981), Auditory processing in aging listeners. *J Speech Hear. Dis.,* 226-240
3. Nordmark, J.O. (1968) Mechanisms of frequency discrimination. *J Acoust Soc Am* 44, 1533-1540

PSYCHOPHYSICAL TUNING IN AUDITORY AM-PROCESSING

STEPHAN D. EWERT AND TORSTEN DAU
Carl von Ossietzky Universität Oldenburg, AG Medizinische Physik,
D-26111 Oldenburg, Germany
E-mail: se@medi.physik.uni-oldenburg.de

1 Introduction

The present study examines effects of frequency selectivity in the modulation frequency domain. Three experiments are presented: The first experiment is adapted from the "classical" band-widening experiment, as suggested by Fletcher (1940), but translated to the modulation-frequency domain. The second experiment directly approaches the modulation-filter shape by varying the spectral distance between the fixed test modulation and the narrowband masker-modulation. In the third experiment, psychophysical tuning curves (PTC) in the modulation-frequency domain are derived. Experimental results are compared with model predictions based on the integrated modulation power in the passband of a hypothetical modulation filter derived in [2].

2 Methods

A 3IFC adaptive procedure (2-down 1-up rule) was used to determine modulation thresholds. A gated bandpass-filtered noise (1000 - 4000 Hz) was used as the carrier. The measurements were performed in the presence of an unmodulated white noise at the same spectrum level, which was complementary bandstop filtered. All signals were 600 ms long including 50 ms $\cos^2$ ramps. The stimuli were presented diotically via headphones (Sennheiser HD25) in a sound-proof booth. The overall signal level was 65 dB SPL. If not stated otherwise, the experimental data are averages from three subjects.

The modulation masker was either bandpass-filtered Gaussian noise (GN) or low-noise noise (LNN). The LNN was generated using an iterative algorithm according to Kohlrausch *et al.* [5], which consecutively compresses the Hilbert-envelope of the signal and restricts the signal to the required bandwidth.

Analytical calculations estimate the modulation-detection threshold by integrating the modulation power of the masker in the passband of the modulation filter tuned to the test-modulation frequency [3].

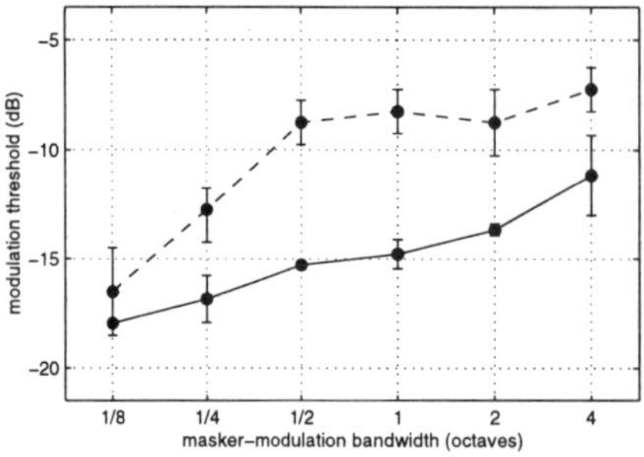

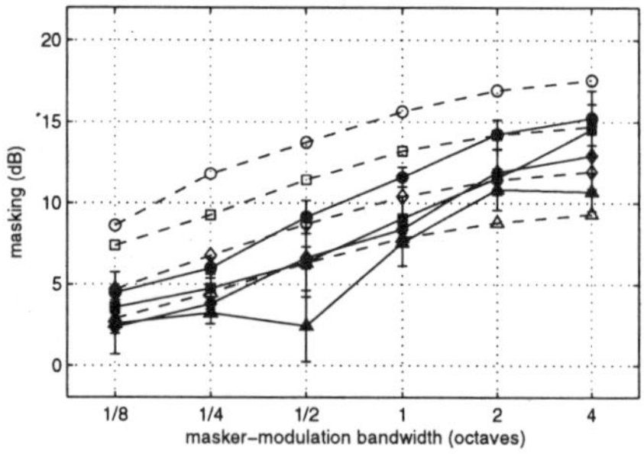

Figure 1. Modulation-detection threshold for 8 Hz as a function of the bandwidth of a GN masker modulation centered at 8 Hz. While varying the masker bandwidth from 1/8 to 4 octaves, the spectral density within the masker passband was kept constant. The upper curve (dashed) is a replot from a study by Houtgast [4]. The lower solid curve shows own averaged data from three subjects for a similar experimental condition. The level of the masker-modulation noise was 7 dB below the level used by Houtgast.

Figure 2. The closed symbols are data from the band-widening experiment with LNN as the modulation masker centered at 8 (●), 16 (■), 32 (◆) and 64 (▲) Hz. The level of the masker modulation was comparable to the level used by Houtgast. In each condition the unmasked modulation-detection threshold (TMTF) is subtracted to express masking. The open symbols are threshold estimates based on the integrated modulation power at the output of the four modulation filters tuned to 8, 16, 32 and 64 Hz, respectively.

3 Experiment 1: Band-widening experiment

For the 8 Hz test-modulation, the effect of varying the bandwidth of the masker modulation was investigated. Fig. 1 shows experimental data obtained with a GN masker (solid curve). The dashed curve represents data from Houtgast at a 7 dB larger masker level. In Fig. 2 experimental data obtained with a LNN masker modulation are shown. Since LNN has a much smaller crest-factor than GN, a larger masker level can be applied without producing over-modulation. Each symbol in the figure represents the masking curve for test-modulation frequencies of 8, 16, 32 and 64 Hz (for details see Fig. 2).

4 Experiment 2: Masked-threshold patterns

Masking patterns were obtained for the test-modulation frequencies 4, 16, 64 and 256 Hz (see Fig. 3). In each condition the spectral position of a half-oct-wide LNN masker was varied in the range from -2 to 2 octaves relative to the test-modulation frequency. The subject's task was to adjust the level of the test-modulation while the masker-modulation level was kept constant.

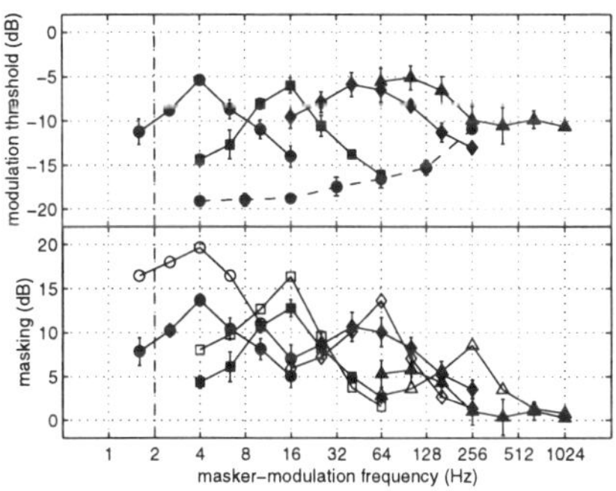

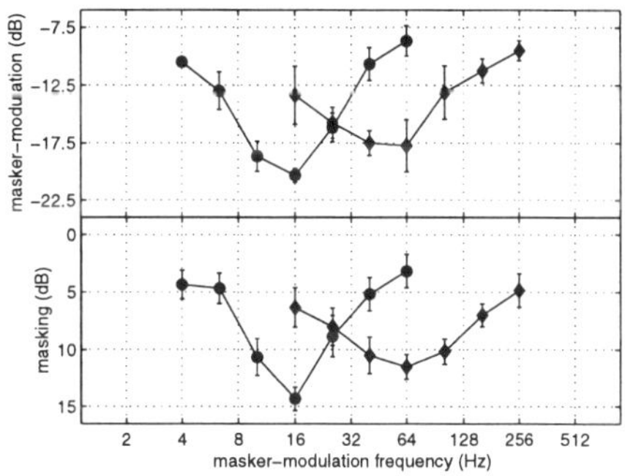

Figure 3. Top panel: Each of the four peaked curves represents the masked-threshold pattern for a certain test-modulation frequency. The lower dashed curve represents the unmasked modulation threshold (TMTF). The bottom panel shows modulation masking patterns (closed symbols) and analytical masking estimates (open symbols). These patterns were derived by subtracting the unmasked threshold from the masked threshold.

Figure 4. Top panel: Psychophysical tuning curves for one subject at a test-modulation frequency of 16 (●) Hz and 64 (◆) Hz. The test-modulation was presented at a fixed near threshold level. The subject's task was to adjust the level of the masker-modulation at threshold. In the bottom panel, the corresponding modulation masking patterns for the same subject are replottet from Fig. 3 in the reverse direction.

5 Experiment 3: Psychophysical tuning curves

Fig. 4 shows psychophysical tuning curves in the modulation-frequency domain. The test-modulation frequencies were 16 and 64 Hz. The method is conceptually related to the derivation of neural tuning curves, where the levels of different input frequencies are adjusted to achieve a constant neural activity. The corresponding psychophysical method requires a fixed test-modulation at a near threshold level and a variable masker-modulation level. This is in contrast to the derivation of the masking patterns from Fig. 3, where the level of the test modulation was varied.

6 Summary and Discussion

Frequency selectivity in AM-detection was investigated by masking paradigms using narrowband Gaussian noise and LNN as the masker-modulation. The detectability of a sinusoidal test modulation was measured in the presence of an additional masker-modulation noise using broadband noise as the carrier. In contrast to the data by Houtgast [4], the results of the band-widening experiment do not allow a reliable derivation of the modulation-filter bandwidth. This is consistent with the analytical masking estimates in Fig. 2, which show that the "classical" band-widening experiment is not appropriate for the derivation of the modulation-filter bandwidth, since the filters are smoothly shaped and relatively broadly tuned. A much higher accuracy is provided by directly deriving the modulation-filter shape with masked-threshold patterns and psychophysical tuning curves as described in Exp. 2 and 3. The bandwidth of the derived modulation filters varies strongly with the modulation frequency. The filter tuned to 4 Hz has a Q-factor of about 1.4 while the filter tuned to 64 Hz shows a much broader tuning (Q=0.7). In comparison to the modulation filter bank supposed by Dau *et al.* [2], which was used for the calculation of the analytical masking estimates, the bandwidth of modulation-filters is about three times larger in the high-frequency region (see Fig. 3).

The use of a stochastic modulation-masker minimizes the influence of beating compared to a sinusoidal masker, but it also limits the available masking power by its large envelope fluctuations. LNN gives a good compromise by providing a maximum amount of masking as well as several advantages of a statistical masker.

Peaked masking patterns were observed for modulation frequencies in the range from 4 Hz to about 100-160 Hz. This demonstrates a spectral decomposition of the signal's temporal envelope over a relatively wide range of modulation frequencies. The fact that no significant differences were found between the modulation-filter-shape estimates obtained with either the "masked threshold" or the "psychophysical tuning" paradigm, gives a strong hint that masked threshold patterns are closely related to the modulation filter shape.

The asymmetry of the masked-threshold pattern at a test-modulation frequency of 256 Hz suggests that the auditory system becomes less sensitive to modulation rates above 100-160 Hz. The detection of high-frequency modulations may be influenced by some "temporal sluggishness", but also beating of modulations may play a role as an additional cue in this condition (e.g., [1]).

References

1. Bacon, S. P., and Grantham, D. W., Modulation masking: Effects of modulation frequency, depth, and phase, *J. Acoust. Soc. Am.* **85** (1989) pp. 2575-2580.
2. Dau, T., Kollmeier, B., Kohlrausch, A., Modeling auditory processing of amplitude modulation. I. Detection and masking with narrow-band carriers, *J. Acoust. Soc. Am.* **102** (1997) pp. 2892-2905.
3. Dau, T., Verhey, J., Kohlrausch, A., Intrinsic envelope fluctuations and modulation-detection thresholds for narrowband noise carriers, *Subm. to J. Acoust. Soc. Am.* (1998).
4. Houtgast, T., Frequency selectivity in amplitude-modulation detection, *J. Acoust. Soc. Am.* **85** (1989) pp. 1676-1680.
5. Kohlrausch, A., Fassel, R., van der Heijden, M., Kortekaas, R., van de Par, S., Oxenham, A., Püschel, D., Detection of tones in low-noise noise: Further evidence for the role of envelope fluctuations, *Acustica united with acta acustica* **83** (1997) pp. 659-669.

INTERFERENCE OF FORMANT-FREQUENCY MODULATION DETECTION (FMDI) BY OFF-FREQUENCY FORMANTS PRESENTED IN SEVERAL BINAURAL SCHEMES

J. LYZENGA, AND R. P. CARLYON

MRC Cognition and Brain Sciences Unit, 15 Chaucer Road, Cambridge CB2 2EF, United Kingdom

The detection of frequency modulation of formants (signals) can be disrupted by simultaneous presentation of off-frequency modulated formants (maskers) to the same ear[1]. The present study examines the effects of various binaural presentation schemes for the masker on this frequency modulation detection interference (FMDI) effect for formants. Signals and maskers were complex tones, bandlimited around 1500 and 3000 Hz, respectively. The signals were always presented to the right ear. The maskers were either stationary or modulated. No interference was found for stationary maskers. The amount of FMDI found for modulated maskers depended strongly on the binaural presentation scheme of the masker. Substantial interference was found for maskers presented to the ipsilateral ear. FMDI was smaller for contralateral maskers. In both cases it increased with increasing masker levels. Substantial interference was also found for binaural (both diotic and dichotic) masker presentations. The perceived location of the maskers, manipulated utilizing inter-aural time and level differences, did not affect FMDI. The amount of interference for binaural maskers was roughly equal to that of the monaural masker with the largest effect. The data were described very well by a simple model in which the amount of FMDI was predicted from a weighted combination of the monaural masker levels.

1 INTRODUCTION

The vowels of natural speech are characterized by the positions of the peaks in their spectral envelopes; the formants. During the pronunciation of vowels and diphthongs, these formants may change their positions. For gaining insight in speech perception, it is interesting to investigate the sensitivity for dynamic changes in formants. Earlier, we have reported frequency modulation detection interference (FMDI) in a monaural study using formant-like complex tones[1]. We presented these "formants" in isolation and simultaneously with unmodulated and modulated off-frequency formants (maskers). We found no FMDI for unmodulated maskers, whereas modulated ones gave rise to FMDI under a number of stimulus conditions.

In the present study we examined this interference effect more closely by introducing inter-aural differences in signals and maskers. We measured listeners' ability to detect a modulation in the center frequency of formants (with a fixed fundamental) presented to the right ear. These stimuli were presented in the absence and in the presence of simultaneous competing formants that were either stationary or modulated. The maskers were presented under different diotic and dichotic conditions. We examined the influence of the ear of masker presentation by presenting the maskers either to the right or to the left ear, and of the perceived location of the masker relative to that of the signal by using inter-aural time and level differences. The various masker conditions are given in Table 1.

TABLE 1. Binaural masker conditions.

Code	Freq. cond.	Ipsi. level	Contra. level	Perceived Location
IS	Stat.	0 dB		Ipsi.
CS	Stat.		0 dB	Contra.
CS+	Stat.		+20 dB	Contra.
I-	Mod.	-20 dB		Ipsi.
I	Mod.	0 dB		Ipsi.
I+	Mod.	+20 dB		Ipsi.
C-	Mod.		-20 dB	Contra.
C	Mod.		0 dB	Contra.
C+	Mod.		+20 dB	Contra.
IC	Mod.	0 dB	0 dB	Central.
ICI	Mod.	0 dB	0 dB	Ipsi., using ITD
ICC	Mod.	0 dB	0 dB	Contra., using ITD
IC+	Mod.	0 dB	+20 dB	Contra.

Level=0 dB signifies that the masker has the same level as the signal.

2 STIMULI AND METHODS

We used a three-interval, three-alternative forced-choice (3I, 3AFC) paradigm to measure frequency modulation detection thresholds (FMTs) for the center frequencies of the formants. When present, the maskers were presented in all three tone intervals. The silent interval between the tone intervals was 400 ms. Four normal hearing subjects were asked to identify the odd tone. Immediate feedback was given.

Both the signals and the maskers were harmonic tones which were band-pass filtered with a triangular spectral envelope (on a log-log scale). The slope of the spectral filter was 50 dB/oct. The signals were located in a MID region (near 1500 Hz) and the maskers in a HIGH region (near 3000 Hz). To reduce mutual influence of the formants at the level of the cochlea, these frequency regions were chosen to be well-separated. The center frequency of the formants was located at a harmonic. Signals and maskers had equal fundamentals of either 80 or 240 Hz. All frequency components of the complexes were in sine phase. Stimulus duration was 250 ms (including 25-ms cosine ramps). The stimuli were presented 30 dB above threshold in a pink noise background. To further reduce peripheral interference between signals and maskers, a narrow-band noise was inserted between them.

For the signals, the center frequency of the triangular filter was modulated at a 5-Hz rate. Two different types of maskers were used: stationary, and modulated.

When modulated, they had a 10-Hz modulation rate and their modulation depth was about four times larger than detection threshold at a d' of 1.

3 RESULTS FOR MONAURAL MASKERS

The results for the monaural maskers are shown by the diamonds in the upper panels of Figure 1. The FMDI was calculated as the ratio between the FMTs with and without a masker. The FMDIs for the two fundamentals were very similar and have been combined in the averages shown in this figure. The error bars are the standard deviations. The three conditions in the left panel are for ipsi- and contralateral stationary maskers. The middle panel shows FMDI for ipsilateral, and the right panel for contralateral modulated maskers.

There is hardly any FMDI for the stationary maskers. For monaural modulated maskers we find substantial FMDI, which is larger for the ipsi- than for the contralateral ear. For both ears, it grows more or less linearly with masker level over the range of levels used. The concept of this linear growth of FMDI with level has been implemented in a simple descriptive model:

$$FMDI = 1 + L_{masker}/C \quad (1),$$

where C equals 32 and 64 for the ipsi- and contralateral maskers, respectively. The purpose of this model was to examine the way ipsi- and contralateral maskers combine when presented simultaneously. The solid lines show the predictions of the model for monaural maskers.

4 RESULTS FOR BINAURAL MASKERS

The bottom panel of Figure 1 shows FMDI for four binaural masker conditions. In comparison with the FMDI for the monaural maskers, it is surprisingly small for the binaural maskers. The amounts of FMDI for the binaural maskers are roughly equal to those for the corresponding monaural maskers. The predictions of the model are shown in the solid line. How well the predictions fit the data depends strongly on the rule that is used to combine the effect of the monaural maskers. The rule used for the current predictions was:

$$FMDI_{total} = Max(FMDI_{ipsi}, FMDI_{contra}) \quad (2),$$

which produces a very good fit to the data. In contrast, direct addition of the effects of both maskers would give predictions that are much larger than the measurements (indicated with the dashed line).

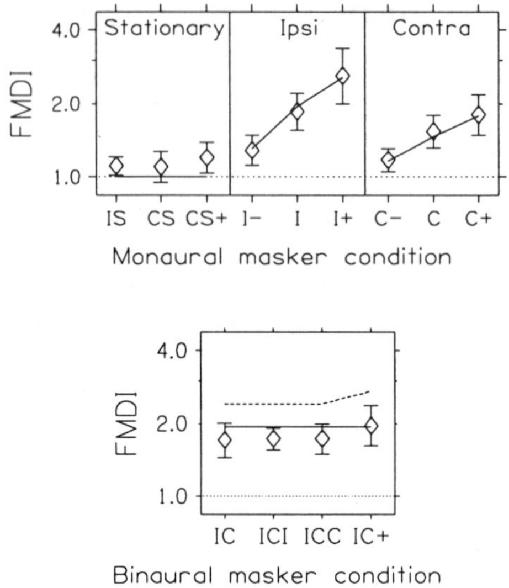

Figure 1. Averaged FMDI

The various perceived locations, as created by inter-aural level and onset-time differences, have no effect on the FMDI. So, we find no evidence for an effect of the perceived location of the maskers on FMDI.

5 CONCLUSIONS

FMDI is larger for ipsi- than for contralateral maskers. Because signals and maskers were well separated and a noise band was inserted between them, it is unlikely that this is due to peripheral effects. For the masker levels used, the amount of FMDI increases roughly linearly with the level of a monaural masker. For binaural maskers, the amount of the FMDI seems to be determined by the monaural masker with the largest effect. The perceived location of binaural maskers has no direct effect on FMDI.

6 REFERENCES

Lyzenga, J., and Carlyon, R. P. "Center frequency modulation detection for harmonic complexes resembling vowel formants and its interference by off-frequency maskers," J. Acoust. Soc. Am. (submitted)

ENVELOPE MODULATION AS A PREREQUISITE FOR TEMPORAL-PITCH PERCEPTION?

LUTZ WIEGREBE AND ROY D. PATTERSON

Centre for the Neural Basis of Hearing, The Physiological Laboratory, Cambridge University, United Kingdom

1 Introduction

Adding a delayed version of a noise sample back to itself generates rippled noise (RN). Iterated rippled noise (IRN) is derived by iterating the delay-and-add process using the output of the previous stage as the input of the next stage [6]. IRN produces a pitch corresponding to 1/IRN delay. The pitch persists even when the sound is high-pass filtered at 12 times the reciprocal of the IRN delay [4]. Typically, when a sound produces a pitch in the absence of resolved harmonics, the pitch is explained in terms of periodic envelope modulation, for example, the pitch of a high-pass filtered sine-phase harmonic complex, or the pitch of sinusoidally amplitude-modulated noise (SAMN). This study presents four experiments designed to search for periodic modulation in the envelope of an IRN stimulus.

2 Experiments

1. IRN and SAMN Masking-period Patterns: Viemeister (1977) measured masking for click signals masked by a SAMN as a function of the temporal position of the click within the modulator period. In Exp. I, we measure masking-period patterns for IRN and SAMN maskers with a 4-kHz tone pip signal to determine whether the masking depended on the position of the tone pip within the IRN cycle. Band-pass filtering alters the envelope of a stimulus so it will be affected by peripheral auditory filtering. We investigated changes of the IRN and SAMN envelope due to changes in filter bandwidth. While leaving the signal frequency constant at 4 kHz, we filtered the IRN and SAMN maskers to bandwidths, which were narrower than the auditory-filter bandwidth at 4 kHz and repeated the masking-period pattern measurements. The results are shown in Fig. 1. Whereas broadband SAMN produces a modulated masking period pattern, the broadband IRN pattern is flat (Col. 1). When the maskers are filtered between 3928 and 4056 Hz, the SAMN masking period pattern (Row 1, Col. 2) becomes much flatter, whereas the IRN pattern (Row 2, Col. 2) develops two peaks. When the maskers are filtered between 3976 and 4008 Hz, the SAMN masking period pattern is almost flat, whereas the IRN pattern reveals modulation (Col. 3).

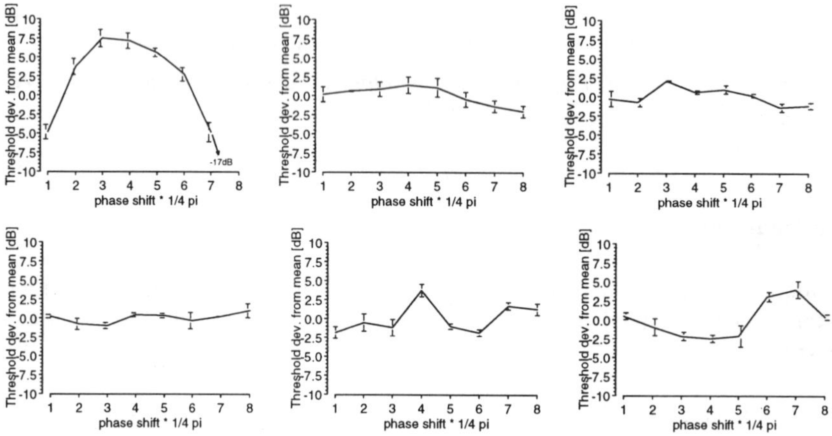

Fig. 1: Masking period patterns for SAMN (Row 1) and IRN (Row 2) maskers unfiltered (Col. 1), and filtered to include either 8 (Col. 2) or 2 spectral peaks (Col. 3).

2. Variations of the IRN Duty Cycle: Replacing a portion of each IRN period with Gaussian noise (GN) generated an IRN stimulus with a variable duty cycle. A square-wave gating function was applied to an IRN to remove a fixed percentage of each IRN period, and then the gaps were filled with GN. The IRN pitch is not altered but the pitch strength decreases with decreasing 'on' portion. The ratio of the on-portion to the IRN delay is the 'IRN duty cycle'.

IRN is distinguishable from GN when the IRN delay is 4 or 8 ms, the gain is 0 dB and the duty cycle is 25%. In this Exp., the on-portion of the IRN duty cycle was phase shifted over the IRN period to determine whether one phase produced a stronger pitch than the other three. The pitch strength varies with IRN gain [6], and so we reduced the gain to determine the minimum required for the IRN-GN discrimination as a function of duty-cycle phase. The IRN-GN discrimination thresholds are plotted as a function of the phase of the IRN gating function for the 25 % duty cycle in Fig. 2. Thresholds for the 4 and 8 ms delays are displayed as open and filled symbols, respectively. Threshold is virtually independent of phase; that is, the deviations from the mean are largely within the range of the standard deviations.

The data in Fig. 2 suggest that the information in the IRN wave that leads to the tonal sensation is uniformly distributed over the IRN period.

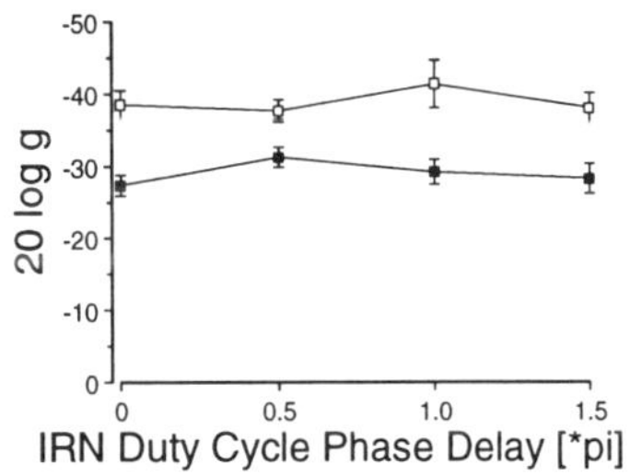

Fig. 2: Detection thresholds for IRN with 25% duty cycle as a function of the phase of the IRN gating function. Thresholds are plotted for an IRN delay of either 4 ms (open symbols) or 8 ms (filled symbols).

3. Detection of an across frequency phase delay for IRN and for cosine-phase harmonic complexes: In the time domain, cosine-phase harmonic complexes show a pronounced transient amplitude modulation of the stimulus as it is represented in the auditory system [2]. Carlyon (1994) has demonstrated that, when listeners are presented with two non-overlapping, narrow spectral bands of cosine-phase harmonic complexes, they can detect a phase delay of one of the complexes relative to the other. We replicated this experiment with two non-overlapping IRNs to determine whether listeners also hear these phase delays for bands of IRN.

The stimuli were either sine-phase harmonic complexes with a repetition period of 16 ms or IRN(16, 1, 32). They were filtered in triangular bands with frequency maxima at 1.5 and 3.5 kHz. The procedure was the same as in [1]. Results are shown in Fig. 3. A phase delay of 45 degrees was just detectable for cosine-phase harmonic complexes. For the IRN(16,1,32) stimulus, performance was at chance for all phase delays up to 180 degrees. Thus, IRN does not produce a repeating feature in its narrow-band envelope sufficient to support the detection of an across-channel delay of this feature.

4. *Detection of a Difference in Fundamental Frequency:* Here, we investigate to which extent two narrow bands of spectrally unresolved IRN support the detection of an across-channel f0 difference. Again, performance is compared to harmonic complexes. Apart from applying an f0 difference, stimuli and procedure were the same as for Exp. III. Results are shown in Fig. 4. Due to the long signal duration (819.2 ms), even a small f0 difference produced a detectable phase delay at some point during the interval for harmonic complexes. Thus, performance is *not* based on the extraction of the f0 of each band. When the stimuli are concurrent IRNs, listeners are confined to genuine f0 processing. The data show that large IRN f0 differences are detectable.

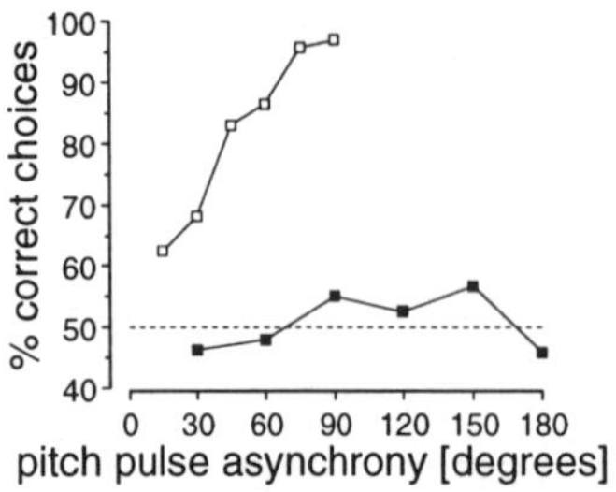

Fig. 3 Across-frequency phase delay detection for narrow-band filters sine-phase harmonic complexes (open symbols) and filtered IRN (filled symbols) both with a period of 16 ms.

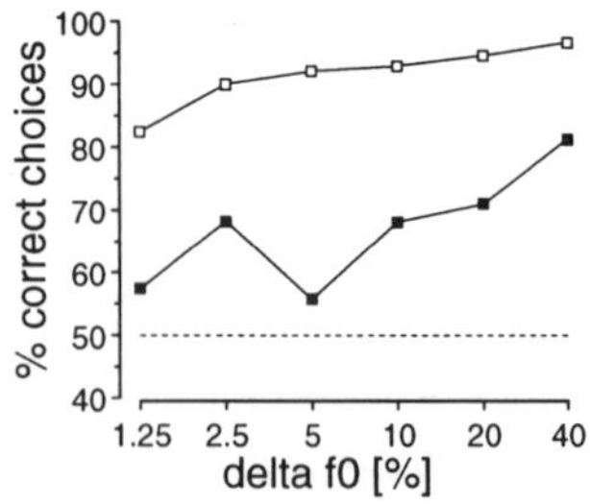

Fig. 4: Detection of a difference in f0 between two narrow-band filtered sine-phase harmonic complexes (open symbols) or between two narrow bands of IRN (filled symbols).

3 Discussion and Conclusions

This study provides three different methods to search for amplitude modulation in the IRN envelope. The negative results strongly suggest that the envelope of spectrally unresolved IRN is flat. Despite this, narrow-band IRN can convey pitch information strong enough to allow for across channel f0 comparisons.

Some perceptual models explain the pitch of unresolved cosine-phase harmonic complexes and SAMN exclusively in terms of first-order inter-spike intervals [3]. For these models, modulation of the stimulus envelope is a prerequisite; multiple peaks per pitch period would lead to multiple spikes per pitch period, which would markedly reduce the number of first-order intervals at the pitch period. The temporal pitch of high-pass filtered IRN is perceived despite the absence of envelope modulation indicating that the auditory system is able to compute higher-order inter-spike intervals.

Supported by the British MRC (G9783469) and the "Deutsche Forschungsgemeinschaft" (Research Grant to the first author).

References:

1. Carlyon, R.P. (1994) J. Acoust. Soc. Am. 95(2), 968-979.
2. Kohlrausch, A. and Sander, A. (1995) J. Acoust. Soc. Am. 97, 1817-1829.
3. Moore, B.C.J. (1997) 4th edition, Academic, London.
4. Patterson, R.D., Handel, S., Yost, W.A. and Datta, A.J. (1996). J. Acoust. Soc. Am. 100(5), 3286-3294.
5. Viemeister, N.F. (1977) in: E.F. Evans and J.P. Wilson (Eds.) Psychophysics and Physiology of Hearing, Academic, New York.
6. Yost, W.A. (1996) J. Acoust. Soc. Am. 100(5), 3329-3335.

PERIOD DIFFERENCE LIMENS FOR HARMONIC COMPLEX TONES IN AND BELOW THE PITCH REGION

KATRIN KRUMBHOLZ, ROY D. PATTERSON AND DANIEL PRESSNITZER

CNBH, Dept. of Physiology, University of Cambridge, U.K.

E-mail: katrin.krumbholz@mrc-cbu.cam.ac.uk

A Introduction

Harmonic complex tones often elicit a pitch, referred to as periodicity pitch (PP), which approximately corresponds to the repetition frequency of the sounds (e.g. de Boer, 1976). The range of repetition frequencies over which the PP is audible is limited. The results of Ritsma (1962) suggest that the existence region of PP depends on the frequency content of the sounds.

The present study investigates the PDLs for bandpass-filtered harmonic complex tones with repetition frequencies both below and above the lower boundary of the existence region of PP. The first experiment examines the influence of the phase spectrum and the frequency content of the harmonic tones on the PDLs; the second experiment investigates the effect of aural distortion products and their dependence on phase spectrum of the stimuli.

B Experiment 1

Stimuli

PDLs for harmonic complex tones were measured for five different filter conditions (FCs). The stimuli were filtered into 1.2 kHz wide frequency bands with rounded spectral edges and a lower cut-off frequency (F_{low}) ranging from 0.2 to 3.2 kHz. The frequency components of the stimuli were added either in cosine phase (COS) or in alternating sine and cosine phase (ALT). PDLs were measured for repetition frequencies (RFs) between 8 and 256 Hz. The difference limens were measured in the presence of a continuous pink noise masker which was lowpass-filtered below the stimuli's passband.

Results

In FCs 1 to 3, the PDLs for the COS stimuli (open symbols in Fig. 1) decrease from about 5% for RFs below 32 Hz to less than 1% at 64 Hz. The transition from high to low PDLs probably reflects the lower boundary of the existence region of PP. For the ALT stimuli, the transition appears to be shifted by one octave to lower RFs (filled symbols in Fig. 1). The slope of the COS PDLs across the transition to the pitch region (between 32 and 64 Hz) becomes increasingly shallow for FCs 4 and 5.

However, the octave shift between COS and ALT stimuli is still visible in FC 4. The exceptionally high PDLs for the ALT stimuli with RFs of 128 and 256 Hz in FC 4 and 5 are probably due to the ambiguity of the pitch elicited by these stimuli (Shackleton and Carlyon, 1994).

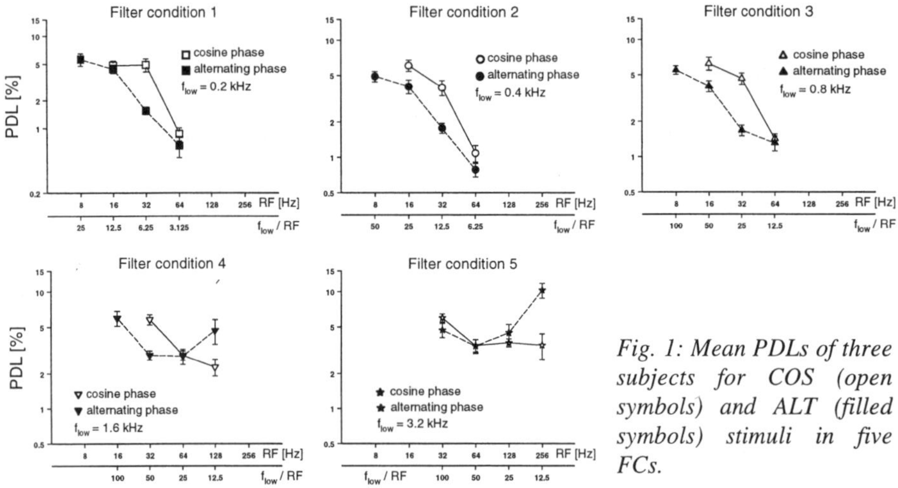

Fig. 1: Mean PDLs of three subjects for COS (open symbols) and ALT (filled symbols) stimuli in five FCs.

Discussion

Ritsma and Hoekstra (1974) measured the period difference limens (PDLs) for bandpass-filtered click trains for filter frequencies between 1 and 8 kHz. They found that discrimination threshold decreased sharply when the harmonic number of the lowest frequency component in the stimuli decreased from about 12 to 8. The decrease of the PDL corresponded approximately to the lower limit of PP found by Ritsma (1962). Shackleton and Carlyon (1994) made a similar observation; they showed that the PDL for harmonic complex tones is much lower when the harmonics are at least partly resolved by the auditory filters than for completely unresolved harmonics. Fig. 2 shows the COS PDLs as a function of RF (left part), and the 10 dB bandwidth of the auditory filter centred at FC (Q_{10dB}) divided by RF (right part). The Fig. shows that, for COS stimuli with component frequencies below about 1.8 kHz, the transition from high to low PDLs which marks the lower boundary of the pitch region occurs at a *constant frequency* between 32 and 64 Hz rather than at a *constant fraction* of the bandwidth of the associated auditory filter.

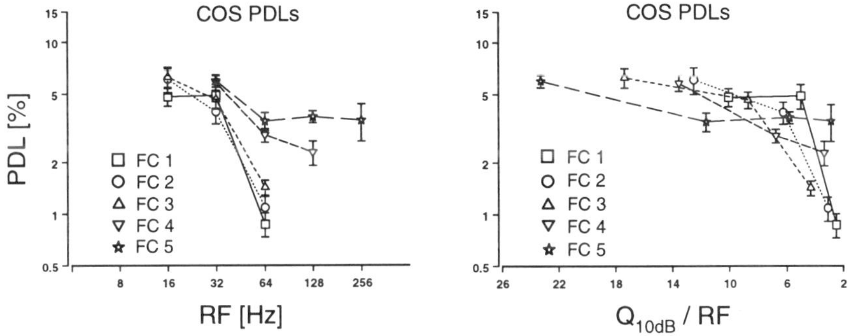

Fig. 2: COS PDLs as a function of the RF of the stimulus (left) and the resolvability of the harmonics (right).

C Experiment 2

For a subset of the filter conditions used in experiment 1 (FCs 3, 4 and 5), the PDLs for COS and Schroeder phase harmonic tones (SCH; Schroeder positive according to Schroeder's (1970) definition) were also measured without the pink noise masker. Fig. 3 shows that the SCH PDLs *without* the masker are approximately equal to the respective COS PDLs measured *with* the masker, indicating that the SCH stimuli produce negligible distortion tones. The COS PDLs, on the other hand, seem to be strongly influenced by the perception of distortion tones well below the signal band when the pink noise masker is omitted (Fig. 4).

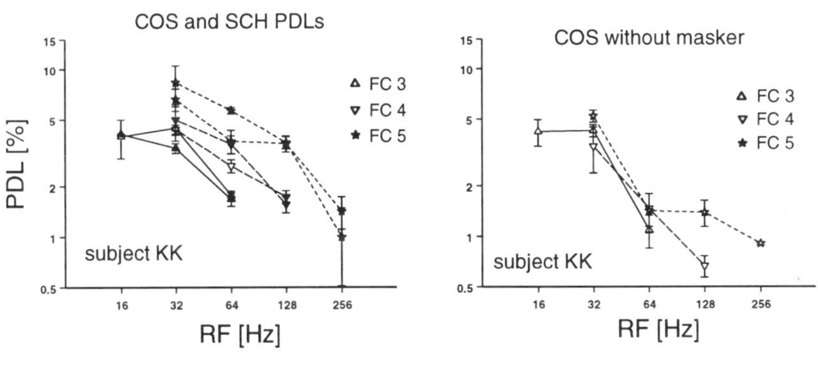

Fig. 3, Left: PDLs for COS stimuli measured with a pink-noise masker (open symbols) and for SCH stimuli without the masker (filled symbols). Right: PDLs for COS stimuli measured without the pink-noise masker.

D Conclusions

The PDLs for harmonic tones with component frequencies below 1.8 kHz reflect the presence of an absolute lower limit of PP. The fact that the lower limit of pitch depends on the phase spectrum of the stimuli suggests that it is determined by limitations in a temporal pitch processing mechanism. The temporal pitch mechanism appears to be unable to handle periods longer than about 30 ms.

The PDLs for cosine-phase harmonic tones with component frequencies above 1.6 kHz are significantly reduced by the perception of distortion tones below the passband of the stimuli when the pink-noise masker is omitted. This is true even though all stimuli were presented at a moderate level (55 dB). Schroeder-phase harmonic tones, on the other hand, do not produce distortion tones strong enough to affect threshold in the period discrimination task.

Acknowledgements

This study was supported by the Medical Research Council (G9703469) and the Fyssen Foundation.

References

1. de Boer E. In: Keidel WD and Neff WD (Eds.), Handbook of Sensory Physiology, New York: Springer, 1976, Vol. 3.
2. Ritsma RJ and Hoekstra A. In: Zwicker E and Terhardt E (Eds.), Psychophysical Models and Psychological Facts in Hearing, Berlin: Springer, 1974.
3. Ritsma RJ. J. Acoust. Soc. Am. 1962; 9: 1224-1229.
4. Shackleton TM and Carlyon RP. J. Acoust. Soc. Am. 1994; 95: 3529-3540.
5. Schroeder MR. IEEE Trans. Information Theory 1970; IT-16 (1): 85-89.

LOCALIZATION EXPERIMENTS WITH SACCADIC RESPONSES IN VIRTUAL AUDITORY ENVIRONMENTS

HEIKE HEUERMANN & HANS COLONIUS

Institut für Kognitionsforschung, FB5 – A6, Universität Oldenburg, 26111Oldenburg, e-mail: mehmood@psychologie.uni-oldenburg.de

In this study, subjects had to localize sound sources in free-field listening conditions and in virtual environments. Under virtual listening conditions, three auditory displays with distinct degrees of individualization were used. In contrast to previous studies using verbal reportexps or pointing techniques, the participants indicated the perceived location of a sound source by turning their eyes towards it. Localization performance under virtual conditions was comparable to free-field conditions for all subjects, at least if individual HRTFs were used. In general, judgments turned out to be more accurate than in previous studies, especially with regard to dispersion of single judgments. This may be due to a close link between the auditory and the oculomotor system leading to a quasi-automatic saccade towards an auditory stimulus, even without a corresponding visual target.

1 Methods

Subjects were seated in a darkened sound-attenuated chamber on a platform in the center of the moving loudspeaker arcs. A dim LED-light served as fixation and reference point. In the free-field experiment, stimuli (1000ms click-trains at 65 dB SPL) were presented by a movable loudspeaker.
Under virtual conditions, stimuli were presented over headphones.

For each subject, three virtual auditory displays were used with distinct degrees of individualization:

1. HRTFs and headphone transfer function measured with a dummy head (Oldenburger Kunstkopf)
2. HRTFs of a dummy head, but subject's own headphone transfer functions
3. Individual HRTFs and headphone transfer function

Stimulus position was varied both in azimuth and elevation with a maximum range of $\pm 30°$.
Eye positions were measured with a video-based eye-tracking system. Head movements during measurement were restricted by a chin rest.
Two experimental blocks under free-field listening conditions were followed by three blocks in virtual environment, each based on a distinct display. Each stimulus was presented 20 times per condition and subject.

2 Results

Experiment 1. Localization in free-field

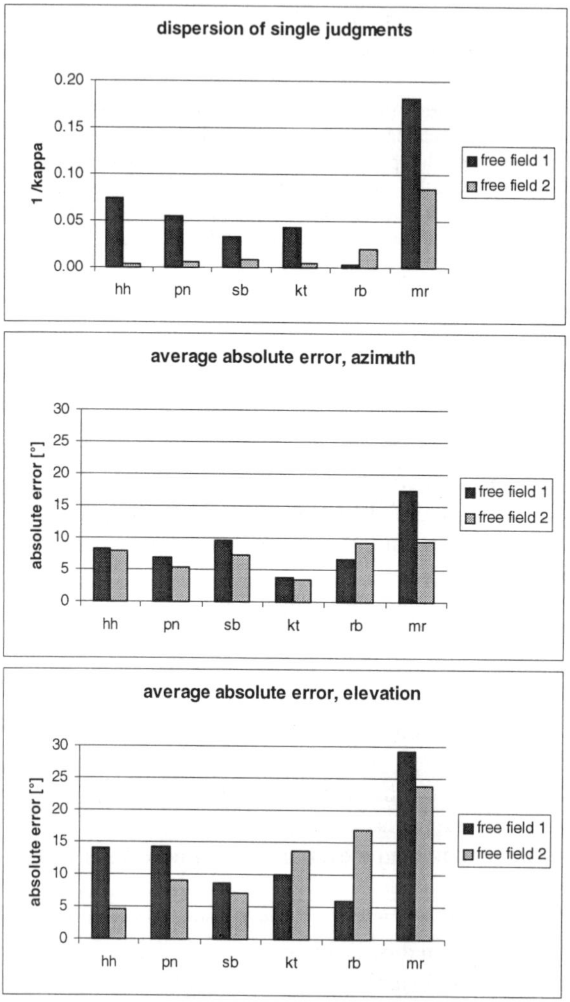

Fig.1: Localization performance of six different subjects in free-field conditions

Figure 1 shows the localization performance of six subjects in terms of average angles of error and dispersion $1/\kappa = N\,(N-R)\,/\,(N-1)2$ (N = number of measurements, R = length of average input vector).

A comparative examination of localization performance in free-field listening conditions shows substantial interindividual differences.

Furthermore, absolute errors of estimation are significantly higher in elevation than in azimuth judgment.

In the second free-field block, four out of six subjects show an improvement of judgment accuracy in terms of decreasing absolute angles of error and remarkbly low dispersion.

Experiment 2. Localization in virtual auditory environments

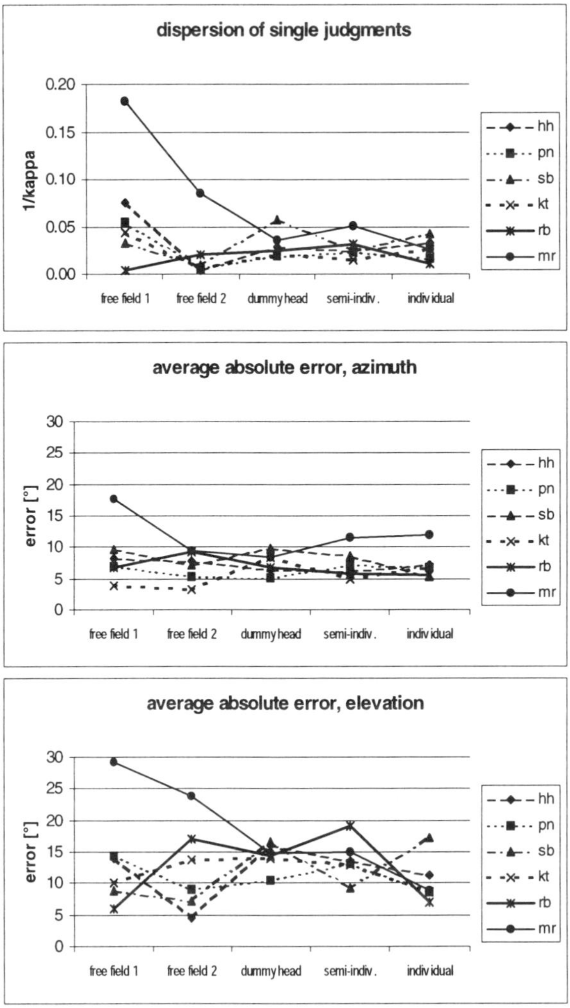

Fig.2: Localization performance of six different subjects in virtual listening conditions

Azimuth estimation accuracy under virtual conditions is almost as good as in corresponding free-field experiments for all subjects, independent of the displays' degree of individualization.

Elevation accuracy, however, seems to be more sensitive to spectral changes caused by using non-individual displays. Under application of individual HRTFs and headphone transfer functions, localization performance is comparable to free-field results for all participants.

Dispersion of single judgments is slightly higher using any virtual display.

Nevertheless, the data show remarkably small angles of error, combined with a generally high precision of estimation towards a given target.

3 Discussion

free field listening	1/kappa	Azim. Error [°]	Elev. Error [°]	Global error [°]
present study	0.008	7.6	11.4	12.3
Wightman & Kistler	0.04	---	---	17.9
Otten	0.044	13.3	19.8	---

virtual environment	1/kappa	Azim. Error [°]	Elev. Error [°]	Global error [°]
present study	0.027	6.4	8.3	12.2
Wightman & Kistler	0.050	---	---	21.5
Otten	0.049	16.15	---	---

Tab.1: comparison of the results presented with some typical data of the literature

A comparison of the results presented with some typical data from previous studies reveals noticeably smaller average angles of error and less dispersion of single judgments in this study. The improvement of estimation accuracy from block 1 to block 2 in the free-field experiment indicates adaptation to unusual environments *in general* (not only to virtual displays) with respect to localization performance. Furthermore, executing a saccade towards an auditory stimulus is a quasi-automatic process minimizing judgment bias. Using eye movements for localization judgments allows rapid collection of data and does not require any special training of the subjects. Therefore, the presented method seems a promising technique in auditory localization minimizing variability and errors in responses towards auditory stimuli.

4 Conclusions

Assessment of localization performance with respect to acoustical stimuli by means of eye-movement measurement turned out to be a suitable technique. Data acquisition is fast and no special training of the subjects is needed. As saccadic responses towards auditory stimuli occur quasi-automatic, variability in the responses to a given target as well as average angles of error in this study are substantially smaller than in previous studies. The results indicate a strong influence of the response measurement technique on performance and suggest that hearing acuity in virtual environments may be better than supposed. Azimuth localization performance in virtual environments was as good as in free field experiments for all participants. Accurate elevation judgments required the use of individual HRTFs for most subjects. In this case, however, elevation estimation accuracy was also comparable to free field results.

Literature:

Wightman, F.L., Kistler, D.J. (1989b). „Headphone simulation of free-field listening II: Psychophysical validation“, J.Acoust.Soc.Am.85, 867-878

Otten, J. (1997). „Psychoakustische Messungen zur Lokalisationsfähigkeit beim Menschen“. Diplomarbeit an der C.v.Ossietzky-Universität Oldenburg

TEMPORAL INTEGRATION OF SHORT TONES IN MASKING NOISE IN LISTENERS WITH NORMAL HEARING AND SENSORINEURAL HEARING LOSS

HENRIK L. OLSEN

Unit of Technical Audiology, Department of Ear and Skin, Karolinska institutet, SE-100 44 STOCKHOLM, Sweden. Henrik.Olsen@ood.ki.se

CAROLINA JOHANSSON

Department of Speech, Music and Hearing, Royal Institute of Technology, SE-100 44 STOCKHOLM, Sweden. carolina@speech.kth.se

1 Introduction

Recently there has been indications that masked thresholds for short tones (less than 20 ms) varies non-monotonically with masker level for normal hearing listeners [1, 2, 4, 6]. Considerably higher masked thresholds for short tones are seen at medium masker levels than for lower and higher masker levels. Since masked thresholds for longer tones varies less with masker level, the result is that temporal integration is higher at these medium levels. This phenomenon has been referred to as a "midlevel hump" or a "severe departure" from Weber's law [1, 7].

There is a lack of studies on masked thresholds and temporal integration at different masker levels for hearing-impaired listeners. A study by Oxenham *et al.* [6] showed no effect of masker level on masked thresholds and temporal integration for four hearing-impaired listeners.

The purposes of the present study was to:

1. investigate this "midlevel hump" for normal hearing listeners using a wide range of signal durations in masked threshold experiments.
2. identify changes, if any, in temporal integration as a function of masker level for hearing-impaired listeners.
3. compare the results obtained with intensity modulated and stationary masking noise.

2 Methods

The target stimulus was a 2 kHz sinusoidal tone. Eight different signal durations where used: 3, 6, 12, 24, 48, 96, 192 and 384 ms. The masker was an octave band noise with centre-frequency 2 kHz. The masker duration was 600 ms and the tone was gated to the noise 200 ms after the onset of the noise. The 3 ms tone was also used together with an intensity modulated noise with the same bandwidth and centre-frequency as mentioned above. The tone was placed in the peaks of the modulation envelope [4].

Signal-to-masker ratios (SMR) at threshold for the different stimulus durations and masker levels were determined. A 3I-3AFC method was employed, estimating the 70.7 % point on the psychometric function [3]. Threshold data for detecting the short tone in the modulated noise were obtained using Békésy-technique.

Five normal hearing and three hearing-impaired listeners were tested. All hearing-impaired listeners had a moderate, sloping sensorineural hearing loss. All subjects listened monaurally through a TDH39 earphone and were seated in a sound-insulated test booth.

3 Results

For the *normal hearing listeners*, the SMR at threshold as a function of masker level is shown in Figure 1a. A follow-up test revealed significantly higher thresholds at masker levels 65 and 85 dB SPL than at 45 dB SPL ($p<0.01$). Temporal integration was calculated as the differences in thresholds between the 3-192 ms tones and the 384 ms tone (Figure 1b). An analysis of the slopes of the temporal integration functions for the short tones, 3-24 ms, showed a significantly steeper slope for masker level 65 dB SPL ($p<0.01$).

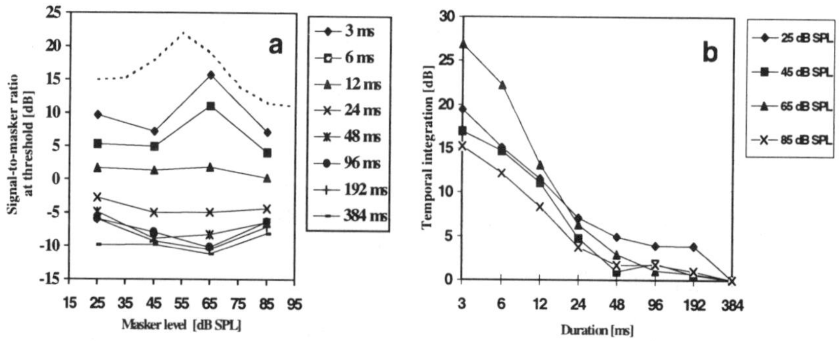

Figure 1. Mean data for the normal hearing listeners. (a) shows the SMR at threshold for the eight different signal durations. Also shown (dashed line) are the SMR for the 3 ms tone in modulated noise, from Lindblad (1992). (b) shows the temporal integration function calculated with the threshold for the 384 ms as reference.

For all the *hearing-impaired listeners*, there was an increase in SMR at threshold with masker level (Figure 2). This was confirmed in the analysis of variance, showing a significant effect of masker level ($p<0.01$). As for the mean SMR, the temporal integration increased monotonically with masker level. An analysis of the slopes of the temporal integration function for the short tone durations, 3 - 24 ms, revealed no significant differences across the four masker levels (p>0.1).

Figure 2 also shows the thresholds for detecting the 3 ms tone in the modulated noise. It can be seen that the results are quite similar in shape compared to the unmodulated condition for the subjects HI-1 and HI-2.

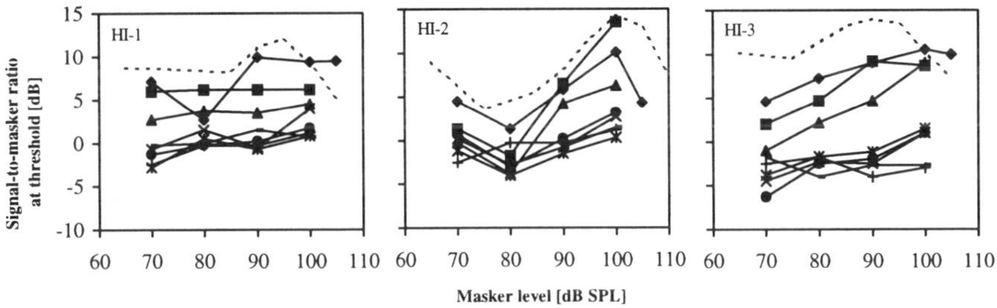

Figure 2. Individual data from the three hearing-impaired listeners. Legends as in figure 1a. The dashed lines show the SMR for the 3 ms tone in the modulated noise.

4 Discussion and conclusions

The results for *normal hearing listeners* are in general agreement with other studies on masked thresholds and temporal integration. The "midlevel hump" was found only for the shortest signal durations at medium masker levels and this will result in a steeper slope of the temporal integration function.

In general, *the hearing-impaired listeners* showed less temporal integration than the normal hearing listeners for all masker levels. In contradiction to the data of Oxenham et al. [6], all our data (unmodulated and modulated masker noise) showed that masked thresholds varies with masker level also for the hearing-impaired listeners. Temporal integration increased with masker level which was a result of the increased SMR for the shortest durations as the masker levels proceeded 80 dB SPL.

A systematic discrepancy was observed for thresholds measured with modulated and unmodulated noise. It is unclear whether these deviations arises from procedural effects (Békésy *vs* 3AFC) or perceptual effects (for modulated *vs* unmodulated noise) or both. Perhaps more important: a more distinct "hump" was seen for all hearing-impaired subjects when the modulated masker was used.

If the data reflect the subject's peripheral compression, then how should we understand our data from the hearing-impaired listeners?

Oxenham *et al.* [6] presented a model that could account for most of their data. The hearing impairment was modelled as a total loss of peripheral compression.

It has been proposed that outer hair cell (OHC) damage causes loss of compression or nonlinearity in the auditory periphery and inner hair cell (IHC) damage causes a linear attenuation of the signal transmission from the cochlea [5].

If we expect our subjects to suffer only from a certain amount of OHC loss, then we would expect to see an increase in masked threshold and in the slope of the temporal integration function in the region of the normal ”midlevel hump”. However, two of our subjects, HI-2 and HI-3, showed some compression above 80 dB SPL which is hard to explain from a residual OHC compression only. Instead, if we assume that the hearing loss of the subjects is caused by damage of both OHC and IHC, then the residual compression can be present at a higher sound pressure level.

5 References

1. Carlyon R. P.and Moore B. C. Detection of tones in noise and the "severe departure" from Weber's law. J Acoust Soc Am 1986;79(2):461-4.
2. Florentine M., Buus S.and Poulsen T. Temporal integration of loudness as a function of level. J Acoust Soc Am 1996;99(3):1633-44.
3. Levitt H. Transformed up-down methods in psychoacoustics. J Acoust Soc Am 1971;49(2):Suppl 2:467.
4. Lindblad A.-C. Temporal and spectral abilities in sensorineural hearing loss. XXI International Congress of Audiology. Morioka, Japan, 1992. Conference proceedings.
5. Moore B. C. Perceptual consequences of cochlear hearing loss and their implications for the design of hearing aids. Ear & Hearing 1996;17(2):133-61.
6. Oxenham A. J., Moore B. C. J., and Vickers D. A. Short-term temporal integration: evidence for the influence of peripheral compression. J Acoust Soc Am 1997;101(6):3676-87..
7. von Klitzing R.and Kohlrausch A. Effect of masker level on overshoot in running- and frozen-noise maskers. J Acoust So12c Am 1994;95(4):2192-201.

TEMPORAL ASPECTS OF LOUDNESS SUMMATION

JESKO L. VERHEY AND BIRGER KOLLMEIER
Graduiertenkolleg Psychoakustik und AG Medizinische Physik, Carl von Ossietzky Universität, D-26111 Oldenburg, Germany
e-mail: jesko@medi.physik.uni-oldenburg.de

1 Introduction

Keeping the sound pressure level of a complex stimulus fixed, the loudness level of the stimulus can significantly increase when the bandwidth exceeds the critical bandwidth (e.g., [6]). This effect was called loudness summation. In most studies, signals with *long* durations (about 1 s) were used to study this effect. There are only few publications, where temporal aspects of the loudness-summation effect were investigated. Most of the data described in the literature [4,5] did not show an effect of duration on loudness summation. On the other hand, Boone [1] found an increasing loudness-summation effect with decreasing signal duration.

The present study investigates the loudness-summation effect for different signal durations at moderate levels. A constant power level was used for the different durations. This is in contrast to other studies about temporal aspects in loudness summation, where the loudness was held constant and both time *and* amplitude were changed.

2 Procedure and Stimuli

Stimuli with different bandwidths were matched in loudness to a reference signal with fixed bandwitdh and level using an adaptive 2–interval, 2–alternative forced–choice procedure. In a trial test and reference signal were presented in random order and with equal a priori probability (silence interval 500 ms). A simple 1–up 1–down procedure was used (50 % point of the psychometric function). Several interleaved adaptive tracks were used [2]. On each trial, the track was chosen randomly from all possible tracks, i.e., from all tracks that had not yet been terminated. To ensure that the interleaved tracks converge at roughly the same time the random choice of tracks is further restricted by the following rule: If none of the tracks is terminated, the number of trials for all different tracks has to be the same before the next trial for all tracks is presented in random order to the listener. If one track is terminated, the rule is applied to the choice of trials from the remaining unterminated tracks.

Bandpass-noise signals geometrically centered at 2 kHz with a flat spectrum were used. The bandwidths of the test signals were 200, 400, 800, 1600, 3200, and 6400 Hz. The reference was a 45-dB 3200-Hz broad bandpass noise. Test signals and reference signal had the same duration. It was either 10, 100 or 1000 ms. Signals were gated with 2.5 ms raised cosine ramps.

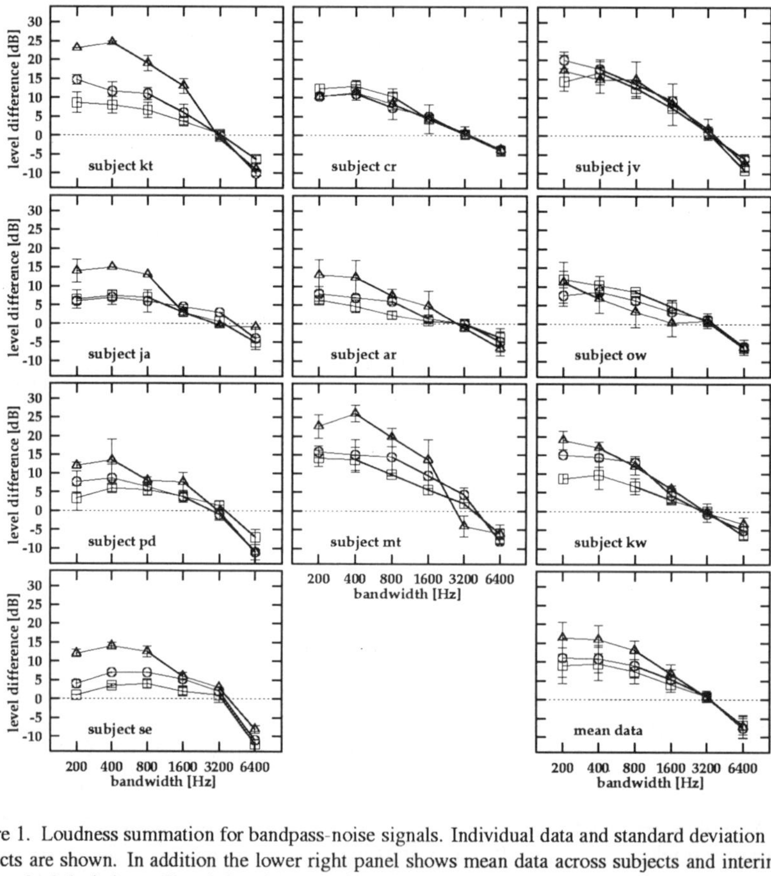

Figure 1. Loudness summation for bandpass-noise signals. Individual data and standard deviation for ten subjects are shown. In addition the lower right panel shows mean data across subjects and interindividual standard deviations. Signal duration was either 1000 msec (squares), 100 msec (circles), or 10 msec (triangles). The level differences between the level of test and reference signal needed to obtain equal loudness are plotted as a function of the bandwidth of the test signal.

3 Results

Figure 1 shows individual data and standard deviations for 10 subjects. In addition, the lower right panel of Fig. 1 shows mean data across subjects and interindividual standard deviations. As a function of the bandwidth, the figure shows the level difference between the test signal and the reference signal necessary to produce the sensation of equal loudness (equal-loudness level difference). A positive level difference means that, to produce the sensation of equal loudness, the test signal must have a higher level than the reference. The figures show equal-loudness level differences for signal durations of 1000 ms (squares), 100 ms (circles), and 10 ms (triangles).

All subjects show positive level differences for test-signal bandwiths < 3200 Hz and no level difference when the test signal and the reference signal have the same bandwidth (3200 Hz). For a bandwidth of 6400 Hz, negative level differences are observed. There are large individual differences among subjects in the amount of loudness summation effect. The level difference between equally loud signals of the same duration with the smallest (200 Hz) and largest (6400 Hz) bandwidth ranges from 10 (subject AR, PD) to 23 dB (subject JV). Although large individual differences are apparent, the average data for signals of 1000-msec duration, (lower right panel, squared symbols) is consistent with the results presented in the literature for long signals (e.g., [6,4]).

The individual data for the different durations can be divided into two different groups. For some subjects (CR, JV, and OW) the loudness summation effect is independent of duration. However, the majority of the subjects (70 %) show an increasing loudness summation effect with decreasing duration, i.e., the slope of the equal-loudness level functions increases with decreasing duration. Averaged across all subjects, the level difference is 15.5 dB for 1000 ms, 18.5 dB for 100 ms, and 23.5 dB for 10 ms.

4 Discussion

In contrast to the present study, most of the data described in the literature [4,5] did not show an effect of duration on loudness summation (see introduction). There is only one study in which a duration-dependent loudness-summation effect was found [1]. Boone's finding that the level difference for short-duration signals is twice the level difference for long-duration signals is in line with the experimental results of the first experiment in the present study.

Loudness models account for the loudness-summation effect for stationary signals by assuming three processing stages: A bank of (critical-band) filters, a compression within each critical band, and a summation across frequency bands. Within these models, the amount of loudness summation is determined by the bandwidth of the critical-band filters and the amount of compression: A higher compression and a smaller filter bandwidth result in a larger loudness summation effect. Most loudness models for time varying sounds assume that the spectral analysis, which is responsible for loudness summation, works nearly instantaneously (e.g. [3]), whereas the overall temporal weighting has longer time constants of about 100 ms. Such a model structure would predict a loudness-summation effect, which is independent of signal duration, in contrast to the present results. A more complex model for calculating loudness of temporally variable sounds was proposed by Zwicker and Fastl (1990). In contrast to the model structure described above, they assumed a duration-dependent lowpass filtering within each critical band before summation across critical bands followed by an overall time weighting with a lowpass filter (τ about 100 ms). However, also this model cannot account for the the data, since only the decay time varies with duration, which is not important for the experiments in the present study.

5 Conclusions

Loudness summation depends on signal duration. It is markedly increased for 10-msec signal, whereas it is almost the same for 100-msec and 1000-msec signals. Present loudness models of time varying sounds cannot account for the experimental data. The reason for this failure is that their spectral-analysis stage acts nearly instantaneously and duration-independent. A model which contains a duration-dependent compression should in principle account for the present data. Further studies should be done to test, how the temporal effects in loudness-summation depend on the level.

References

1. Boone, M.M. (1973) Loudness measurements on pure tone and broad band impulsive sounds. Acustica 29, 198–204
2. Florentine, M., Buus, S., and Poulsen, T. (1996) Temporal integration of loudness as a function of level. J. Acoust. Soc. Am. 99, 1633–1644
3. Ogura, Y., Suzuki, Y., and Sone, T. (1993) “A new method for loudness evaluation of noises with impulsive componente” Noise Control Engineering Journal 40, 231–240
4. Port, E. (1963) “Über die Lautst"arke einzelner kurzer Schallimpulse” Acustica 13, 212-223.
5. Zwicker, E. (1965) Temporal effects in simultaneous masking and loudness. J. Acoust. Soc. Am. 37, 132–141
6. E. Zwicker, G. Flottorp, and S. S. Stevens, (1957) “Critical bandwidth in loudness summation” JASA 29, 548-557.

THE EFFECT OF LATERALIZED TEMPORAL FRINGES ON FUNDAMENTAL FREQUENCY DISCRIMINATION

H. GOCKEL, R.P. CARLYON AND C. MICHEYL

MRC Cognition and Brain Sciences Unit, 15 Chaucer Road, Cambridge CB2 2EF, UK

E-Mail: hedwig.gockel@mrc-cbu.cam.ac.uk

Fundamental frequency (F0) discrimination can be impaired substantially by the presence of another complex (the "fringe") immediately before and after the target complex. It has been shown [2] that for the impairment to occur (i) target and fringes have to be in the same frequency region; (ii) if all harmonics of target and fringes are unresolved then they may differ in F0; otherwise, they have to be similar. The present experiments investigated the effect of fringes lateralized differently from the target. In a 2I-2AFC procedure, difference limens for F0 for a 100-ms harmonic target complex were measured in the presence and absence of 200-ms harmonic fringes. The nominal F0 was 88 Hz or 250 Hz. Stimuli were bandpass filtered between 125-625 Hz, 1375-1875 Hz, or 3900-5400 Hz. The target was presented monaurally, while the fringes were: (i) contralateral with the same level as the target, (ii) contralateral with a level increased by 20 dB, (iii) dichotic (ILD of +20 dB contralateral), (iv) diotic, and (v) ipsilateral. The effect of the fringes was reduced when their subjective location differed from that of the target. This reduction depended on the resolvability of both the fringes and the target. The results are consistent with the idea that the fringes produce interference when the listeners have difficulty segregating the target from the fringes.

1 Introduction

Many sounds that we hear in everyday life contain components whose frequencies correspond to integer multiples of a fundamental frequency (F0). Such harmonic tones generally elicit a pitch sensation corresponding to the F0, even when there is no physical energy at that frequency. The present study is part of a series investigating the mechanisms of F0 encoding and how this mechanism is affected by the presence of other sounds. Carlyon [1] showed that F0 discrimination can be impaired substantially by the presence of another harmonic complex (the "fringe") immediately before and after the target complex. This has been attributed to listeners over-integrating information such that the estimate of the target-F0 is influenced by F0 information from the fringe. More recent experiments [2] showed that for the impairment to occur (i) the target and fringes have to be in the same frequency region; (ii) if all harmonics of the target and fringes are unresolved then they may differ in F0; otherwise, they have to be similar.

The present experiments investigated the effect of temporal fringes lateralized differently from the target, compared to monaural stimulation. Specifically, it was asked whether contralateral presentation and perceived contralateral location of the fringes relative to the target would reduce their deleterious effect. This would provide some information concerning the possible "site" of the affected mechanism and its susceptibility to perceptual segregation.

2 General Method and Stimuli

The difference limen for F0 (DLF0) for a 100-ms harmonic target complex was measured in the presence and absence of 200-ms harmonic fringes. In a 2I-2AFC task, subjects were required to indicate the interval with the higher target F0. The F0s of the two successive targets differed by an amount $\Delta F0$ and were geometrically centred on a nominal F0 of either 88 or 250 Hz. The fringes (when present) were presented immediately before and after the target, but not simultaneously. The F0 of the fringes was either 88 or 250 Hz. The target was always presented monaurally to the left ear of the subjects.

The variable of main interest was the lateralization of the fringes relative to the target, which was manipulated using the following conditions (all levels are specified as the level per component before any filtering was applied): (i) contralateral at a level of 45 dB SPL ("c45"), (ii) contralateral at a level of 65 dB SPL ("c65"), (iii) dichotic at a level of 45 dB ipsilateral, and 65 dB SPL contralateral: PERceived CONtra ("percon"), (iv) diotic at a level of 45 dB SPL: PERceived MID("permid"), and (v) ipsilateral at a level of 45 dB SPL ("ipsi"). Both the target and the fringes consisted of a large number of harmonics bandpass-filtered into the same frequency region, which, in three different experiments, could be the LOW, MID, or HIGH region; these regions had passbands (3-dB down points) of: 125-625 Hz, 1375-1875 Hz, and 3900-5400 Hz, respectively. According to Shackleton and Carlyon's [3] definition of resolvability, harmonics of a 88-Hz F0 are resolved only in the LOW region, whereas harmonics of a 250-Hz F0 are resolved in the LOW and MID region but unresolved in the HIGH region. The number of harmonics presented varied with F0 and region, but always "filled" the passband and skirts down to a level of 0 dB SPL. All stimuli were gated with 20-ms raised-cosine ramps. The inter-interval delay was 500 ms. The stimuli were presented in a background of pink noise (low pass filtered at 20 kHz, with a spectrum level of 10 dB at 1 kHz), over Sennheiser HD 414 headphones. The adaptive procedure tracked thresholds, corresponding to 71% correct responses. Feedback was provided following each response. At least four subjects participated in all experiments.

3 Results

3.1 Experiments I and II: LOW and MID region

Results were very similar in the LOW and MID frequency region, where harmonics for at least one F0 were resolved. The bars in Fig. 1 show the geometric mean ratios between thresholds in conditions with and without fringes for five subjects in the MID frequency region. If the target and fringes differed in F0, then the presence of the fringes did not significantly impair F0 discrimination of the target. However, if the target and fringes had the same nominal F0, then the presence of the fringes significantly impaired F0 discrimination of the target. The effect of the fringes was significantly lower when the fringes were lateralized differently from the target. However, even for purely contralateral presentation (c45, c65) the fringes still had a significant effect. Hence, the effect of the fringes in the ipsilateral ear could be reduced by moving their perceived location away from that of the target.

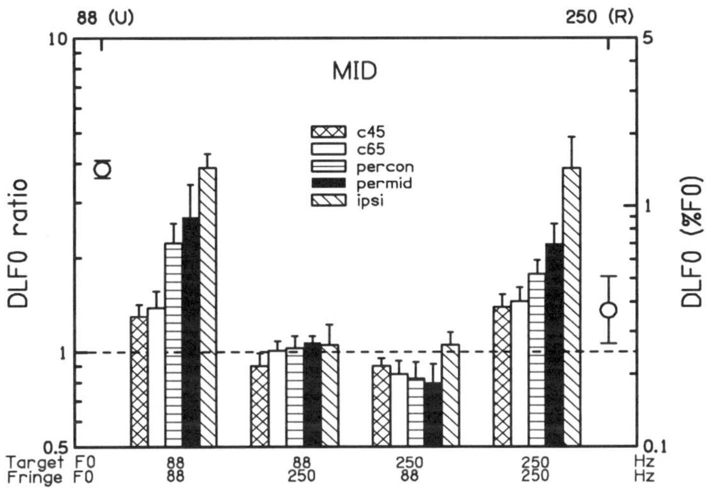

Figure 1. Geometric mean thresholds and the corresponding standard errors for five subjects in the MID frequency region (Exp.II). The open circles correspond to DLF0s in the absence of any fringe, and are plotted with respect to the right-hand ordinate. The bars, plotted with respect to the left-hand ordinate, show the DLF0s for the 88-Hz and 250-Hz targets in the presence of 88-Hz or 250-Hz fringes, divided by the DLF0 for each target when presented alone.

3.2 Experiment III: HIGH region

In the HIGH frequency region, where harmonics for both F0s were unresolved, results differ from those for the LOW and MID frequency regions in two ways: (i) The presence of the fringes significantly impaired F0 discrimination for the target even if target and fringes differed in F0; (ii) Only when target and fringes differed in F0, was the effect of the fringes significantly reduced in conditions "percon" and "permid" compared to the ipsilateral condition. Moving the perceived location of the fringes away from the target did not help (enough) when target and fringes had equal F0s.

4 Summary and Conclusions

In the LOW and MID frequency regions, where harmonics were resolved for at least one F0, ipsilaterally presented fringes only raised thresholds when they had the same nominal F0 as the target. In contrast, they raised thresholds in the HIGH region even when their F0 differed markedly from that of the target. In all regions, the amount of the impairment could be reduced by presenting the fringes only contralaterally to the target. Moving the perceived location of binaural fringes away from that of the target reduced their effect in all conditions except in the HIGH frequency region when target and fringes had the same F0. A possible explanation is that fringes only impair performance when they are not perceptually segregated from the target. Initial segregation might depend on F0 differences and might then be enhanced by a difference in perceived location. In the LOW and MID regions, a big difference in F0 (mixed nominal F0 conditions) is sufficient to induce segregation and eliminate the interference of the fringes. Even when the nominal F0s are the same, there were small F0 differences between the target and fringes introduced by the measurement procedure. The initial segregation produced by these differences in F0 might then be enhanced by a difference in lateralization. In the HIGH region, where F0 differences may be much less effective at triggering an initial segregation, the lateralization cues might exert an effect only when the F0 difference between the targets and fringes is very large, as occured when they had different nominal F0s.

5 Acknowledgement

Work supported by the Wellcome Trust.

References

1. Carlyon R. P., Masker asynchrony impairs the fundamental-frequency discrimination of unresolved harmonics, *J. Acoust. Soc. Amer.* **99** (1996) pp. 525-533.
2. Micheyl C. and Carlyon R. P., Effects of temporal fringes on fundamental-frequency discrimination, *J. Acoust. Soc. Amer.* (1998) in press.
3. Shackleton T. M. and Carlyon R. P., The role of resolved and unresolved harmonics in pitch perception and frequency modulation discrimination, *J. Acoust. Soc. Amer.* **95** (1994) pp. 3529-3540.

AN EFFECT OF THE COHERENCE BETWEEN ENVELOPES ACROSS FREQUENCY REGIONS ON THE PERCEPTION OF ROUGHNESS

DANIEL PRESSNITZER[1], STEPHEN MCADAMS[2,3]
(1) CNBH, Dept. of Physiology, University of Cambridge, U.K.
(2) IRCAM, Paris, France
(3) Laboratoire de Psychologie Expérimentale (CNRS), Université René Descartes, Paris, France

1 Introduction

1.1 Roughness and envelope coherence: within- and cross-channel cues

Auditory roughness arises when rapid beats can be perceived[1]. It is considered to be linked to the amount of amplitude fluctuation within a given auditory filter, around a "roughest" modulation frequency of 70 Hz[2].

The way roughness builds up if sounds have energy spread over a large frequency region is unclear. When two sinusoidally amplitude-modulated (SAM) tones with different carrier frequencies but a same modulation frequency are added, conditions with "co-phasic" envelopes are rougher than those with "anti-phasic" envelopes[2]. This could be accounted for by cross-channel comparisons of envelope coherence.

However, in places where the two SAM tones can interact along the basilar membrane, the resulting envelope in the anti-phasic conditions will be almost flat. The difference between co- and anti-phasic modulations could then be explained by within-channel cues only[3]. Some limited cross-channel processing had to be introduced in models in order to reproduce the low roughness of wide-band noise[3,4], but the hypothesised mechanism has not been directly tested. New stimuli were designed to address this issue.

The stimuli were derived from SAM-tones by introducing a time-jitter in their envelope on a period-to-period basis. A jAM tone can be described by its average modulation frequency $\overline{f_m}$, its centre frequency f_c, its amount of jitter (that controls the bandwidth), and its modulation depth m (that precisely sets the envelope *rms* value). All jAM tones with a same set of parameters should thus produce exactly the same roughness. It is however possible to obtain incoherent envelopes within a set of parameters by using different random samples for the jitter. Two jAM tones can then be added, without producing as important within-channel cues as anti-phasic SAM-tones.

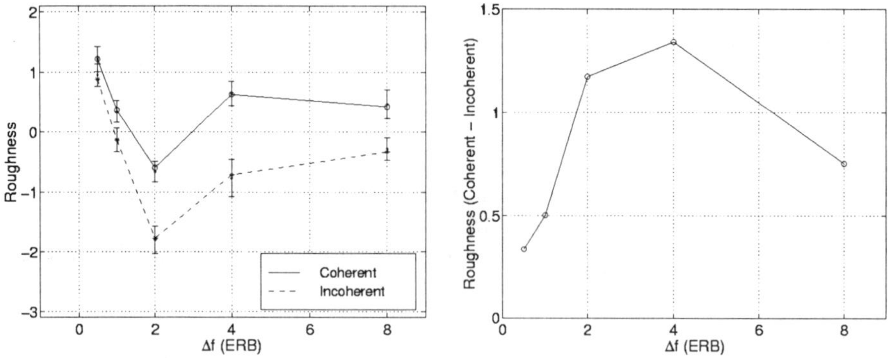

Figure 1. Experimental results. Left panel shows the relative roughness for the 12 stimuli. Right panel shows the coherence effect.

2 Experiment

2.1 Method

Each stimulus consists of the addition of two jAM tones with $\overline{f_m} = 50$ Hz and $m = 0.7$. The maximum amount of envelope period jitter was such that the range of possible frequencies for each period was between 33 and 100 Hz. One of the tones had a carrier of 4000 Hz and the second carrier took values between 4233 Hz and 9772 Hz. Values of the frequency difference between carriers were $\Delta_f = [1/2, 1, 2, 4, 8]$ on the ERB-rate scale. For half of the conditions, both jAM tones had the same envelope ("coherent"). For the other half of the conditions, two different random samples were used to compute the envelope jitter ("incoherent"). In this latter case, the correlation between envelopes was close to 0. Paired comparisons were used to collect roughness judgements. The experimental procedure and analysis methods have been described elsewhere[5]. Fifteen subjects took part in the experiment.

2.2 Results

The shape of the relative roughness values shown in Figure 1 is rather complex. For the purpose of the present paper, we will only focus on the "coherence effect" (the roughness difference between coherent and incoherent conditions). Overall,coherent envelopes yielded more roughness than the incoherent ones. This effect is low for a frequency separation of $\Delta_f = 1/2$ ERB, increases to a maximum at 4 ERB and then decreases but is still present at 8 ERB.

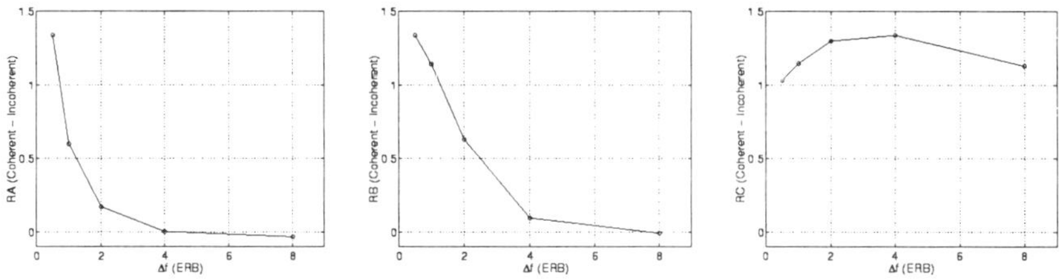

Figure 2. A comparison between the coherence effect predicted by the three roughness combination methods RA, RB, and RC described in the text.

3 Comparison of computational summation methods

Let us consider the way classic models of roughness would handle the stimuli. The signals would first be decomposed into i frequency channels and partial roughnesses r_i would be computed[2,3,4]. The next step is then to combine the partial roughnesses.

A straight-forward method is to add up the partial roughnesses across channels: $RA = \frac{1}{N}\sum_{i=1}^{N} r_i$. Although this method is surely wrong (because of the excessive roughness it would predict for white noise), it describes the overall influence of within-channel cues. A second method takes into account the correlation between the envelopes on which the roughness estimation is based[3,4]. Roughnesses in adjacent channels only add up if their envelopes are correlated: $RB = \frac{1}{N}\sum_{i=2}^{N-1}(c_{i,i-1}c_{i,i+1})^2 r_i$. Finally, all possible pairs of channels can be considered and roughness in any two channels only add up if the corresponding envelopes are correlated: $RC = \frac{1}{N(N-1)}\sum_{\substack{i,j=1 \\ j\geq i}}^{N} c_{i,j}(r_i + r_j)$.

The prediction of these algorithms are displayed in Figure 2, using a common front end to compute the partial roughnesses. The method RA only predicts a coherence effect if the excitation pattern of the sounds overlap. This is in clear opposition with the experimental results. For the same reasons, the method RB also fails, as it predicts an effect only if adjacent channels are involved. The last method is the only one to produce the correct trend, although the effect is systematically over-estimated to a substantial degree. It predicts a coherence effect proportional to the number of incoherent channels. As $\Delta_f$increases, this number first increases as the two jAM tones become separately resolved, and the coherence effect gets bigger. When the jAM tones do not overlap anymore, the number of incoherent channels decreases as the auditory filters widen in the frequency region of the higher jAM tone, and the coherence effect diminishes. This is in line with the experiment.

4 Discussion

The roughness of complex sounds not only depends on the partial roughnesses in different frequency regions but also on the coherence between the modulations present in these regions: incoherent modulations are less efficient than coherent ones. In the experiment we presented, this effect was clearly the result of cross-channel cues as it was small for small Δ_f where potential within-channel cues were maximal, and remained significant for large Δ_f where within-channel cues would be negligible.

Computational simulations indicated that a simple correlation between envelopes across all channels could qualitatively account for the trend in the results. However, we do not wish to imply that the auditory system performs such a correlation. Firstly, the same kind of effect could be achieved by some kind of cancellation across channels. Secondly, the coherence effect observed for a complex-sound attribute such as roughness could reflect the simultaneous action of cross-channel mechanisms of various kinds[6].

Finally, we only focused on one aspect of the data in this paper. There are actually other questions raised by the complex evolution of roughness with Δ_f and by measures made at $\Delta_f = 0$. Some of these issues are currently being investigated and they point to the need for a revision of the current "within-channel" roughness models before proceeding any further in trying to quantitatively adjust the cross-channel combination algorithm. [Work supported by the Fyssen Foundation.]

References

1. Hermann L. F. von Helmholtz. *On the Sensations of Tone as the Physiological Basis for the Theory of Music.* 1877. 2nd. Ed. trans. A. J. Ellis from German 4th Ed., 1885, Dover, 1954.
2. E. Terhardt. On the perception of periodic sound fluctuation (roughness). *Acustica*, 30:201–212, 1974.
3. W. Aures. Ein Berechnungsverfahren der Rauhigkeit [A roughness calculation method]. *Acustica*, 58:268–281, 1985.
4. P. Daniel and R. Weber. Psychoacoustical roughness: Implementation of an optimized model. *Acustica*, 83:113–123, 1997.
5. D. Pressnitzer and S. McAdams. Phase effects in roughness perception. In A. Rees, A. Q. Summerfield, and R. Meddis, editors, *Psychophysical and Physiological Advances in Hearing*, Whurr Publishers, 1998.
6. J. W. III Hall, J. H. Grose, and L. Mendoza. Across-channel processes in masking. In B. C. J. Moore, editor, *Hearing*, Academic Press, 1995.

A NEW APPROACH FOR THE EVALUATION OF TONAL NOISE (TONALITY)

MATTHIAS VORMANN, MARKUS MEIS, VOLKER MELLERT, AUGUST SCHICK
Graduiertenkolleg Psychoakustik, Carl von Ossietzky-Universität
26111 Oldenburg
E-mail: matthias@aku.physik.uni-oldenburg.de

1 Introduction:

Tonal noise (DIN 45681 [1]) is an important parameter affecting sound quality and is substantial for the assessment of the annoyance of tonal components in environmental noises.

For the measurement of time varying weak tonal components of complex technical and natural sounds none of the existing methods correspond to subjective ratings [2].

This is due to the fact that conventional methods neglect the time and frequency resolution as well as masking properties of the human hearing [2]. Furthermore a possible frequency dependence of tonal noise components is disregarded.

The aim of this study is to investigate the aspects of frequency for tonal noise.

2 Method:

In the following experiments 36 sounds are generated by adding 6 sinusoidal tones with different frequencies f (equally shifted on a bark-scale) to 6 uniform masking noises with decreasing steps of sound level. The highest level $L_{N,1}$ for each individual tone is chosen as the noise-level L_G which is the masked threshold of the tone (with sound level L_T) in the critical band around the tone-frequency plus the masking index a_v ($a_v =: L_T - L_G$). The following 4 levels L_N of the uniform masking noise are successively decreased in 4 steps of 12 dB. The sixth level $L_{N,6}$ is chosen as infinity. All 36 modifications of the sounds are displayed in figure 1, showing the different frequencies in horizontal and the different levels L_N in vertical direction. All of the sounds are scaled to the same loudness (12 sone GF).

3 Experiments:

1. Subjective scaling of the 18 sounds with f = 100, 920 and 2700 Hz from figure 1 is done by forced choice paired comparison in random order.

2. Subjective scaling of the 18 sounds with different levels of L_N = 12, 36 and infinity (pure tone) dB from figure 1 is done by forced choice paired comparison in random order.

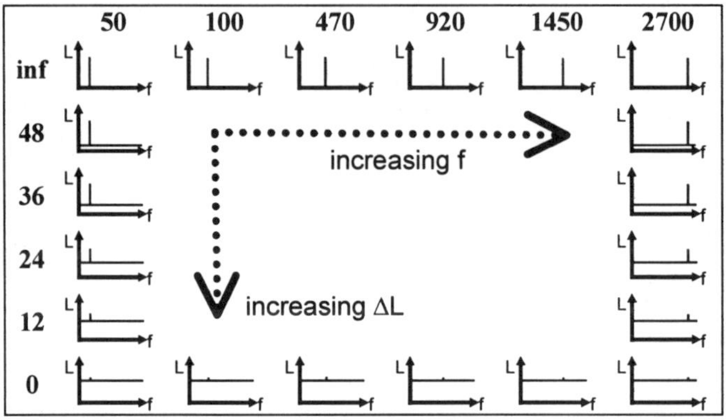

Figure 1. The 36 sounds used for experiments: horizontal axis: frequency of test tone [Hz], vertical axis: noise-level $L_N = L_T - L_G - a_v$ [dB].

4 Results:

Both experiments show very good (individual) consistency of the subjects. The results are reliable and reproducible. The overall consistency among all subjects ('Akkordanz' [3]) is relatively high and very significant. The results are shown in tables 1 and 2: The numbers n in the tables account for the quantity how often (how many times) the signal is rated more tonal as n other signals in the same experiment. Because there are 18 different sounds the maximal number is $n = 17$. All figures show averaged data of the corresponding groups.

4.1 Experiment 1:

- 11 Subjects: 2 female, 9 male, age 22 to 30, normal-hearing
- individual consistency [3]of subjects: between 0.73 and 0.99, average 0.89
- cyclic-triads [3] of paired comparison: between 66 and 3, average 27 (204 possible)
- significance-level [3] (SL): >0.9995 for all subjects
- overall consistency ('Akkordanz' [3]): over all subjects: 0.51, SL >0.999
- the subjects show in general 2 different behaviours: group1: a pure tone is the maximum possible tonal noise, group2: a pure tone is not the max. tonal noise
- group1: 8 subjects, overall consistency ('Akkordanz' [3]): 0.70 SL >0.999
- group2: 3 subjects, overall consistency ('Akkordanz' [3]): 0.55 SL >0.999

All	50	100	470	920	1450	2700
inf		10,7		13,2		12,8
48		9,4		13,4		12,8
36		7,5		11,9		12,0
24		5,5		10,8		10,5
12		3,7		7,4		6,4
0		0,6		2,4		1,9

Table 1. Results for experiment 1: the upper left table contains data averaged over all 11 subjects, the lower left shows the averaged data for group 1, in the lower right table data is averaged over the subjects of group 2.
The numbers in the tables account for the quantity how often the signal is rated more tonal as all other signals.

Gr.1	50	100	470	920	1450	2700
inf		13,5		16,6		15,4
48		9,3		13,4		12,4
36		7,1		11,4		11,3
24		5,1		9,5		9,3
12		3,3		6,4		5,3
0		0,5		1,8		1,8

Gr.2	50	100	470	920	1450	2700
inf		3,3		4,0		6,0
48		9,7		13,3		14,0
36		8,7		13,3		14,0
24		6,3		14,3		13,7
12		5,0		10,0		9,3
0		1,7		4,0		2,3

For this experiment the subjects split up into 2 groups: 8 of 11 subjects perceive a pure tone as the most possible tonal noise, the remaining 3 stated the pure tones to be a different perception "something else but a tonal noise". The first group rates pure tones with a frequency about 900 Hz as most tonal (table 1, lower left panel). For every constant L_N the 900 Hz-tone is most, the 50 Hz-tone rated least tonal (group 1). All subjects are experienced in psychoacoustic experiments, but it is remarkable, that group 1 consists only of natural scientists, whereas the subjects of group 2 are non-natural scientists.

4.2 *Experiment 2:*

- 9 Subjects: 1 female, 8 male, age 22 to 30, normal-hearing
- individual consistency [3] of subjects: between 0.67 and 0.94, average 0.85
- cyclic-triads [3] of paired comparison: between 14 and 80, average 35 (204 possible)
- significance-level [3] (SL): >0.9995 for all subjects
- overall consistency ('Akkordanz' [3]): over all subjects: 0.46, SL >0.999

All	50	100	470	920	1450	2700
inf	8,2	9,1	12,4	13,9	13,8	12,7
48						
36	4,6	6,4	9,8	11,2	11,9	10,9
24						
12	0,3	1,9	5,8	6,7	6,9	6,6
0						

Table 2. Results for experiment 2: averaged data over all 9 subjects.
The numbers in the tables account for the quantity how often the signal is rated more tonal as all other signals.
A frequency dependency with a maximum tonality in the range of 920 to 1450 Hz is obvious.

In this experiment a similar frequency-dependence for tonality is obtained as in experiment 1: the maximum tonality for the test-tone is perceived in the frequency region between 920 and 1450 Hz. The pure tone with a frequency of 50 Hz is more tonal as any sound with a tonal component and L_N = 12 dB. This is remarkable due to the fact, that frequencies lower than about 100 Hz are generally suggested not to evoke the sensation of tonality [1].

5 Discussion

- The experiments show that frequency has a distinct effect on tonal-noise perception, which is contradictory to the DIN45681 [1].
- The fact that sounds with a tone of 50 Hz and high L_N values are rated more tonal than tones with higher frequencies but smaller L_N values is contradictory to the DIN45681 [1], too.
  $\Rightarrow$ both points requires the DIN45681 to be modified
- It seems that different subjects have different perceptions of tonal noise, especially the sense of pure tones is still not clear.

6 Future Research:

Additional experiments with many more subjects for both experiments have to be done to verify these results and to be able to use different statistic methods such as analysis of variance. Especially the difference between natural scientists and 'other' subjects has be further investigated.

The aim of this is to modify and improve the DIN45681 [1]. In the following step the DIN45681 will be implemented on a computer to receive data which is in line with subjective ratings.

Therefore the influence of the selection of subjects must be explored and taken into consideration adequately.

References

1. Deutsche Norm, DIN 45681 (draft), Januar 1992, Detection of tonal components of noise and determination of a tone adjustment for the assessment of noise immissions.
2. Daniel, P., Vormann, M. and Ellermeier, W., Berechnung der Tonhaltigkeit von Reifengeräuschen mit einem Modell der Tonhöhenwahrnehmung, Fortschritte der Akustik – DAGA 98, Oldenburg: DEGA, in print
3. Bortz, J., Lienert, G.-A. and Boehnke, K., Verteilungsfreie Methoden in der Biostatistik, Springer-Verlag Berlin Heidelberg 1990.

FACTORS INFLUENCING EQUAL- LOUDNESS LEVEL CONTOURS

C. RECKHARDT AND V. MELLERT
Physik Akustik Universität Oldenburg,
D- 26111 Oldenburg
E-mail: carsten@aku.physik.uni-oldenburg.de

B. KOLLMEIER
Medizinische Physik Universität Oldenburg
D- 26111 Oldenburg

This work investigates the influence of the experimental setup on the resulting equal-loudness levels contours (ELLC) obtained with an adaptive 2 alternative forced choice procedure. The amount of change in the resulting contours is measured for different parameters. For a simple adaptive procedure the influence of the starting level is smaller than the range effect when using the method of constant stimuli. The effect of the starting level is reduced when using a large initial step size in the adaptive procedure. However the effect can still amount 6 dB. This effect diminishes when using an interleaved procedure. The maximum measurement error for the different procedures is estimated. In addition, it is shown that interindividual variance is reduced when the individual threshold is taken into account.

1 Introduction

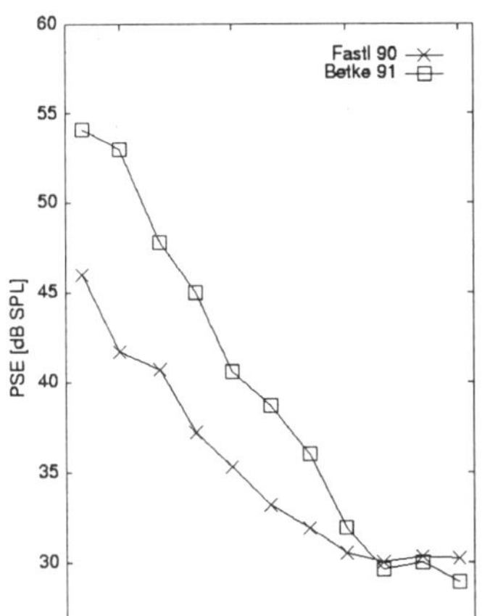

In the literature considerable differences between data on equal loudness level contours (ELLC) exist. These differences are partly explained by the experimental context, e.g. when using the method of constant stimuli for the determination of ELLC the point of subjective equality (PSE) is shifted up to 13 dB when changing the absolute position of the presented test tone levels [1]. Some of these differences can be attributed to the experimental context and described by a simple adaptation model [1]. Applied to data from literature the model can not minimize variances between different laboratories [2].

Figure 1. 30 phon curve from Betke and Fastl obtained with the method of constant stimuli.

2 Experiments

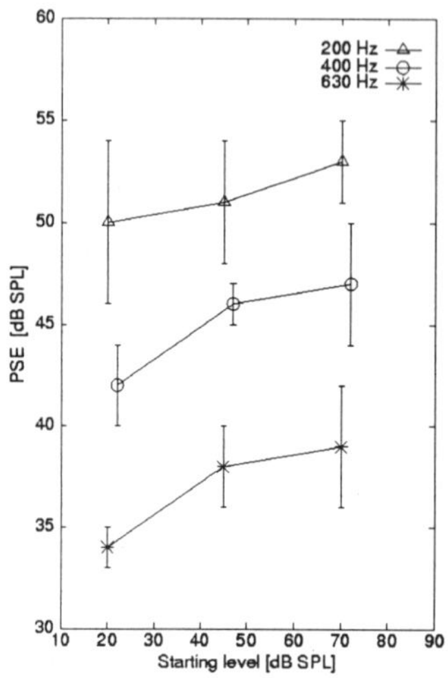

The influence of different parameters on the results obtained with an adaptive 2-AFC 1- step procedure is tested. These parameters are the starting level of the test- tone and the initial step size for the adaptive procedure.

6 subjects participated in the experiments. The test- tone frequencies are 200, 400, 630 and 1000 Hz. The frequency of the reference is 1000 Hz and the level is 30 and 50 dB. Duration for both signals is 1 s with a pause of 500 ms between signals. Test and reference signal were presented in random order via headphone.

Figure 2. Mean PSE and standard deviation for 4 test- tone frequencies averaged over 6 subjects. Level of the reference 30 dB. Initial step- size 8 dB. X-axis starting level for the adaptive procedure.

The largest influence of the starting level is 5 dB for the 630 Hz test-tone. The effect of increasing PSE with increasing starting level occurs for all frequencies.

3 Individual differences and threshold of hearing

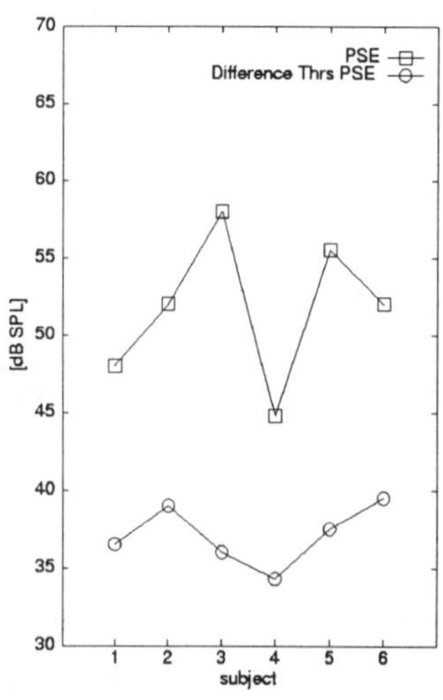

Measured PSE and data from the literature show considerable differences especially at low levels and test tone frequencies far away from the reference frequency. To explain these differences the individual threshold of hearing is measured for every subject at the test- tone frequencies. Taken into account the individual threshold of hearing standard deviation between subjects is halved. The obtained contours could be interpreted as ELLC sensation level (ELLC -SL).

Figure 3. Individual PSE (starting level 45 dB) and values subtracted by individual threshold of hearing.

4 Influence of initial step- size

The influence of the initial step- size is tested for starting levels of: 30, 50, 70 dB SPL. In addition for 200 Hz a starting level of 90 dB SPL is presented. The initial step sizes are 8, 16 and 4 dB. The smallest step- size is tested last in order to reject the hypothesis training effects could minimize the influence of the starting point.

f [Hz]	30 – 50 dB	50 –70 dB
200	0,9	2,8
400	0,7	2,0
630	1,8	2,4
1000	0,5	-0,1

Table 1. Difference between starting level 30 and 50 dB and 50 and 70 dB for 4 test- tone frequencies. Initial step- size 8 dB.

f [Hz]	30 – 50 dB	50 –70 dB	70 –90 dB
200	1,5	0,5	4,6
400	-0,1	0,5	-
630	-0,1	1,8	-
1000	0,3	-0,4	-

Table 2. Initial step- size 16 dB

f [Hz]	30 – 50 dB	50 –70 dB	70 –90 dB
200	1,3	5,1	4
400	1,3	2,6	-
630	2,7	3,9	-
1000	0,5	0,1	-

Table 3. Initial step- size 4 dB.

Increasing the initial step -size minimizes the effect of the starting level. Extreme starting levels shift the PSE also for large initial step- sizes, e.g. 60 dB change of the starting level leads for the initial step- size of 16 dB with an adaptive procedure to a difference of 6,6 dB.

The largest effect using an adaptive procedure is found with an initial step- size of 4 dB. In this case a 60 dB change of the starting level shifts the PSE 10,4 dB. In comparison to these results using the method of constant stimuli a 30 dB change of the absolute level shifts the PSE about 13 dB [1].

5 Interleaved Procedure

Instead of measuring equal loudness level contours for each frequency in a different run, in the interleaved procedure the test- tone frequencies are presented in a random order [3].

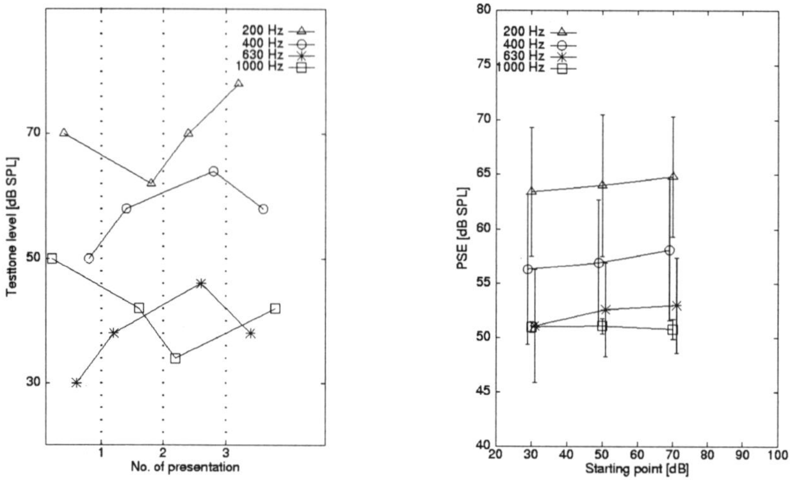

Figure 4. Left: Example for the presentation of the test signals in the interleaved procedure. Initial step-size 8 dB. Right: Mean PSE and standard deviation for 4 test-tone frequencies averaged over 6 subjects. X-axis starting level for the adaptive procedure.

The difference between the starting level is less than 2 dB. In addition the same experiment is done with starting levels of 50, 70 and 90 dB. The difference between the obtained PSE is smaller than 2.5 dB.

6 Summary and conclusion

The influence of the presented test- tone levels on the obtained ELLC is smaller with an adaptive procedure than using the method of constant stimuli. Large initial step-sizes reduce the effect of the starting point but extreme starting levels can shift the PSE, too. The influence of the starting point is minimized when an interleaved procedure is used. Interindividual variances are larger than intraindividual. These differences can be explained by taking into account the individual threshold of hearing .The obtained curves can be interpreted as "equal- loudness level contours sensation level" (ELLC-SL).

References

1. B.Gabriel: Equal-loudness Level Contours: Procedures, Factors and Models. Shaker Verlag Aachen 1996 ISBN 3-8265-2049-1
2. V.Mellert, C.Reckhardt: Modellierung von Rangeeffekten bei der Bestimmung von Ispophonen, In: Fortschritte der Akustik, S. 498 + 499 DAGA '97 Oldenburg, DEGA e.V.1997
3. J.Verhey, et al.: Temporal aspects of loudness summation, this proceedings

SOUND AND VIBRATION AT LOW FREQUENCIES

MICHAEL A. BELLMANN, VOLKER MELLERT, CARSTEN RECKHARDT AND HERMANN REMMERS*

Physik Akustik, Universität Oldenburg

**Institut für Technische und Angewandte Physik GmbH*

D-26111 Germany

E-mail: michael@aku.physik.uni-oldenburg.de

1 Introduction

Equal-loudness level contours are assumed to reflect the frequency sensitivity of the auditory system. Studies on equal-loudness level contours (ELLC) show considerable variations, even when the same measurement method is used. Recent studies [1, 6] report about the influence of measurement method, choice of test tone level and threshold of hearing. Moreover, in literature only few data at low frequencies exist [e.g. 5]. Since hearing of low frequency sound at high levels is often combined with the perception of vibration. Additional investigations are carried out to determine ELLC in the frequency range of 16 to 160 Hz at 60 phon. In this study an adaptive 2-AFC procedure [6, 7] is used, which minimises the influence of the experimental procedure, e.g. the measured data are not influenced by starting levels of test tone. In a study of Reckhardt et al. is shown, that the interindividual variance is decreased by taking the individual threshold of hearing into account. Therefore, the threshold of hearing in a frequency range of 40 to 160 Hz is measured as well.

Additionally the influence of vibration on loudness perception is examined. First, the perception threshold of vibration is determined in a frequency range of 16 to 80 Hz. Then the influence of vibration on loudness perception is measured in presence of distinct vibration levels above threshold.

2 Experiment 1: Equal-loudness level contours

First, a “new reference level” at 100 Hz is measured with an adaptive 2-AFC method with a reference frequency of 1 kHz. Then ELLCs are measured at 60 phon with an adaptive 2-AFC interleaved method, with an initial step size of 8 dB and an ending step size of 2 dB. Test and reference signals are sinusoidal tones with a duration of 1 s and 500 ms pause. The test frequencies are in 1/3-octave steps from 16 to 160 Hz, except 50 and 80 Hz. The starting level of the test tone is varied 25 dB (ISO value -10 dB, +15 dB) around the ISO 226 [2]. 6 normal hearing subjects

(3 female and 3 male) take part in this measurement in a pressure chamber. The results are higher than the standard curve specified in ISO 226 (Fig. 1). No systematically dependence of the starting levels of test tone can be observed.

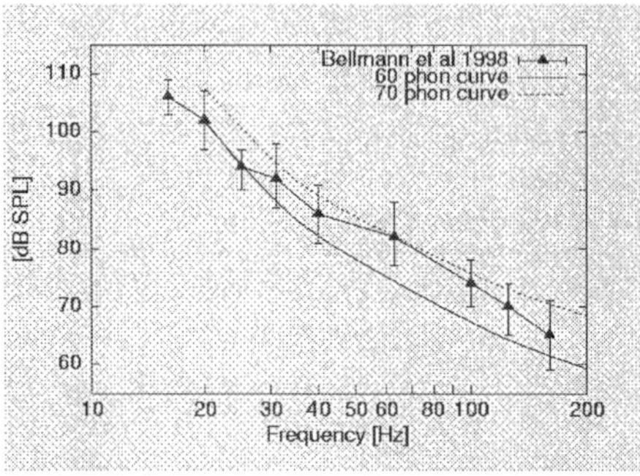

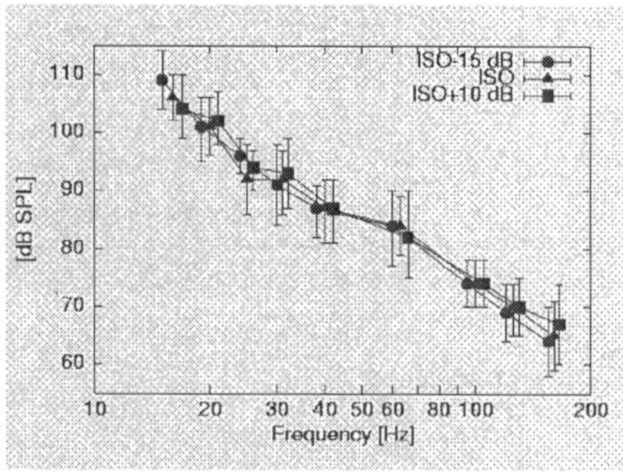

Figure 1 and 2: Equal-loudness level contours of 60 phon in a frequency range of 16 to 160 Hz and the standard 60 and 70 phon curves specified in ISO 226 [2] are shown in the left figure. The mean values and deviations of ELLC for different starting levels of test tone are in the right figure. The starting levels are a) ISO-level minus 15 dB, b) ISO-level and c) ISO-level plus 10 dB. For better illustration, the frequencies for different start levels are slightly shifted.

3 Experiment 2: Auditory threshold

To investigate the influence of the individual threshold of hearing on the interindividual variance, the auditory threshold is measured in a frequency range of 40 to 160 Hz for 5 normal hearing subjects (2 female and 3 male). An adaptive 3-AFC method is used. The duration of signals is the same as in Exp. 1 but the ending step size is 1 dB. Test frequencies are in 1/3-octave steps from 40 to 160 Hz. The measured data are like the standard data of ISO 389-7 [4]. For the test frequency of 100 Hz, the measured data of ELLC and ELLC SL (ELLC minus individual threshold) are shown in figure 5. The interindividual variance for the other test frequencies looks similar. The results from low frequencies support the hypothesis of Reckhardt [6], that the interindividual variance of ELLC decreases by taking the individual threshold into account.

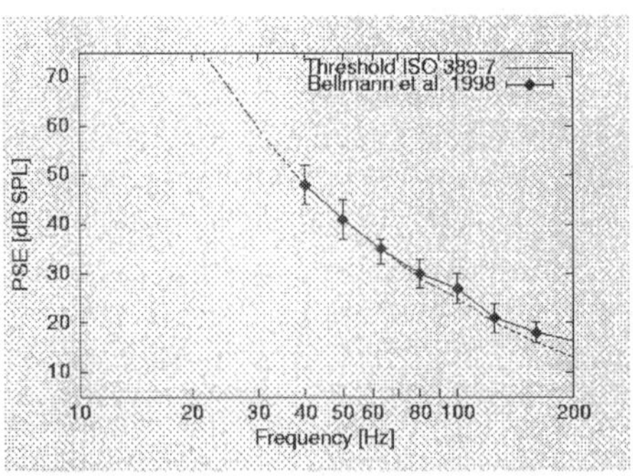

Figure 3: Auditory threshold of 5 normal hearing subjects and the standard threshold specified in ISO 389-7 [4].

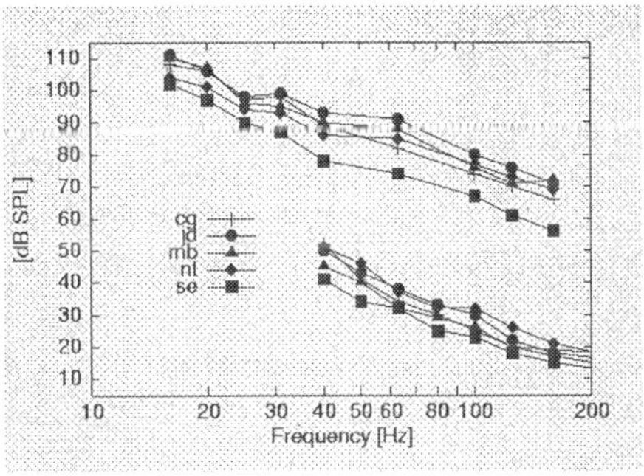

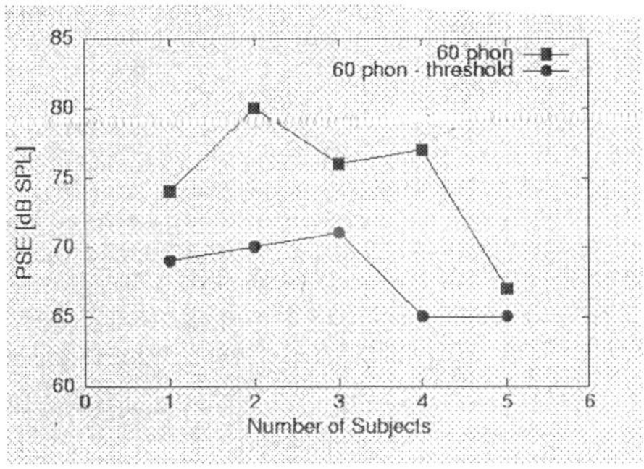

Figure 4 and 5: In the left figure are results of 5 normal hearing subjects for the 60 phon curve and the auditory threshold. Measured data of ELLC and ELLC SL (20 dB shifted for better illustration) for the test tone frequency of 100 Hz for 5 subjects are shown in the right figure.

4 Experiment 3: Perception of vibration at frequencies lower than 80 Hz

Hearing of low frequency sound at high levels is often combined with the perception of vibration. Therefore the influence of vibration on loudness perception is investigated in the following experiment. First, the perception threshold of vibration is determined in a frequency range of 16 to 80 Hz. The Sound and Vibration Reproduction System (SVRS) developed at itap, called „vibrationpad", produces vibrations with accelerators in three direction, which are fixed on a board (100 x 130 x 4 cm). Sound can be reproduced by earphone or loudspeaker. The frequency range of each accelerator is 10 to 125 Hz. The perception threshold of vibration is measured for 6 normal subjects (2 female and 4 male) with the same measurement method as in Exp. 2 and the test frequencies are in a range of 16 to 80 Hz in 1/3-octave steps. Measured threshold increases 2,5 dB from 16 to 31,5 Hz and decreases for higher frequencies. In this set up the perception of vibration is limited at 40 Hz, because the emitted sound is above the auditory threshold.

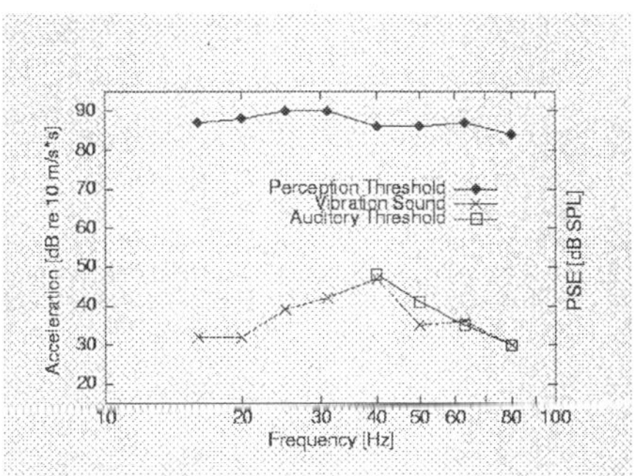

Figure 6: The perception threshold of vibration in z-direction (Vibration threshold) is in this figure. Additionally the auditory threshold of the subjects is shown, because the board emits sound due to the vibrations (Vibration sound).

5 Summary and conclusion

In this study equal-loudness level contours (ELLC) are measured in a frequency range of 16 to 160 Hz at 60 phon with an adaptive 2-AFC interleaved method. The measured data are higher than the standard curves specified in ISO 226 [2]. Influence of the starting test tone level cannot be observed. Additionally the auditory threshold is measured in a frequency range of 40 to 160 Hz, measured data look like the standard curve specified in ISO 389-7 [4]. The results from low frequency measurements support the hypothesis of Reckhardt et al. [6], that the interindividual variance of ELLC decreases, if the individual auditory threshold is taken into account (ELLC SL). The variance of ELLC SL is 6 to 10 dB in comparison to 10 to 15 dB at ELLC or auditory threshold. Another aspect of this study is to investigate the influence of vibration on loudness perception. Therefore the perception threshold of vibration in a frequency range of 16 to 80 Hz in z-direction is determined. This threshold increases frequency dependent from 16 to 31.5 Hz approximately 2,5 dB and decreases for higher frequencies with 3 dB/octave. The perception of vibration is limited in this set up by the emitted sound of the "vibrationpad". In the next step the influence of vibration to ELLC will be examined in presence of distinct vibration levels above the perception threshold of vibration in a frequency range of 16 to 80 Hz.

6 References

1. Gabriel B. et al., Modellierung von Kontexteffekten bei der Messung von Isophonen, *Fortschritte der Akustik,* DAGA 97 (1997) pp. 338-339
2. ISO /R 226, Normal equal-loudness contours for pure tones and normal threshold of hearing, (1987)
3. ISO 2631-1, Evaluation of human exposure to whole-body vibration Part 1: General requirements, (1997)
4. ISO 389, Acoustics-Reference zero for calibration of audiometric equipment - Part 7: Reference threshold of hearing under free-field and diffuse-field listening conditions, (1996)
5. Lydolf M. and Møller H., New Measurement of the Threshold of Hearing and Equal Loudness Level Contours at Low Frequencies, *8th i. Meeting on Low Frequency Noise and Vibration* (1997), p. 76-84
6. Reckhardt C., Mellert V. and Kollmeier B., Quantifizierung der Einflußgrößen bei der Bestimmung von Isophonen, *Fortschritte der Akustik,* DAGA 98 (1998), (in press)
7. Verhey J. L. and Kollmeier B., Messungen zur zeitabhängigen Lautheitssummation, *Fortschritte der Akustik,* DAGA 98 (1998), (in press)

III. Recent Advances in Models of Auditory Processing

PECULIARITIES IN TRANSIENT RESPONSES IN LINEAR AND NONLINEAR COCHLEA RESPONSES

H. DUIFHUIS

Department of Neurobiophysics, Research School for Behavioral and Cognitive Neurosciences (BCN), University of Groningen, Nijenborgh 4, 9747 AG Groningen, Netherlands.

This study presents results from our time domain nonlinear model of the cochlea. After a general presentation of the model structure, some noticeable aspects of the cochlear time response are shown, in particular for a periodic pulse stimulus.

1 Introduction

Cochlear nonlinearity now is generally assumed to originate within the cochlea, and more in particular at intact outer hair cells. Although this might be considered the final answer to the question: where are distortion products generated, apparently this is not yet the case. The solution is complicated by the fact that the phenomenon has been observed at a limited number of sites, which has led to the 'concept' of *generation site*. Although such a concept may be helpful in a thought experiment, the real system does not deal with one (or a few) generation site(s); it rather provides a generation area, or many distributed sites. Moreover, the distortion products generated at these areas are subject to reflections at places where coupling of energy flux is not optimal. This will occur at the middle ear interfaces, but also at the helicotrema. The basic effects of those reflections are already visible in simplified models of the cochlea.

One of the basic aspects of biophysical models of the cochlea is that they treat the coupling of the different elements in a proper way. That coupling involves two-way traffic, which is essential for any nonlinear structure. [In a linear structure it can be effectively hidden by the impedance concept, which in fact involves summation of all interaction, resulting in a net amplitude and phase behavior.] The involvement of the nonlinearity in principle invalidates the normal systems approach in the frequency domain. It is more useful to analyze nonlinear systems in the time domain. The increase of computer power over the last years now has made such an approach feasible, even on PC's or notebooks, if one can accept a processing speed that is about 100× real time.

2 Some basic specifications

A traditional 1-dimensional model is used for the computation of the presented results. That implies that simplifications include the assumptions of: rigid walls of the cochlea, open at round and oval windows only; inviscid, linear fluid; ignored coiling; 2-channel model (ignore Reissner's membrane); rectangular channels; properties of elements of the structure of the organ of Corti compiled in a cochlear 'partition'; etc.

An approximation of the 3-D rectangular cochlea can be formulated in terms of a tridiagonal matrix equation, where the off-diagonal elements determine the longitudinal (fluid-) coupling. The equations are constrained by physical conservation laws. One of the most important is the conservation of energy and energy flow. The energy balance involves all sources. In case of distortion, the spectral distribution of energy is not conserved. Obviously, energy involves amplitude and velocity. The ratio between these two is no longer essentially constant, and therefor no longer expressible as impedance [further refs: 1-5].

Because of the feasibility to use regular PC's for the model study, we are in the process of presenting a general-purpose time domain cochlea model, with the option of several tunable features. The general setup is shown in Figure 1. The basic element is the kernel that solves the matrix equation of cochlea and middle ear as fast as possible, with a reasonable and reliable number of spatial elements and time steps. The matrix equation is solved using Gaussian elimination; the time equation uses a Runge-Kutta-4 method. This kernel interacts with the input signal and with the partition characteristics. The latter are adjustable. Although the kernel operates most efficiently under Fortran 90, interfaces for input and

Model Structure

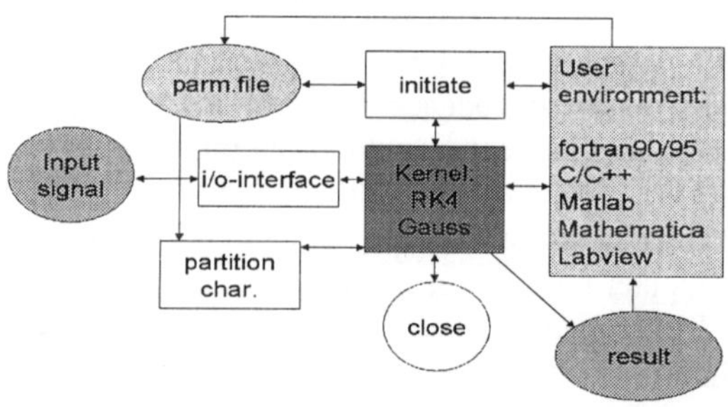

Figure 1. Structure of the cochlea model. The kernel executes the basic computations, which solve the matrix equation. The program is started with an initiation block, which interprets and sets options and parameters specified in the parameter file. The kernel interacts continuously with the i/o-interface and with the partition characteristics. Results are provided in data files and/or they are visualized using, e.g., MATLAB.

output with other platforms are provided. Thus, input signals as well as results can be connected in several formats.

Since the interaction between kernel and partition characteristics is continuous, it enables inclusion of feedback, thereby modeling efferent adaptation effects.

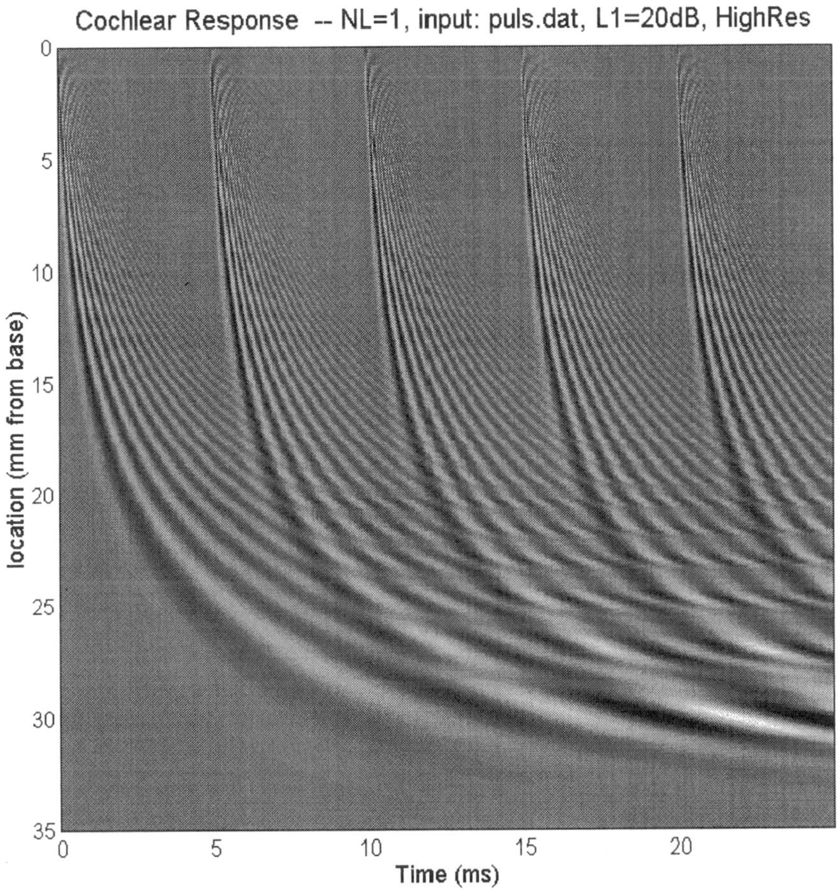

Figure 2a. Cochlear partition response profile: first 25 ms of the velocity response to a periodic pulse with a repetition frequency of 200 Hz (period 5 ms). Gray scale provides a 'linear' code of velocity. Note the increasing interaction with increase of time. Obviously, the responses to the lower harmonics build up relatively slowly. At the apical end of the cochlea we observe changing gradients over at least 100 ms.

3 Time response

Figures 2a-b present the response to a 200 Hz periodic pulse at 20 dB SPL. Figure 2a gives the first 25 ms of this response, Fig. 2b give the response from 75-100 ms. [The artefact around x_{BM}= 5 ms displays graphical interference in the presentation; it does not reflect interference in the actual cochlear response.]

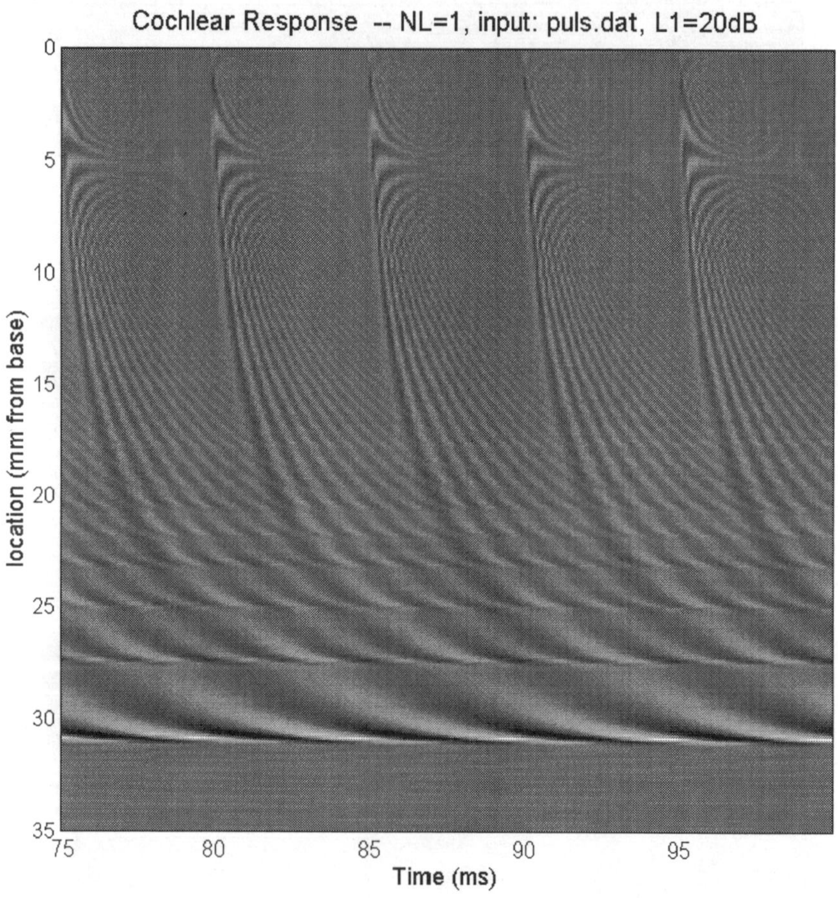

Figure 2b. Continuation of the response (75-100ms). Although the response is relatively stable, the basal end of the cochlea still shows onset effects. The 'group delay' shows clear discontinuities near the peak responses to the lower harmonics [at approx. 31, 27, 24.5, 23, etc. mm from the base].

In this case the cochlea parameters were linear. Hence, the presented response profile (obtained at 20 dB) is not sensitive to level. The Figures used automatic amplitude scaling. The profile, however, does depend on the termination of the auditory canal (open in Fig. 2). Figure 3 show a response profile determined from 50-75 ms after stimulus onset, for open (3a) and plugged (3b) cochlea. [The interference artefacts mentioned before are present in all pictures. Figure 2a was computed for a higher spatial resolution than the others; it shows less interference.]

Note that the response profile is more evenly distributed in the nonlinear cochlea, and that the response envelope depends on the presence of the ear canal plug.

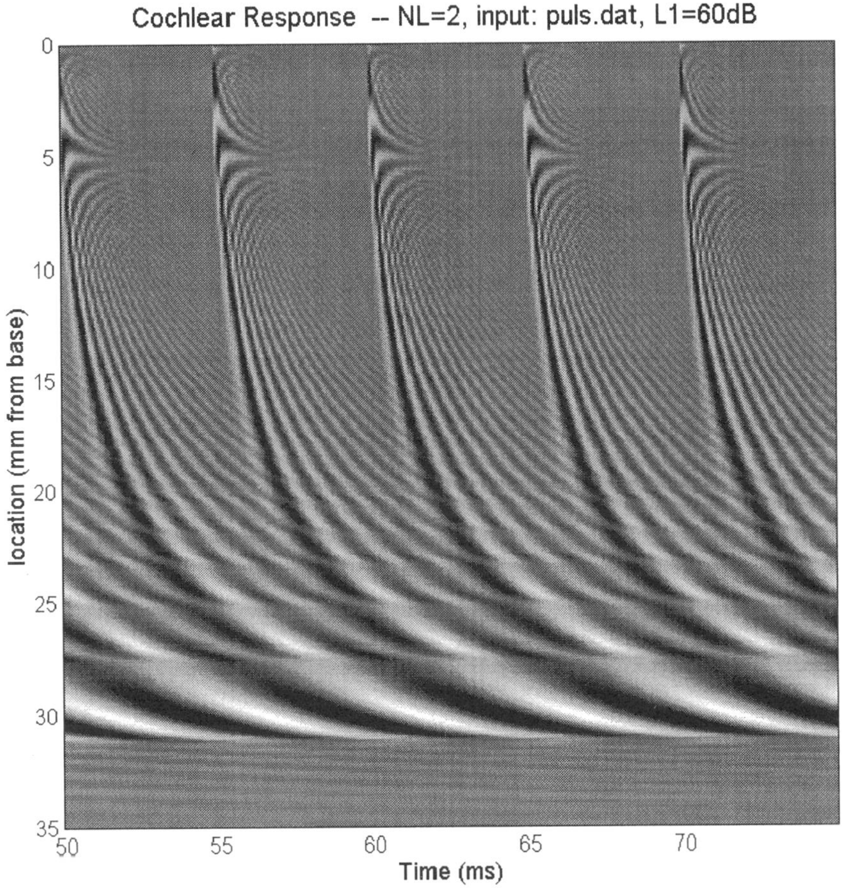

Figure3a. Response of a nonlinear (passive) cochlea to a periodic pulse at 60 dB SPL. Time interval: 50-75 ms after stimulus onset. Ear canal open.

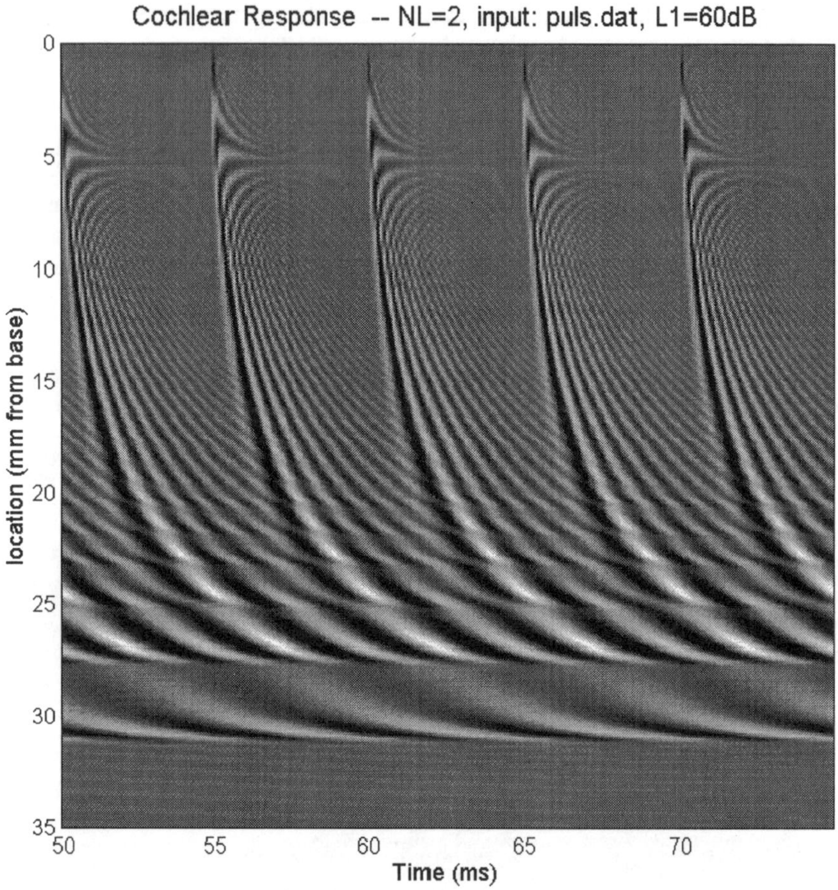

Figure 3b. As Fig. 3a, but ear canal plugged.

4 Discussion

The results shown here are from a linear and from a passive nonlinear 1-D cochlea model. In the nonlinear version the nonlinearity was determined by an exponentially increasing damping. Although many aspects of the responses were to be expected, the discontinuities, e.g., at x_{BM} = 25 mm, indicate significant but unexpected gradients, both in place and in time. These are partly due to the fact that spatial resolution is limited, and that transition from 2nd to 3rd to harmonic does predict temporal discontinuity. However, the phase changes (and related amplitude changes) over this transition region are larger than expected.

We also find that it takes a long time for the response to stabilize, in particular at the low-frequency end, near the helicotrema. It takes at least 100 ms for the apical response to settle. Obviously, this is not only determined by the local properties of the partition, but also by reflections at the apex. Although the apical coupling between the scalae may be quite optimal for a limited frequency range, it is virtually impossible that his coupling would be optimal over the entire audio range, and that for all ears.

In conclusion: the time domain model of the cochlea can now be used effectively and reliably for the analysis of the behavior of the peripheral ear. On the one hand this can unravel traditional questions, on the other, it will provide new predictions and new questions.

References

1. Duifhuis, H., Hoogstraten, H. W., van Netten, S. M., Diependaal, R. J., and Bialek, W. (1985). Modelling the cochlear partition with coupled Van der Pol oscillators, in *Peripheral Auditory Mechanisms,* edited by J.B. Allen, J.L. Hall, A. E. Hubbard, S. T. Neely, and A. Tubis (Springer, New York), pp.290-297.
2. Diependaal, R. J., Duifhuis, H., Hoogstraten, H. W., and Viergever, M. A. (1987). Numerical methods for solving one-dimensional cochlear models in the time domain, *J. Acoust. Soc. Am.* 82, 1655-1666.
3. Duifhuis, H. (1988). Cochlear macromechanics, in: *Auditory Function,* edited by G. M. Edelman, W. E. Gall, and W. M. Cowan (Wiley, New York), pp. 189211.
4. Van den Raadt, M. P. M. G., and Duifhuis, H. (1990). A generalized Van der Pol-oscillator cochlea model, in *The Mechanics and Biophysics of Hearing,* edited by P. Dallos, C. D. Geisler, J. W. Matthews, M. Ruggero, and C. R. Steele (Springer, Berlin), pp.227-234.
5. van Hengel, P. W. J., Duifhuis, H., and van den Raadt, M. P. M. G. (1996). Spatial periodicity in the cochlea: The result of interaction of spontaneous emissions? *J. Acoust. Soc. Amer.* 99, 3566-3571.

THE AUDITORY PERIPHERY AS A SIGNAL PROCESSOR

RAY MEDDIS

Centre for the Neural Basis of Behaviour at Essex, Department of Psychology, Essex University, Colchester, UK, CO4 3SQ

email rmeddis@essex.ac.uk

Auditory signal processing begins long before the signal reaches the brainstem. The waveform is first split by the head into two waveforms. The pinnae introduce further spectral transformations that assist with estimating the elevation of the source. The ear canal and the middle ear action of the malleus incus and stapes attenuate potentially interfering very low and high frequency signals. Mechanical filtering of the signal in the cochlea can help to identify the spectral profile of the sound but may also have far reaching implications for segregating simultaneously present sounds sources through subsequent channel selection. The nonlinear response of the cochlea compresses the signal into a more manageable narrow dynamic range. It also cunningly fails to compress attenuated signals remote from best frequency with the effect of smearing information usefully across the array of mechanical filters. Electrical transduction at the inner hair cell transforms the signal in a number of ways. The process of adaptation makes it easier to identify the onset of a sound (or its modulation). Half-wave rectification and low pass filtering isolate the envelope of the signal for high frequency carriers while leaving the fine structure in place for low frequency signals. All this sophisticated signal processing occurs before the nervous system can become involved. As a consequence auditory modelling needs to replicate all of these processes if psychophysical phenomena are to be explained and predicted with quantitative accuracy. However, each one of these stages could easily occupy a lifetime of research. An important problem in modelling, therefore, is knowing how much detail is required and how much can be omitted. The answer lies in an understanding of what each stage contributes to the listening experience.

1 Introduction

Models come in all shapes and sizes, serving a range of different purposes. When producing a model it helps to specify from the beginning what kind of model is intended and what purpose is to be served. Below we shall be discussing computational models that accept an arbitrary acoustic signal and generate a prediction of the response of the mammalian auditory nerve (AN) bundle as a function of time since the onset of the signal. The computations are performed in the time domain. The purpose of the models is to provide an input into other models and systems such as speech recognisers, auditory prostheses, assessments of the quality of transmission channels, models of auditory brainstem functioning, etc.. This kind of model of the auditory periphery must therefore be judged by its ability to emulate the response of individual nerve fibres and the practicality of using such a model in the desired context.

In principle, the model could consist of a simple transform accepting acoustic input and generating acoustic output. Such a transform might be inferred from measurements of AN fibre activity in response to a range of sound stimuli. Unfortunately, the signal processing activity taking place in the auditory periphery is too subtle for such an approach. AN fibre response to sound has many interesting features that must be reproduced in a useful model. For example, the response in a given fibre is typically limited to a narrow range of frequencies; its strength may or may not be nonlinear with respect to signal amplitude; it adapts in strength as the stimulus continues, it reflects stimulus fine-structure at low frequencies but stimulus envelope at high frequencies; it is stochastic in nature; it depends on the location of the source of the sound. Moreover, there are two AN bundles (left and right), each producing different responses to the same sound. Altogether, the response is so complex that a simple transform is unlikely to be found and another approach is required.

The most popular approach so far has involved the idea that the auditory periphery can be simulated using a cascade of separate processes each contributing one or more of the properties described above. Such a system might involve a head-related transfer function (HRTF), compression, a filter bank, half-wave rectification, low-pass filtering, automatic gain control and random spike generation. A refinement of this approach is to identify each sub-transform with a known physiological process occurring in the auditory periphery such as reflection and refraction in the pinnae, vibration response of the organ of Corti, mechanical to electrical transduction at the inner hair cell, and generation of AN spike activity at the hair-cell/ AN synapse. The advantage of this latter approach is that it allows each stage of the process to be tuned using data from animal measurements. The disadvantage is that the model becomes highly detailed and risks becoming computationally too slow for practical purposes. Fortunately, computer technology (including DSP and VLSI technology) has developed quickly enough for this not to have become an issue for most models so far.

Below we shall discuss some aspects of each stage of the sequence. Space does not permit a comprehensive review, but the basic principles will be illustrated using the models in use at the Centre for the Neural Basis of Behaviour at Essex. These models are readily available in the form of a computer routine library embedded in a development system for auditory modelling (DSAM) accessible over the internet at http://www.essex.ac.uk/hearinglab. The model will be presented as if it were a simple cascade (feed-forward system), even though we know that feedback loops do exist both within the cochlea itself and also involving efferent AN fibres.

2 The 'two-ear' transform

The sound source generates a single wave front but the ear uses two receptors located on either side of the head deep inside an ear canal whose entrance is adorned with a pinna and whose shape in humans is best described as complex. What began as a single wave front becomes two different waveforms at the left and right eardrums. The differences are shaped by the pinnae but are also crucially influenced by the effect of the head's position between the two receptors; an effect that can be represented using low-pass filtering to simulate head-shadow and the introduction of a time-delay to simulate the extra travel time around the head.

It is safe to say that we do not yet have successful models based on physical principles (and in a form appropriate for rapid computational evaluation) that generate two location-dependent signals. Modellers either avoid this problem altogether by concentrating on problems involving monaural stimuli presented using headphones or adopt simple strategies such as the introduction of fixed left-right time delays related to stimulus azimuth and low-pass filtering to simulate head-shadow effects.

Efforts in our laboratory to simulate mathematically the effect of the pinnae did result in an expression that might be used to predict changes in the spectral profile as a function of elevation (Fig. 1). However, this method has not yet been used as a component in a larger system. In this example, the model assumes that the concha takes the shape of an open spiral and predicts the peaks and valleys in the spectrum as a function of sound source location. It works moderately well for the main

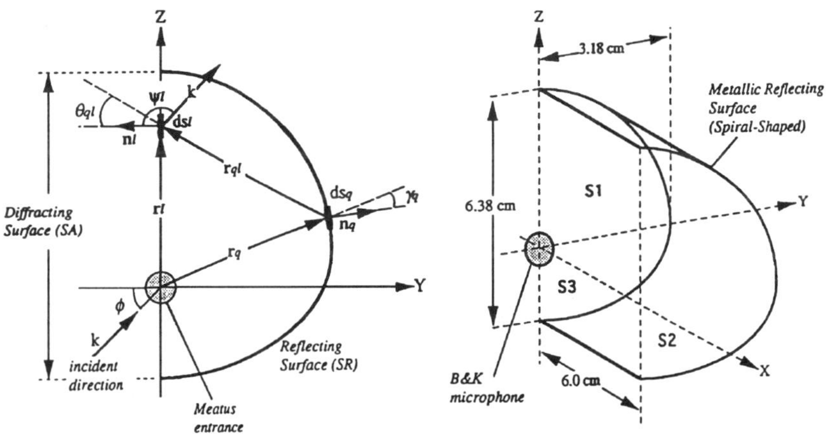

$$\frac{p_T}{p_o} = 1 + \iint_{S_R}\iint_{S_A} \frac{R(r_q,\gamma_q)K(\theta_{ql},\psi_l,r_{ql})}{r_q r_{ql}} \exp\{j[k(r_q + r_{ql}) - \pi/2] + \varphi_l\} ds_l ds_q$$

The transfer function of the concha

Figure. 1. One solution to the problem of modelling the reflections and refractions in the area of the concha. The formula is given purely to illustrate the fact that the problem is non trivial and probably not yet useful for everyday computational use. (Taken from Lopez-Poveda and Meddis, 1996).

spectral peak but is too cumbersome for most applications.

There are two solutions to this problem in current use; the first uses mannequin recordings and the second uses HRTFs. Obviously, the first method depends on the availability of the necessary equipment and may be limited by the artificial pinnae supplied by the manufacturer. At Essex, we have made rubber models of the pinnae of individual subjects to fit on the mannequin. There is no doubt that there are considerable differences between pinnae. Whether these are of practical significance remains an open question. We have noted some benefit from using appropriate pinnae but the effect is typically small. The alternative approach is the use of HRTFs. These functions are discovered by recording broadband signals using probe microphones placed in a listener's ear canal close to the eardrum. The resulting impulse response can then be convolved with free-field stimuli to emulate the wave front at the eardrum. This is more appropriate to a completely computational environment although it also suffers from the problem of requiring a different set of HRTFs for each individual modelled. For most purposes, the use of HRTFs must be the best solution at present for modellers wishing to study binaural phenomena. Beginners might like to examine a Masters thesis by Frank Filipanits Jr. which is available on the internet as a quick introduction (Filipanits 1994).

3 Cochlear filtering

All modern models of the auditory periphery use a bank of filters to simulate the response of the basilar membrane to stimulation. For the most part these are linear filters and are inspired either by psychophysical tuning curves (Patterson and Moore, 1986) or by physiologically measured tips of auditory nerve tuning curves (Evans, 1977). Fortunately, both approaches give similar results in terms of the width of the filters and their gammatone impulse responses. The availability of software linear gammatone filters means that they have rapidly come to dominate the auditory modelling scene. Once again, however, it is important to remain vigilant when using psychophysically-derived filter information to simulate mechanical selectivity on the basilar membrane. This approach implies that there is no other contribution to psychophysical selectivity other than that supplied by the linear aspects of BM functioning. Nonlinear effects such as two tone suppression as well as central effects such as might arise from the use of lateral inhibition may also be at work. Ideally we should work from the physiological to explanations of the psychophysical. In practice, of course, we have no direct knowledge of human cochlear functioning and only limited data on animal cochlear response. Pragmatism is therefore the driving principle in establishing useful filter banks.

Because of anatomical constraints, we only have data on the input/output (I/O) characteristics of BM response at a small number of sites along the membrane whose best frequencies are roughly 0.8, 8, 18, and 36 kHz. I/O functions for first three of these sites are given in Fig. 2.

What is immediately clear in the figures is that the response is decidedly nonlinear for tones near best frequency. BF here means the frequency provoking the largest response near 0 dB SPL. However, the response is clearly linear for tones remote from BF. It is reasonable to require that auditory models are similarly nonlinear in their response.

Figure 3 offers a general scheme that combines simple model components into a larger system that generates the necessary nonlinear I/O functions. It is a phenomenological rather than a mechanical model; that is, it aims to generate the

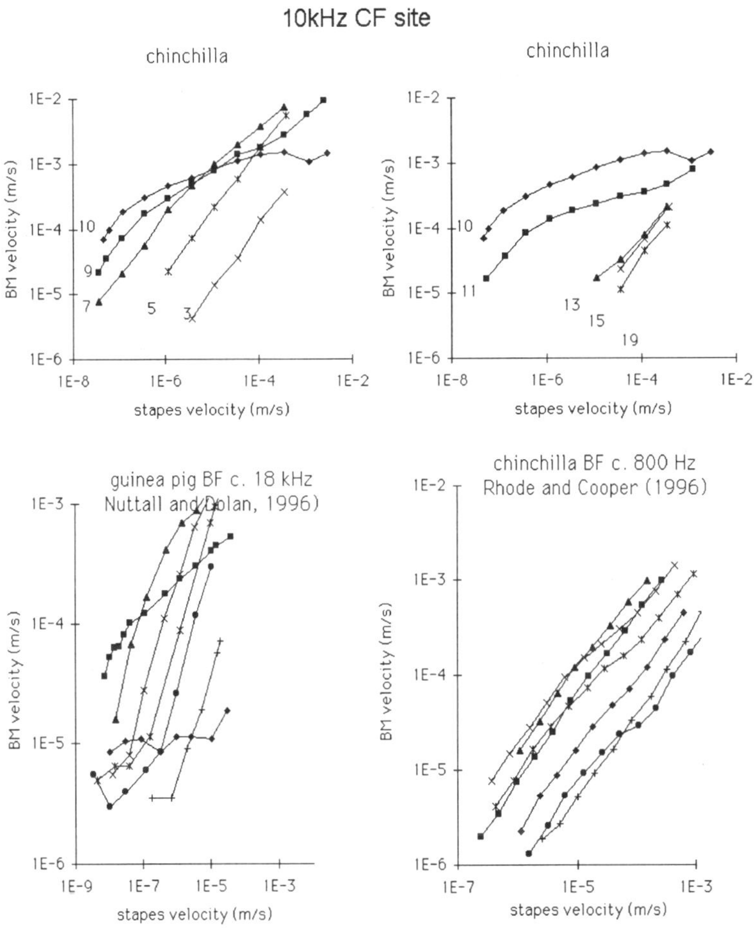

Figure. 2. Input output functions at three sites on the basilar membrane measured using laser velocimetry. The 10 kHz site is based on data from Ruggero et al. (1997) and is represented in two figures where the numerals refer to the frequency of the tone.

right behaviour rather than explain the origin of that behaviour. Figure 4 shows that it is capable of generating I/O functions that are close to those measured using chinchillas and guinea pigs. Of course, different parameters are required at each site but the basic model is unchanged despite the substantial change in the shape of the I/O functions between sites.

The model is a reworking of a number of ideas already present in the literature (e.g. Goldstein, 1990; Pfeiffer, 1970). The model consists of two signal processing paths whose individual outputs are summed to generate a prediction of the resulting motion of the BM. The simpler path consists of an attenuator followed by a wide linear gammatone filter (cf. Goldstein's 'tail filter'). The other path splices a compressive nonlinearity between two third-order gammatone filters (cf. Pfeiffer's bandpass nonlinearity). This corresponds to the sensitive tuning curve tip of AN measurements. The benefits of the system include appropriate distortion products to

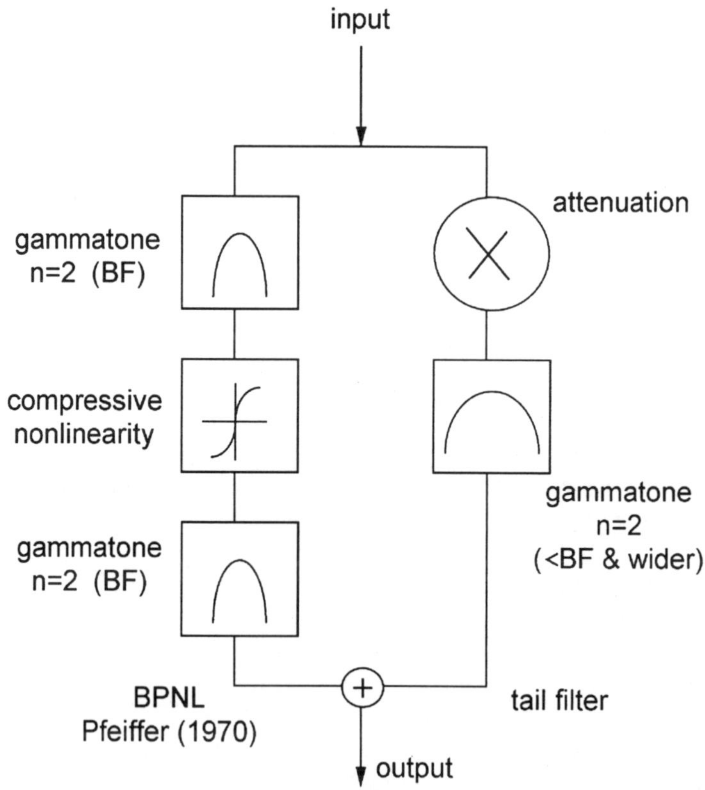

Figure. 3. Dual resonance nonlinear filter model for simulating responses on the basilar membrane (work in progress)

two-tone stimuli and a realistic response to intense stimuli (greater filter bandwidths and a shift in the best frequency).

While a useful fit to the data can be obtained at the three sites studied, we still have the problem of establishing parameters at intermediate sites if we are to generate a useful filterbank. Most studies of human hearing involve stimuli characterised by frequencies below 5 kHz. The gap in the animal data stretches from 0.8 kHz to 8 kHz and it is not clear that we can safely interpolate between the parameters for these two points. A way forward is to study AN data for intermediate sites and a project to do this is already planned.

4 Inner hair-cell transduction

The inner hair cell (IHC) is normally credited with a series of signal-processing functions including adaptation, half-wave rectification, low-pass filtering and some

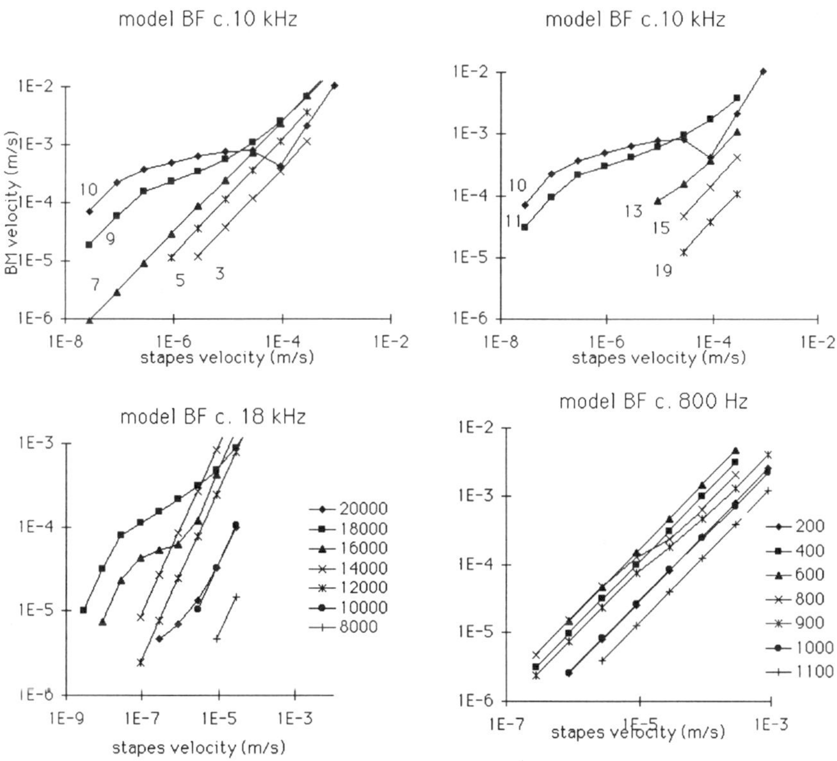

Figure. 4. Input/output functions for the model shown in Fig. 4 for three different sites on the basilar membrane. These data should be compared with Fig. 3. These results are 'work in progress'

further compression. In fact, it is best to separate the IHC function into two separate stages; the generation of the receptor potential (RP) and the synaptic response. The RP is measurable (e.g. Cody and Russell, 1978) and can be shown to be responsible for the half-wave rectification, the low-pass filtering and some nonlinear compression at very high signal levels. The synapse is then responsible for adaptation and possibly some further low-pass filtering. Despite being otherwise very useful, a hair cell model from our lab (Meddis, 86) did not distinguish clearly between these two stages and located all of the low-pass filtering in the synapse. This is clearly a mistake as the phenomenon can already be measured in the preceding RP stage. This is a useful lesson that producing the right behaviour is no guarantee that the model explains that behaviour.

There are now many models of IHCs (Hewitt et al, 1991) to choose from and most do a reasonable job. We have recently been working on an updated version of our original model that incorporates the two separate stages of RP generation and synaptic response (Lopez-Poveda et al., 1997 see Fig 5).

The first stage follows that of Shamma et al. (1986) and uses the nonlinear output of the BM model to control the generation of the RP. The RP is then used to drive the release of transmitter substance into the synaptic cleft. The probability of release of an available quantum of transmitter, k(t), is an exponential function of the RP

k(t)= Z exp*(h. V(t)) where V(t) is the receptor potential (see also Schoonhoven, 1998). Z determines the spontaneous rate of firing (i.e. when V(t) =0). When quanta are released, the amount of available transmitter is reduced and the system response is weakened until the transmitter store can be replenished. This

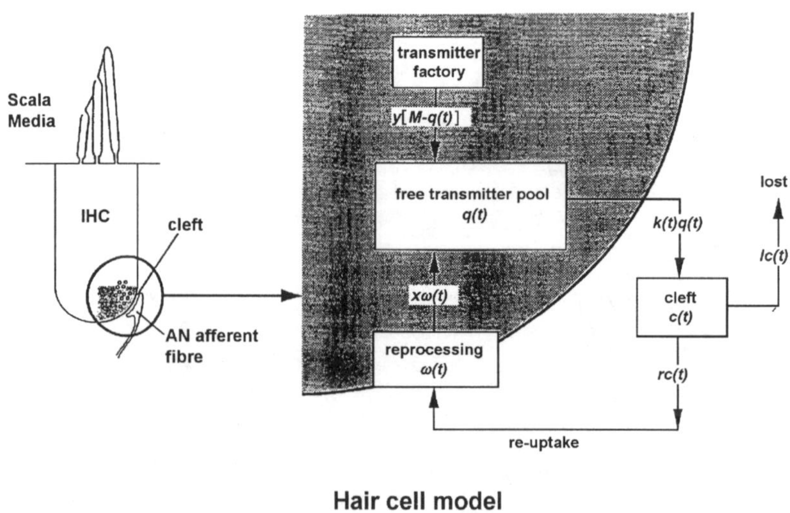

Figure. 5. Schematic of IHC/AN synapse as implemented in Meddis (1986)

is a plausible (though not proven) account of adaptation.

If the input to the IHC model is from a nonlinear model of BM response, we obtain some special benefits. Firstly it allows us to generate a theoretically respectable model of the differences among low medium and high spontaneous AN fibres. Physiological measurements have made it clear that the difference between HSR and LSR fibres is more complex than can be simply explained by different rates of release of transmitter at the synapse (Winter et al., 1990). The LSR rate-intensity function is generated in response to a compressed input and has a correspondingly wide dynamic range. The HSR response rises quickly during the initially linear section of the BM I/O function and saturates before the nonlinear response stage (Fig. 6). It has a narrow dynamic range as a result.

Nonlinear BM responding offers us the opportunity to solve another problem of existing hair cell models which give a good account at CF but are unrealistic for frequencies above and below CF (Fig. 7). This is because the response of the BM may be nonlinear at CF but is clearly linear for remote frequencies. A good nonlinear model solves this problem and gives sensible responses for frequencies remote from CF.

A novel feature of the new model is the use of small numbers (<20) of individual quanta of transmitter to simulate events. It is assumed that the release of a single quantum is enough to produce a post-synaptic spike in the AN fibre (see, for example, Siegel and Dallos, 1992. As described above, the RP is used to generate a probability of release that is applied to each available quantum. A random number

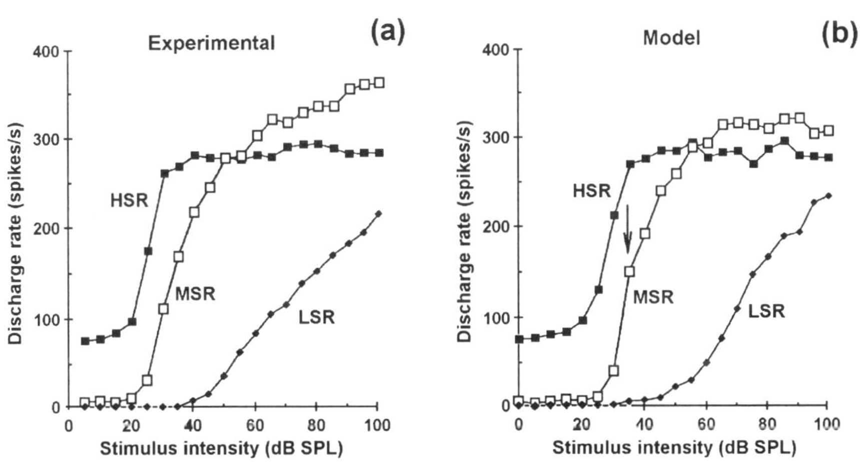

Figure. 6. Different response shapes for AN rate/intensity functions that can be achieved when nonlinear filtering is used. The rate of firing below 0 dB SPL is determined by the parameter Z in the model.

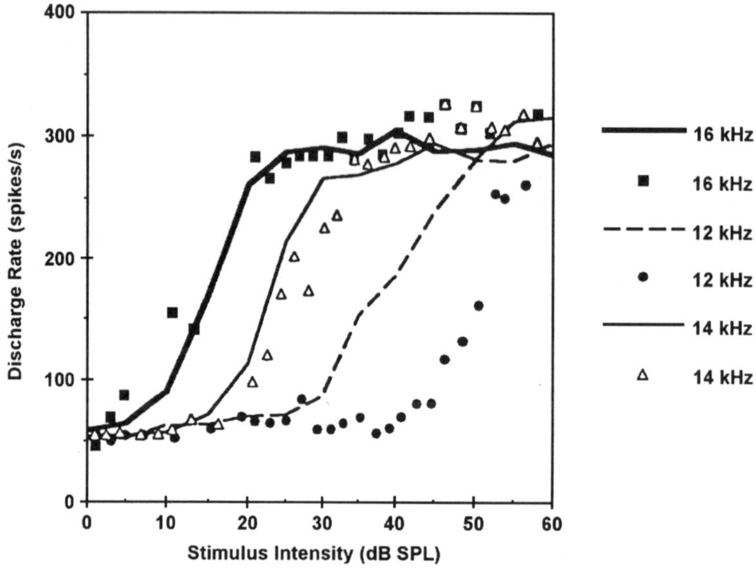

Figure. 7. Rate intensity functions for frequencies remote from a fibre's best frequency as generated using a model with a nonlinear filter to represent the basilar membrane. Dots are animal data. Clearly, we do not yet have the right parameters for the filter. Model results are from Lopez-Poveda et al. (1997).

generator decides whether each quantum is released or not at that time. If at least one is released, a spike occurs, assuming that the fibre is not already in a refractory state. This is a new approach to the stochastic nature of the response of the AN fibre. Previously, we regarded the release of transmitter as a continuous function and the generation of a spike as determined post-synaptically on a probabilistic basis. The new approach suggests that the main events determining AN fibre response are essentially pre-synaptic. At present this remains purely speculative, but it gives a useful account of the behaviour of AN fibres in response to sound in the form of realistic post-stimulus time histograms (PSTH).

5 System considerations

Putting the whole system together requires many decisions, not least of which may be the availability and compatibility of software. The DSAM system is intended as a platform for developers of auditory models. It is not a model in itself but is a set of conventions for writing modules that are mutually compatible. One important use of DSAM is to compare alternative models once they have been written according to the conventions of the system. It is our hope that these conventions will be adopted widely so that modellers can share software and make their models open to independent scrutiny. As it stands at present, the DSAM environment includes modules from many different models along with routines for generating and handling sound files and routines for analysing the output of the system (e.g. FFT, PSTH, autocorrelation, cross-correlation).

For many purposes, quite simple models of the auditory periphery can be used to good effect. In much of our own work on pitch and auditory streaming, a simple linear gammatone filter bank and basic hair cell transduction were enough to permit a useful exploration of the topics and good quantitative explanation of a number of important psychophysical results. In these models we also largely ignored the stochastic nature of AN fibre response without great penalty. It seems unlikely that we can continue to ignore stochasticity indefinitely, however. Serious psychophysical modelling means that we shall have to predict thresholds and this implies some description of system noise. The randomness of AN spike activity is the most obvious source of such noise and imposes severe constraints on information transmission between the ear and the brain. There are about 17000 nerve fibres operating in the region below 5 kHz in human listeners. If we use 100 channels, that gives us only 170 fibres per channel, some of which will be refractory. If we routinely work with probabilities, we are operating with the assumption that we have an infinite number of fibres per channel. This is clearly not justified and each channel will be very noisy; noisy enough, perhaps, to provide a quantitative basis for predicting psychophysical thresholds.

Ahead of us lies the quantum leap in auditory modelling implied in the introduction of nonlinear and stochastic models of BM response. Much needs to be done to perfect these models and establish the appropriate parameters for their use. However, I have no doubt that they will be soon be in routine use and that we shall become so convinced of their value that we shall wonder how we ever managed to work without them.

References

1. Cody, A.R., Russell, I. J., The Responses of hair cells in the basal turn of the guinea-pig cochlear to tones. *Journal of Physiology* **383**, (1987) pp. 551-569.
2. Evans, E. F., Frequency selectivity at high signal levels of single units in cochlear nerve and nucleus. In *Psychophysica and Physiology of Hearing* ed. by E. F. Evans and J. P. Wilson (Academic press, London, 1977) pp. 185-192.
3. Filipanits, F. Jr., Design and implementation of an auralization system with a spectrum based temporal processing optimization, Master's Research Project, University of Miami May 1994 (http://alumni.caltech.edu/~franko/thesis/thesis.html#Index).

4. Goldstein, J. L., Modeling rapid waveform compression on the basilar membrane as multiple-bandpass-nonlinearity filtering. *Hearing Research* **49**, (1990) pp. 39-60.
5. Hewitt, M J., and Meddis R., An evaluation of eight computer models of mammalian inner hair-cell function. *Journal of the Acoustical Society of America* **90**, (1991) 904-917.
6. Lopez-Poveda, E. A., O'Mard, L. P. and Meddis, R., A revised computational inner-hair cell model. *Psychophysical and Physiological Advances in Hearing* edited by A. R. Palmer, A. Rees, A. Q. Summerfield, and R. Meddis. (Whurr, London, 1998) pp. 112-121.
7. Lopez-Poveda, E. and Meddis, R., A physical model of sound diffraction and reflections in the human cochlea. *Journal of the Acoustical Society of America* **100**, (1996) pp. 3248-3259.
8. Meddis, R., Simulation of Mechanical to Neural Transduction in the Auditory Receptor. *Journal of the Acoustical Society of America* **79**, (1986). Pp. 702-711.
9. Nuttall, A.L. and Dolan, D.F. Steady state sinusoidal velocity responses of the basilar membrane in guinea pig. *Journal Acoustical Society America* **99**, (1996) pp. 1556-1565.
10. Patterson, R. D., and Moore, B. C. J., Auditory filters and excition patterns as representations of frequency resolution. In *Frequency Selectivity in Hearing.* edited by B. C. J. Moore (Academic Press, 1986) pp.123-177.
11. Pfeiffer R. R., A model for two-tone inhibition of single cochlear-nerve fibers. *Journal of the Acoustical. Society of America* **48**, (1970) 1373-1378.
12. Rhode W. S. and Cooper N. P., Nonlinear mechanics in the apical turn of the chinchilla cochlea. *Auditory Neuroscience* **3**, (1996) pp. 101-121.
13. Ruggero, M.A., Rich, N.C., Recio, A., Narayan, S. S. and Robles, L., Basilar membrane responses to tones at the base of the chinchilla cochlea. *Journal of the Acoustical. Society of America* **101**, (1997) pp 2151-2163.
14. Schoonhoven, R., Prijs, V F. and Frijns, J. H. M., Transmitter release in inner hair cell synapses: a model analysis of spontaneous and driven rate properties of cochlear nerve fibres. *Hearing Research*, (1998) pp. 247-260.
15. Shamma, S., Chadwick, R. S., Wilbur, W. J., Morrish K, A., and Rinzel J., A Biophysical model of Cochlear Processing: Intensity dependence of pure tone responses. *Journal Acoustical Society America* **80**, (1986) pp. 33-145.
16. Siegel, J. H., and Dallos, P. Spike activity recorded from the organ of Corti *Hearing Research* **22**, (1992) pp. 245-248.
17. Winter, I M., Robertson, D., and Yates, G K., Diversity of characteristic frequency rate-intensity functions in guinea pig auditory nerve fibres. *Hearing Research* **45**, (1990) pp. 191-202.

MODELLING CHANGES IN FREQUENCY SELECTIVITY WITH LEVEL

BRIAN R. GLASBERG, BRIAN C. J. MOORE AND MICHAEL A. STONE
Department of Experimental Psychology, University of Cambridge, Downing Street, Cambridge CB2 3EB, England
E-mail: bg12@cus.cam.ac.uk, bcjm@cus.cam.ac.uk, mas19@cus.cam.ac.uk

It is widely believed that the frequency selectivity measured behaviourally in masking experiments is largely determined by the filtering that occurs on the basilar membrane. The tuning of a specific point of the basilar membrane is determined by two components: a passive component reflecting the hydromechanical properties of the basilar membrane and surrounding fluids and structures; and an active component reflecting the influence of the outer hair cells on basilar membrane motion. We have modelled this as the sum of two rounded-exponential functions, following Patterson *et al.* [17]. One of these functions represents the "tip" filter determined by the active mechanism. It is assumed to have a gain at the tip which varies with level in the same way as the basilar-membrane gain, for a tone close to the characteristic frequency; the gain is greater at low levels than at high levels. The low-frequency side of the tip filter is also assumed to vary in sharpness with level. The second rounded-exponential function represents the "tail" filter determined by the passive mechanism. Its gain and shape are assumed to be invariant with level. The model was used to fit a large body of data obtained with the notched-noise method. The model took into account the modification of the spectrum of the stimuli produced by the middle ear. Generally, the model fitted the data well using only a small number of free parameters. The filter shapes derived from the notched-noise masking data show a level dependence that strongly resembles the level dependence of filtering on the basilar membrane.

1 Introduction

A commonly used method for measuring frequency selectivity is the notched-noise method [3,4,15,16,17,18]. This method has several advantages compared to methods such as psychophysical tuning curves [1,26]. Specifically:

1. It limits the amount of off-frequency listening (the use of a filter that is not centred at the signal frequency) that is possible and off-frequency listening can be taken into account in the analysis of the results.
2. It is not influenced by combination tones produced by the interaction of the signal and masker [7], although combination products produced by the interaction of components within the upper band of noise can have a small masking effect [13].
3. The detection cues available to the subject do not change markedly as the notch width is altered (except, perhaps, for the special case when no notch is present, where "profile analysis" [5] may play a role).

Auditory filter shapes estimated using the notched-noise method have been characterised as having a sharply tuned "tip" and a broader "tail" [4,17]. Patterson *et al.* [17] proposed that each side of the auditory filter can be described as the sum of two rounded-exponential functions:

$$W(g) = (1\text{-}w)(1 + pg)\exp(\text{-}pg) + w(1 + tg)\exp(\text{-}tg)$$

where g is normalised deviation from the centre frequency of the filter (deviation divided by centre frequency), p is a parameter determining the sharpness of the main passband of the filter, t is a parameter determining the sharpness of the tail of the filter, and w is a parameter determining the point at which the response becomes dominated by the tail rather than the tip. The filter has unity gain at its tip.

Usually, auditory filter shapes have been measured using a fixed noise level. When the noise level is varied, the derived filter tends to become broader on the low-frequency side as the level increases [3,8,11,16]. Recently, Rosen and his co-workers [21,22] have proposed a method for deriving auditory filter shapes from notched-noise data in which several different levels are used within the same experiment. In their work, thresholds have been measured as a function of notch width both at several fixed noise spectrum levels and at several fixed signal levels; all of the data were fitted simultaneously. The analysis of the results was based on the assumption that each side of the filter could be described as the sum of two rounded exponential functions (as described above), but the parameters describing the filter shape were made functions of level. They reported that the data were fitted better when the parameters were made to depend on the signal level than when they were made to depend on the masker spectrum level. As found in earlier work, they reported that the low-frequency side of the filter broadened with increasing level. The broadening could be well described by a change in the gain of the tail filter relative to that of the tip filter; the value of the parameter w increased with increasing level.

It is widely believed that the frequency selectivity measured behaviourally in masking experiments is largely determined by the filtering that occurs on the basilar membrane. Recent work measuring the tuning of single points on the basilar membrane [23,24,25] has also been interpreted as reflecting the combined effect of two filters: a filter reflecting the "passive" hydromechanical properties of the basilar membrane and surrounding fluids and structures; and a filter reflecting the active influence of the outer hair cells on basilar membrane motion. At low sound levels, the filter associated with the active mechanism has a strong influence, and sharp tuning is observed. As the level is increased, the gain of this filter decreases, so that the "passive" filter plays a greater role. Thus, the tuning becomes broader with increasing level, especially for frequencies below the characteristic frequency of the place on the BM whose response is being measured.

In the present paper, we assume that the "active" filter corresponds to the tip filter measured in notched-noise experiments, while the "passive" filter corresponds to the tail filter. We describe a new way of analysing the data from notched-noise experiments, based on the assumption that the gain of the tip filter varies with level in a similar way to the basilar-membrane input-output function for a frequency close to CF. We show that filter shapes derived using this method vary with level in a similar way to filtering characteristics measured on the basilar membrane.

2 Variation of the gain of the tip filter with level

Our method of analysis is based on the assumption that the low-frequency side of the auditory filter can be represented as the sum of a tip filter and a tail filter. The gain of the tail filter is assumed to be invariant with level. The gain of the tip filter is assumed to vary with (input) level in a similar way to the basilar-membrane input-output function for a frequency close to CF. A schematic family of such input-output functions is shown in Figure 1. The functions have shapes resembling those measured on the basilar membrane in animals [19,23,24]. The functions are nearly linear for very low sound levels (i.e. the slopes are close to 1 on a dB versus dB scale), but are compressive (with slopes less than 1) for mid-range levels. The curves have been drawn so that an input level of 100 dB leads to a response of 100 dB, based on the assumption that the gain provided by the active mechanism becomes negligible at very high sound levels. The curves vary in the maximum gain, *Gmax*, that is assumed for the active mechanism. For the uppermost curve, the output level is 60 dB for an input level of 0 dB, so *Gmax* is 60 dB. For the lowest

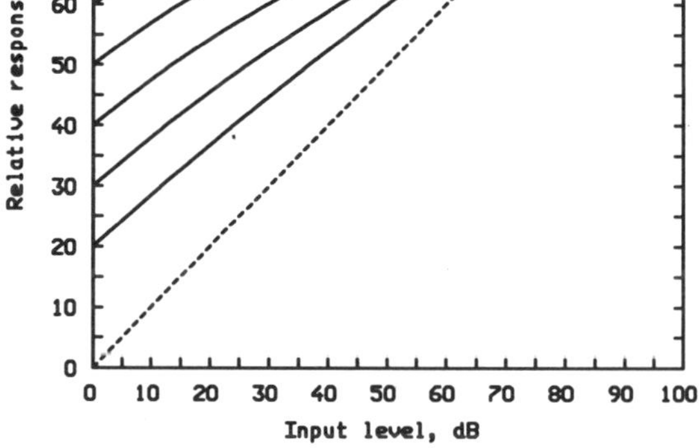

Figure 1. Schematic input-output functions on the basilar membrane for values of the maximum gain, *Gmax*, ranging from 20 to 60 dB.

curve, *Gmax* is only 20 dB. It is widely believed that the active mechanism provides more gain at medium and high frequencies than at low frequencies, so one aim of our modelling is to find the input-output function that gives the best fit to the data at each centre frequency.

3 Characterising the auditory filter shape as a function of level

The upper side of the filter is modelled as a single rounded-exponential filter:

$$W(g) = (1 + p_u g)\exp(-p_u g)$$

where g is normalised deviation from the centre frequency and p_u is a parameter determining the sharpness of the filter. The filter has unity gain at its tip. The value of p_u is assumed to be independent of the input level and is therefore described by one parameter. The lower side of the filter is modelled as the sum of two rounded-exponential filters, for the tip and the tail:

$$W(g) = [G(1 + p_l g)\exp(-p_l g) + (1 + tg)\exp(-tg)]/(G + 1)$$

The value of G specifies the gain of the tip filter relative to that of the tail filter. The term $(G + 1)$ normalises the filter gain to unity at its tip, so that the upper and lower halves of the filter have the same relative gain at the tip. The value of G is assumed to vary with the input level in a similar way to the basilar-membrane input-output function; the exact function (based on the value of *Gmax*) is chosen to give the best fit to the data, as described in the next section. We assume that the gain of the tip filter equals the gain of the tail filter for an input level of 100 dB. The input level is specified as the level per ERB, where ERB stands for the equivalent rectangular bandwidth of the auditory filter. Notice that the ERB here is that around the frequency components of interest. For example, suppose that we wish to determine the response of an auditory filter centred at 2000 Hz to the noise components centred around 1000 Hz. The input level per ERB for the ERB centred at 1000 Hz is obtained by summing the powers of all components in the stimulus over the appropriate frequency range around 1000 Hz (roughly from 935 to 1065 Hz). It is this input level that is used to determine the gain of the tip filter, and hence the response of the filter to the components around 1000 Hz. In practice, the level per ERB is calculated using a rounded-exponential weighting function rather than a rectangular function. We are aware that this procedure is somewhat problematic; for a discussion of the issues involved, see Moore [9], and section 6.

The value of p_l is assumed to be a linear function of the input level per ERB and is therefore described by two parameters. The value of t is assumed to be independent of level. The absolute threshold is modelled by an internal noise, and is also described by one free parameter. This gives a total of five free parameters.

4 Derivation of auditory filter shapes from notched-noise data

The procedure for deriving auditory filter shapes using the new method is as follows:

1. The effective spectrum reaching the cochlea is calculated for each notch width, taking into account the response of the headphone at the eardrum and the transfer through the middle ear. The transfer function assumed for the latter is as described by Moore *et al.* [12]
2. Reasonable starting values are chosen for the five parameters.
3. In addition, a fixed value is chosen for the maximum gain of the tip filter, *Gmax*, for an input level of 0 dB SPL. Levels are effective levels at the cochlea, assuming that the middle ear has a gain of 0 dB at 1000 Hz.
4. The input level per ERB is calculated across the frequency range covered by the noise. The current parameter values determine the ERB at the centre frequency of interest. The relative values of the ERB at other centre frequencies are assumed to be as described by Glasberg and Moore [3].
5. For each notch width, excitation patterns are calculated for the noise alone and the noise plus the signal at a specified level. The excitation pattern is defined as the output of each auditory filter as a function of filter centre frequency [3,10]. The excitation patterns include the effect of the assumed internal noise related to the absolute threshold.
6. The maximum difference in excitation level between the two excitation patterns (noise alone and noise plus signal), ΔL, is calculated. The signal level is adjusted until the value of ΔL corresponds to a fixed criterion value, C. This gives a predicted signal threshold for that notch width.
7. This is repeated for each notch width. The value of C is adjusted so that the mean of the predicted thresholds equals the mean of the obtained thresholds. This requires an iterative procedure, i.e. step 6 has to be repeated for each value of C, until the criterion in step 7 is satisfied.
8. The sum of the squared deviations between the obtained and predicted thresholds (in dB) is calculated.
9. The resulting sum of squares is entered into a multi-dimensional minimisation routine to find the values of the five parameters that give the best fit to the data, i.e. that minimise the sum of squares.

This procedure gives the best-fitting parameters for the filter for the given assumed value of *Gmax*. The whole procedure is then repeated for a series of values of *Gmax*. The value of *Gmax* giving the lowest overall sum of squares is taken as the final value. The best-fitting parameters obtained with that value of *Gmax* define the filter shape for any given input level. In practice, for some data sets, the sum of squares does not vary markedly with *Gmax*, so its value is not well defined. In such cases, we have chosen a value of *Gmax* within the range found for other data sets at the same centre frequency.

5 Filter shapes derived using the new method

We have analysed a large body of data in the literature, as well as unpublished data obtained in our own laboratory. We give here a sample of the results of the analyses.

5.1 Data of Rosen et al. [21] at 2000 Hz

Rosen *et al.* [21] presented a comprehensive set of notched-noise data for a centre frequency of 2000 Hz, using fixed noise spectrum levels of 20, 30, 40, 50 and 60 dB, and fixed signal levels of 30, 40, 50, 60 and 70 dB SPL. We present here an analysis of the data set that they described as "mean of three". However, data points were excluded in cases where the overall noise level exceeded 85 dB SPL; we suspected that the data in these cases might have been influenced by activation of the acoustic reflex. The analysis was based on the remaining 127 data points. No correction was made for the response of earphone used (Etymōtic Research ER2), as this was designed to have a flat response at the eardrum. The value of *Gmax* was taken as 55 dB. The sum of the squared deviations of the data from the fitted values was 297 dB^2. The root-mean-square (rms) deviation of the data from the fitted values was 1.5 dB ($\sqrt{}(297/127)$. Figure 2 shows the filter shapes derived from these data for input levels ranging from 20 to 90 dB.

In the upper panel, filter responses are plotted assuming a gain of unity (0 dB) at the tips; thus, at the tips, the output level equals the input level. The filters for input levels of 80 and 90 dB are shown as dashed lines, as the data do not define the characteristics of the filter well for these input levels (recall that data were excluded when the noise level exceeded 85 dB SPL). The lower panel shows the same filters plotted in terms of gain. Dashed lines indicate regions where the filter characteristic was not well defined by the data; for example, the filter for an input level of 20 dB is not well defined more than 20 dB down from its tip. When plotted in terms of gain, it is clear that the gain at the tip increases progressively with decreasing input level, while the gain on the low-frequency tail is invariant with level. This is consistent with basilar-membrane responses [20,24,25]. The gain at the tip is close to 0 dB for a 90-dB input, and increases to about 47 dB for a 20-dB input. The gain would be 55 dB for a 0-dB input. Note that all levels are specified as the effective level at the input to the cochlea, after allowing for the transfer function of the middle ear (assuming an effective transfer of 0 dB at 1000 Hz).

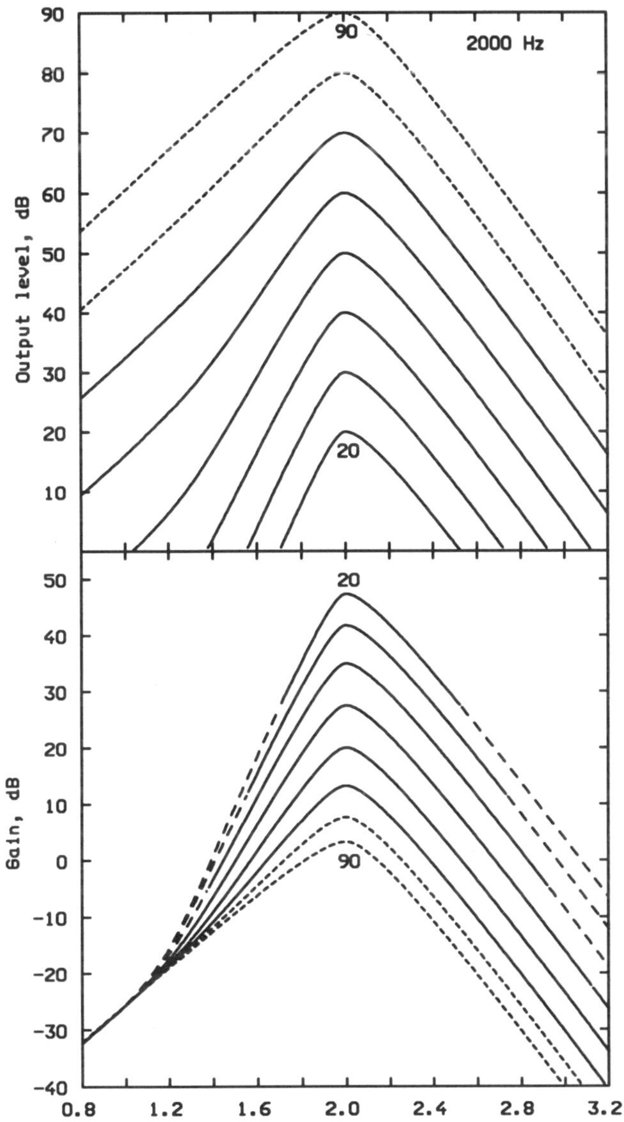

Figure 2. Auditory filter shapes derived from the data of Rosen *et al.* [21] for input levels ranging from 20 to 90 dB in 10-dB steps. In the upper panel, filter responses are plotted assuming a gain of unity (0 dB) at the tips; thus, at the tips, the output level equals the input level. The lower panel shows the same filters plotted in terms of gain.

5.2 Data of Lutfi and Patterson [8] at 1000 Hz

Lutfi and Patterson [8] presented data obtained using the notched-noise method for signal frequencies of 1000 and 4000 Hz. We present here an analysis of the data obtained at 1000 Hz. They used fixed noise spectrum levels of 20, 35 and 50 dB, and fixed signal levels of 20 and 35 dB SPL. Although the published paper presented data only for conditions where the notch was placed asymmetrically about the signal frequency, data were gathered for symmetrically placed notches as well. These data were supplied to us by Roy Patterson, via Stuart Rosen, and were included in our analysis. The analysis took into account the response of the earphone used (TDH 49). The value *Gmax* was taken as 40 dB. The root-mean-square (rms) deviation of the data from the fitted values was 1.4 dB.

Figure 3 shows the filter shapes derived from these data for input levels ranging from 20 to 90 dB. As for Figure 2, the upper panel shows filter responses plotted assuming a gain of unity (0 dB) at the tips, while the lower panel shows responses plotted in terms of filter gain. As for the filters shown in the lower panel of Figure 2, the gain at the tip increases progressively with decreasing input level, while the gain on the low-frequency tail is invariant with level. However, the gain at the tip changes less with level at 1000 Hz than at 2000 Hz, as the value of *Gmax* is different for the two levels.

6 Discussion

One disappointing aspect of our analysis is that the value of *Gmax* was not well defined by the data. For many of the data sets examined, the goodness of fit, indicated by the rms deviation of the data points from the fitted values, did not vary markedly with changes in the assumed value of *Gmax*. This is partly a consequence of the choice of model fitted to the data. In our analysis, the lower skirt of the tip filter, determined by the parameter p_l, was made a function of the input level per ERB. This meant that much of the level dependence of the filter shape was determined by the variation of p_l with level, rather than by the varying tip-to-tail ratio, as determined by the variation of G with level. An alternative approach, advocated by Rosen *et al.* [21], is to make p_l invariant with level. In this case, the variation in filter shape with level is forced to depend more on the tip-to-tail ratio, and the value of *Gmax* is better defined. We have tried fitting the data with p_l invariant with level. We found that this did, indeed, give better defined values of *Gmax*. For the data of Rosen *et al.* [21] at 2000 Hz, the best fitting value of *Gmax* was 55 dB. However, the rms deviation of the data from the fitted values increased from 1.5 dB (with p_l varying with level) to 1.8 dB (with p_l invariant with level). For the data of Lutfi and Patterson [8] at 1000 Hz, the best fitting value of *Gmax* was 40 dB. Again, there was an increase in the rms deviation of the data from the fitted values, but in this case it was smaller, from 1.4 to 1.5 dB.

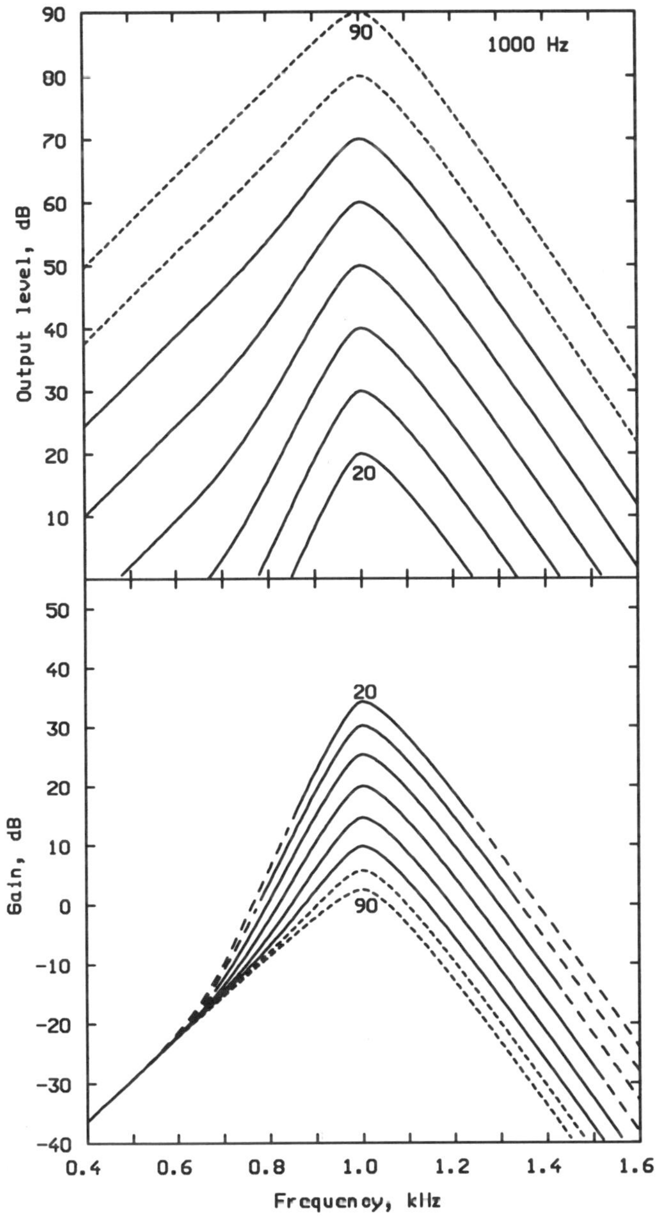

Figure 3. As Figure 2, but showing auditory filter shapes derived from the data of Lutfi and Patterson [8] for a centre frequency of 1000 Hz.

As described earlier, our analysis was based on the assumption that the filter shape in response to a given group of components within the noise masker was determined by the power of those components summed over a frequency range of one ERB (using a rounded-exponential weighting function). For large notch widths this meant that the gain of the tip filter depended on the levels of components that were remote from the centre frequency of the filter. It is not clear whether this is realistic. An alternative approach is to assume that the filter shape is determined by the output of the filter rather than its input [21]. It seems likely that neither approach is correct. The filtering on the basilar membrane probably involves a feedback mechanism which means that it is unrealistic to characterise the shape as dependent only on the input or the output. In addition, simultaneous masking may be partly influenced by suppression of the signal by the masker, when the masking is determined mainly by frequency components well below the signal frequency [2,14]. Such an effect is not readily incorporated in a model based on filters that are level-dependent, but are quasi-linear at a given level. However, our assumption that the gain of the tip filter can be reduced by components of the masker that are well below the signal frequency is consistent with the effects of suppression, since suppression is often described as an effective attenuation of the suppressed component [6].

In summary, the present approach to the derivation of auditory filter shapes from notched noise data is more closely related than previous approaches to the physiological properties of the filtering measured on the basilar membrane. The approach is clearly over-simplified, and needs further refinement. Nevertheless, auditory filter shapes derived using our new method do vary with level in a way that closely resembles the level variation of tuning on the basilar membrane. This encourages us to believe that the basic properties of the auditory filters are determined in the cochlea.

7 Acknowledgements

This work was supported by the Medical Research Council (UK). We thank Stuart Rosen and Roy Patterson for helpful discussions.

References

1. Chistovich L. A., Frequency characteristics of masking effect, *Biophys.* **2** (1957) pp. 743-755.
2. Delgutte B., Physiological mechanisms of psychophysical masking: Observations from auditory-nerve fibers, *J. Acoust. Soc. Am.* **87** (1990) pp. 791-809.
3. Glasberg B. R. and Moore, B. C. J., Derivation of auditory filter shapes from notched-noise data, *Hear. Res.* **47** (1990) pp. 103-138.

4. Glasberg B. R., Moore, B. C. J., Patterson, R. D. and Nimmo-Smith, I., Dynamic range and asymmetry of the auditory filter, *J. Acoust. Soc. Am.* **76** (1984) pp. 419-427.
5. Green D. M., *Profile Analysis* (Oxford University Press, Oxford, 1988).
6. Javel E., Suppression of auditory nerve responses: I. Temporal analysis, intensity effects and suppression contours, *J. Acoust. Soc. Am.* **69** (1981) pp. 1735-1745.
7. Lutfi R. A. and Patterson, R. D., Combination bands and the measurement of the auditory filter, *J. Acoust. Soc. Am.* **71** (1982) pp. 421-423.
8. Lutfi R. A. and Patterson, R. D., On the growth of masking asymmetry with stimulus intensity, *J. Acoust. Soc. Am.* **76** (1984) pp. 739-745.
9. Moore B. C. J., Frequency analysis and masking. In *Hearing*, ed. by B. C. J. Moore (Academic Press, Orlando, Florida, 1995) pp. 161-205.
10. Moore B. C. J. and Glasberg, B. R., Suggested formulae for calculating auditory-filter bandwidths and excitation patterns, *J. Acoust. Soc. Am.* **74** (1983) pp. 750-753.
11. Moore B. C. J. and Glasberg, B. R., Formulae describing frequency selectivity as a function of frequency and level and their use in calculating excitation patterns, *Hear. Res.* **28** (1987) pp. 209-225.
12. Moore B. C. J., Glasberg, B. R. and Baer, T., A model for the prediction of thresholds, loudness and partial loudness, *J. Audio Eng. Soc.* **45** (1997) pp. 224-240.
13. Moore B. C. J., Glasberg, B. R., van der Heijden, M., Houtsma, A. J. M. and Kohlrausch, A., Comparison of auditory filter shapes obtained with notched-noise and noise-tone maskers, *J. Acoust. Soc. Am.* **97** (1995) pp. 1175-1182.
14. Moore B. C. J. and Vickers, D. A., The role of spread of excitation and suppression in simultaneous masking, *J. Acoust. Soc. Am.* **102** (1997) pp. 2284-2290.
15. Patterson R. D., Auditory filter shapes derived with noise stimuli, *J. Acoust. Soc. Am.* **59** (1976) pp. 640-654.
16. Patterson R. D. and Moore, B. C. J., Auditory filters and excitation patterns as representations of frequency resolution. In *Frequency Selectivity in Hearing*, ed. by B. C. J. Moore (Academic, London, 1986) pp. 123-177.
17. Patterson R. D., Nimmo-Smith, I., Weber, D. L. and Milroy, R., The deterioration of hearing with age: frequency selectivity, the critical ratio, the audiogram, and speech threshold, *J. Acoust. Soc. Am.* **72** (1982) pp. 1788-1803.
18. Patterson R. D., Peters, R. W. and Milroy, R., Threshold duration for melodic pitch. In *Hearing - Physiological Bases and Psychophysics*, ed. by R. Klinke and R. Hartmann (Springer, Berlin, 1983) pp. 321-325.
19. Robles L., Ruggero, M. A. and Rich, N. C., Basilar membrane mechanics at the base of the chinchilla cochlea. I. Input-output functions, tuning curves, and response phases, *J. Acoust. Soc. Am.* **80** (1986) pp. 1364-1374.

20. Robles L., Ruggero, M. A. and Rich, N. C., Two-tone distortion in the basilar membrane of the cochlea, *Nature* **349** (1991) pp. 413-414.
21. Rosen S., Baker, R. J. and Darling, A., Auditory filter nonlinearity at 2 kHz in normal hearing listeners, *J. Acoust. Soc. Am.* **103** (1998) pp. 2539-2550.
22. Rosen S., Baker, R. J. and Kramer, S., Characterizing changes in auditory filter bandwidth as a function of level. In *Auditory Physiology and Perception*, ed. by Y. Cazals, K. Horner and L. Demany (Pergamon Press, Oxford, 1992) pp. 171-177.
23. Ruggero M. A., Responses to sound of the basilar membrane of the mammalian cochlea, *Curr. Opin. Neurobiol.* **2** (1992) pp. 449-456.
24. Ruggero M. A., Rich, N. C., Recio, A., Narayan, S. S. and Robles, L., Basilar-membrane responses to tones at the base of the chinchilla cochlea, *J. Acoust. Soc. Am.* **101** (1997) pp. 2151-2163.
25. Sellick P. M., Patuzzi, R. and Johnstone, B. M., Measurement of basilar membrane motion in the guinea pig using the Mössbauer technique, *J. Acoust. Soc. Am.* **72** (1982) pp. 131-141.
26. Small A. M., Pure-tone masking, *J. Acoust. Soc. Am.* **31** (1959) pp. 1619-1625.

COMPUTATIONAL MODELING OF OUTER HAIR CELL DAMAGE: IMPLICATIONS FOR HEARING AID SIGNAL PROCESSING

CHRISTIAN GIGUÈRE

Audiology/Speech-Language Pathology Program, School of Rehabilitation Sciences, University of Ottawa, Ottawa, Ontario K1N 6N5, Canada.
E-mail: cgiguere@uottawa.ca

AND GUIDO F. SMOORENBURG

Laboratory of Experimental Audiology, Department of Otorhinolaryngology, University Hospital Utrecht,3508GA Utrecht, The Netherlands.
E-mail: g.smoorenburg@kmb.azu.nl

This paper describes a number of computer simulations illustrating the main effects of outer hair loss upon the representation of sounds on the basilar membrane. These include (1) a basalward shift of the place of maximum vibration, (2) a decreased gain near the place of maximum vibration, (3) a broadening of the tuning curves, (4) an altered summation of activity across frequency components, and (5) an altered temporal response. It is argued that, for large classes of sounds, the basilar membrane patterns resulting from a loss of outer hair cells cannot, in principle, be compensated for by hearing aid signal processing techniques, such as multi-band amplitude compression and spectral sharpening.

1 Introduction

Sensorineural hearing-impaired listeners suffer from a variety of auditory deficits such as elevated threshold, reduced frequency selectivity and temporal resolution, abnormal growth of loudness, and poor speech perception. Outer hair cell (OHC) damage is the underlying cause in many cases, which should indicate the type of compensating signal processing required. Yet, despite rapid advances in hearing aid technology, the optimal solution for a given hearing configuration is still a matter of debate and research [7,11].

Computational models have been proposed [4,5] to help understand and predict the effects of sensorineural hearing loss on speech perception. However, while OHC damage is a condition affecting peripheral processing, such models must include additional assumptions about the relevant signal processing stages and decision processes in the central auditory system in order to make speech intelligibility predictions. This complicates the interpretation of results.

In contrast, this modeling study focuses on the effects of OHC damage directly at the level of the auditory periphery. A main question is to which extent the basilar membrane (BM) response from a damaged cochlea can be restored to that of a normal cochlea by processing of the input signal, such as in a hearing aid.

While restoration of the BM response is not in itself the goal of a hearing aid, perfect compensation is a sufficient condition to insure normal hearing in the case of a pure OHC loss. Thus, the question has important theoretical implications.

2 Model of the normal cochlea

The computational model of the human auditory periphery described in [2,3] is used in the present study. The model consists of a series of cross-coupled outer ear, middle ear and cochlear stages. The latter stage is a nonlinear 1-D transmission line simulation of BM motion. The cochlea is segmented into N = 320 channels from base (n = 1) to apex (n = 320), or about 10 channels per mm along the BM. In each channel, a source $V_n^{ohc}(t)$ produces a nonlinear and saturating pressure, assumed to originate from the OHCs, as follows:

$$V_n^{ohc}(t) = -K * G * (1 + |I_n(t)|/I_o)^{-0.5} * R_n * I_n(t)$$

where n is the channel number, R_n is the effective passive resistance associated with the BM motion, $I_n(t)$ is the BM volume velocity, I_o is a constant (3.6×10^{-2} cm/s times the BM channel cross-section area), G is the maximum OHC feedback gain (0.99), and K is the OHC loss factor. Functionally, $V_n^{ohc}(t)$ compensates for the pressure loss $R_n*I_n(t)$ across the BM passive resistance. This reduction in damping depends on the instantaneous motion of the BM through a compressing function of $I_n(t)$. The choice of the nonlinearity is based on previous work [3].

Figure 1a shows the BM excitation patterns at 1000 Hz for a „normal" operation of the OHC pressure source (K=1). The curves express the time-average BM velocity in each channel following half-wave rectification. They are normalized along the vertical axis to emphasize the level-dependent effects of the OHCs. Below and above the characteristic place (CP), defined as the place of maximum vibration at low input levels, the curves are superimposed, indicating linear behavior. Near the CP, the gain and width of excitation depend on level. At low levels ($|I_n| << I_o$), the BM is sharply tuned and the gain is maximal. At high levels ($|I_n| >> I_o$), the BM is broadly tuned and the gain near the CP is gradually reduced to that of the passive case. Also, the place of maximum vibration moves towards the base by an amount equivalent to about ½ octave in characteristic frequency. Overall, the BM model shows about 30 dB of amplitude compression near the CP. The 10-dB quality factor (Q_{10}) of the BM model decreases from 6.4 to 0.9 as the input level increases from 20 to 80 dB SPL. Figure 1b shows the input-output (I/O) function of the model at 1000 Hz measured at the low-level characteristic place (n = 184) and along the locus of maximum BM vibration irrespective of place. At low and high input levels, the growth is linear, and at intermediate levels, the growth is compressive. The difference between the two curves in Fig. 1b reflects the shift of the place of maximum vibration with input level.

The physiological data reported by Johnstone *et al.* [6] in the basal end of the guinea pig's cochlea indicate a compression factor of about 36 dB and a decrease in Q_{10} from about 7.0 to 0.7 for input levels from 20 to 80 dB SPL. The place of maximum vibration also shifts by an amount equivalent to ½ octave. More recent physiological data reported by Ruggero [10] near the basal end of „an exceptionally sensitive" chinchilla cochlea show almost 60 dB of compression and a decrease in Q_{10} from about 4.5 to 1.5 over the same input range. The physiological data also indicate a gradual reduction of compression from base to apex. At the apex, it is typically only about 10-20 dB [9]. In the model, the feedback gain G of the OHCs does not depend on place, i.e. the BM has the same level-dependent characteristics from base to apex. Thus, the model is fairly representative of the physiological data in the central portion of the mammalian cochlea, while it shows more compression than the physiological data near the

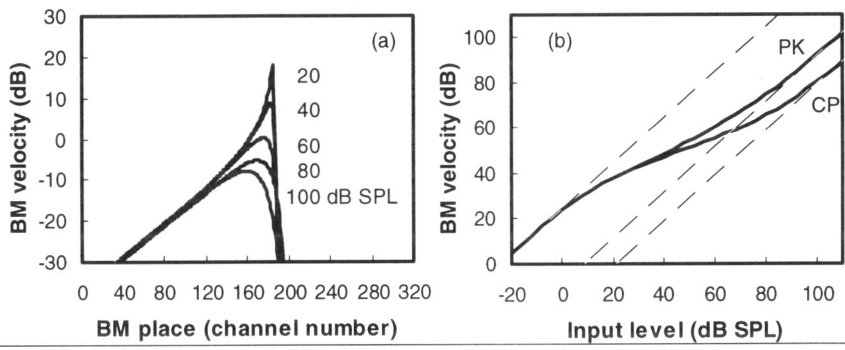

Figure 1. (a) BM excitation patterns for a pure tone of 1000 Hz at several input levels in the model of the normal cochlea. Curves are shifted down by the amount equal to the input level. (b) I/0 functions at 1000 Hz at the low-level characteristic place (CP) and at the peak (PK) BM vibration irrespective of place. Dashed curves have slopes of 1dB/dB.

3 Model of OHC damage

3.1 Pure-tone stimulation

OHC damage is simulated in the model by multiplying the feedback gain G by a loss factor K ranging from 0 to 1. Figure 2a shows the excitation patterns for a 1000 Hz tone at 40 dB SPL for the „normal" cochlea (K = 1) and for three cases of

OHC damage (K = 0.75, 0.50 and 0). The latter cases correspond to 25%, 50% and 100% OHC losses, respectively. In all cases, the loss is uniformly distributed along the cochlea. As the OHC loss increases, the gain gradually decreases near the CP, and the place of maximum vibration moves towards the base. There is no effect further away from the CP. As a result, there is an increasing broadening of the excitation patterns with OHC loss. In the 50% OHC loss case, for example, the width of the excitation pattern has increased by a factor 2.6 compared to the normal cochlea.

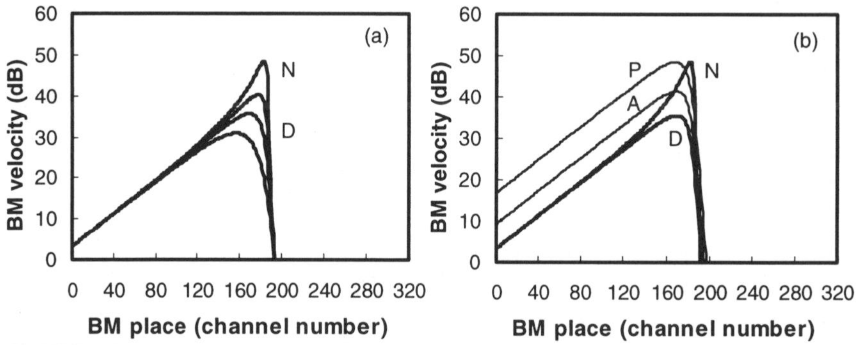

Figure 2. (a) BM excitation patterns for a pure tone of 1000 Hz at 40 dB SPL in the normal (N) cochlea and for three cases of damaged (D) cochlea: 25, 50 and 100% OHC loss. (b) Two attempts at restoring the BM excitation pattern of the damaged cochlea (50% OHC loss) to that of the normal cochlea: BM-peak (P) and BM-average (A) compensation.

Figure 2b shows the BM excitation patterns after two different amplification schemes aimed at restoring the response of the damaged cochlea (50% OHC loss) to that of the normal cochlea. Since the input, a pure tone, is already the narrowest of signals, the BM pattern in the damaged cochlea cannot be sharpened by signal processing. The loss in gain near the CP can be compensated for by amplification. One possibility is to set the gain so that the peak of the excitation pattern just matches that of the normal cochlea. This is referred to here as „BM-peak compensation". As shown in Fig. 2b, this scheme restores the maximum BM vibration to that of the normal cochlea, but there are marked differences in the widths of excitation and in the exact places of maximum vibration. Amplification also elevates the tails of the excitation pattern. In particular, the basal tail is lifted far above that of the normal cochlea. Thus, the benefits of amplification are offset by a substantial increase in basalward (upward) spread of masking.

Another possibility is to set the gain so that BM vibration, averaged across the entire cochlea, is matched to that of the normal cochlea. This is referred to here as „BM-average compensation". It mimics loudness compensation. As shown in Fig. 2b, this leads to a compromise between restoring the loss of gain near the CP, and minimizing the deleterious effect of amplification on the basalward spread of masking. Thus, the place-dependent effects of OHC loss are such that, even for pure tones, it is not possible to restore the BM excitation patterns to that of the normal cochlea.

3.2 Complex-tone stimulation

Figure 3 illustrates an additional effect of OHC loss, namely, an altered summation of BM activity across frequency components. Figure 3a shows the BM pattern in the normal cochlea for three pure tones (875, 1000, 1125 Hz) presented at equal levels of 40 dB SPL. The three tones correspond to the 7th, 8th and 9th harmonics of a fundamental frequency of 125 Hz, which falls within the range of voice fundamentals for male speakers. When presented together, the three tones lead to three very close but distinct peaks along the BM. The height of each peak is slightly smaller (by 1-2 dB) than the peak of the corresponding tone when presented alone, because of tone-on-tone suppression effects [3] across the three components. Also, additional peaks are seen on the apical side of the three primary peaks due to the generation of combination tones [3]. In contrast, the BM pattern in the case of the damaged cochlea (50% OHC loss) in Fig. 3b shows only one broad peak, and this peak is higher by about 3 dB than the peak arising from any of the three tones when presented alone. The frequency spacing between harmonics has become smaller than the residual frequency resolution of the cochlea. BM activity sums differently across frequency components in the damaged and normal cochleae. A compensation scheme adjusted for pure tones or narrow-band signals will overcompensate, in general, for complex sounds or broadband signals. Thus, the gain required for compensation depends on the bandwidth of the signal.

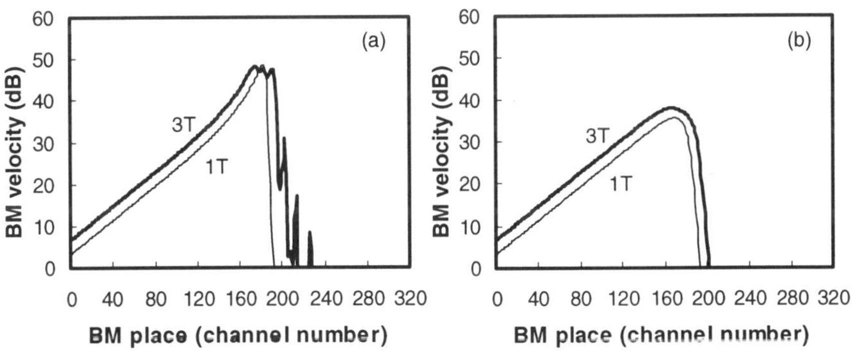

Figure 3. BM excitation patterns for a complex tone (3T) composed of 3 pure tones (875, 1000, 1125 Hz) at 40 dB SPL in (a) the normal cochlea and (b) the damaged cochlea (50% OHC loss). For comparison purposes, the excitation pattern of the middle tone (1T) presented alone is also shown.

3.3 Vowel representation

Figure 4a shows BM excitation patterns for a synthetic vowel /ae/ presented at 55 dB SPL. The fundamental frequency (F0) is 125 Hz. The vowel consists of this fundamental and all harmonics up to 4000 Hz. The height of each harmonic was taken as the height of the vocal tract transmission for this vowel [8]. The harmonics were added in cosine phase.

In Fig. 4a, the fundamental and the three following harmonics are clearly resolved on the BM pattern of the normal cochlea. The first formant frequency (F1 ≅ 700 Hz) falls between the 5th and 6th harmonics of F0. The second (F2 ≅ 1750 Hz) and third (F3 ≅ 2375 Hz) formants fall on the 14th and 19th harmonics of F0. These three formants are clearly resolved. In contrast, the BM pattern for the damaged cochlea (50% OHC loss) is lower and shows much broader features. In particular, F0 and next three harmonics are much less sharp and severely attenuated. The two harmonics defining F1 have fused to produce a broader and lower peak, and the BM peaks for F2 and F3 have themselves fused into one broader formant that is barely resolved. The BM pattern of the damaged cochlea after the „BM-average compensation" scheme, applied independently to each harmonic of the vowel, shows a general restoration of the overall BM vibration to that of the normal cochlea, but there are marked differences. Near the apex, where harmonics are individually resolved, the amplification scheme under-compensates the normal cochlea. In the base, where harmonics are very closely spaced on the BM, the amplification scheme leads to overcompensation.

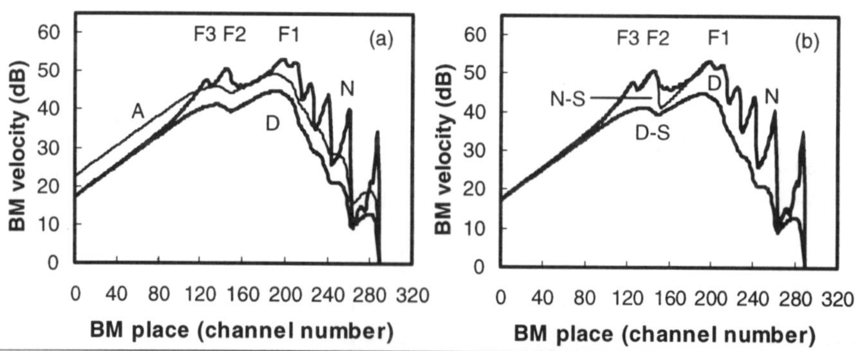

Figure 4. (a) BM excitation patterns for a synthetic vowel /ae/ presented at 55 dB SPL in the normal (N) cochlea, the damaged (D) cochlea (50% OHC loss), and the damaged cochlea after BM-average (A) compensation. (b) BM excitation patterns for a „spectrally-enhanced" version of vowel /ae/ in the normal (N-S) cochlea and the damaged (D-S) cochlea (50% OHC loss). For comparison purposes, the excitation patterns of the original vowel are also shown.

Figure 4b presents the results of an attempt to increase the resolution of the formant peaks in the excitation pattern for vowel /ae/ by sharpening the spectral representation of the formants in the input signal. Only 8 of the 32 harmonics of the vowel are retained: the fundamental (F0) and the three following harmonics, the two harmonics (5th, 6th) defining the first formant, the harmonic (14th) defining the second formant, and the harmonic (19th) defining the third formant. When processed through the model of the normal cochlea, this „spectrally-enhanced" vowel shows a better peak-to-valley ratio between F1 and F2 than the original synthetic vowel by about 7 dB. When processed through the model of the damaged cochlea (50% OHC loss), there are virtually no changes to the excitation pattern, and the increment in peak-to-valley ratio between F1 and F2 is only 0.2 dB. Since there are only one or two harmonics per formant in the above example and that the excitation pattern of a pure tone cannot by itself be sharpened (Fig.2b), it follows that vowel formants cannot be sharpened on the BM of the damaged cochlea by manipulation of the input signal.

3.4 Temporal processing

The previous simulations illustrated the effects of OHC damage on the steady-state excitation patterns of the BM. Figure 5 shows the temporal evolution of the BM vibration in response to the vowel /ae/ for three glottal periods. In the case of the normal cochlea in Fig. 5a, the second and third formants appear in the top half of the plot as cones of vibration that ring over the entire length of a glottal period at the frequency of the 14th and 19th harmonics, respectively. The first formant appears in the bottom half of the plot, where harmonics are individually resolved.

The temporal evolution of the BM in the case of the damaged cochlea (50% OHC loss) after „BM-average compensation" is shown in Fig. 5b. The general level of BM activity has been restored, but there are marked differences with the normal cochlea. Over the entire length of the BM, the damaged cochlea behaves as a broadly tuned band-pass filter with a short impulse response. In particular, there is no clear ringing at any particular frequency in the F2-F3 formant region, but rather a temporal modulation of vibration indicative of a mixture of harmonics. Also, BM vibration is distributed throughout the glottal period in Fig. 5a, whereas it is concentrated at the time of the glottal pulse in Fig. 5b. The compensation scheme has led to another compromise whereby the vibration located in-between the glottal pulses in the normal cochlea has been traded for a large impulse at the glottal moment in the damaged cochlea. Thus, the firing patterns of the vowel in the auditory nerve of the damaged cochlea following inner hair cell processing will likely consist of an oversized onset response (and refractory effects) at each glottal pulse followed by a depressed ongoing response. Extraction of the formant frequencies from temporal processing may therefore be significantly affected in the damaged cochlea, even after amplitude compensation. In the normal cochlea, temporal processing was found to be a very effective way to extract stable vowel representations from auditory-nerve firing patterns [12].

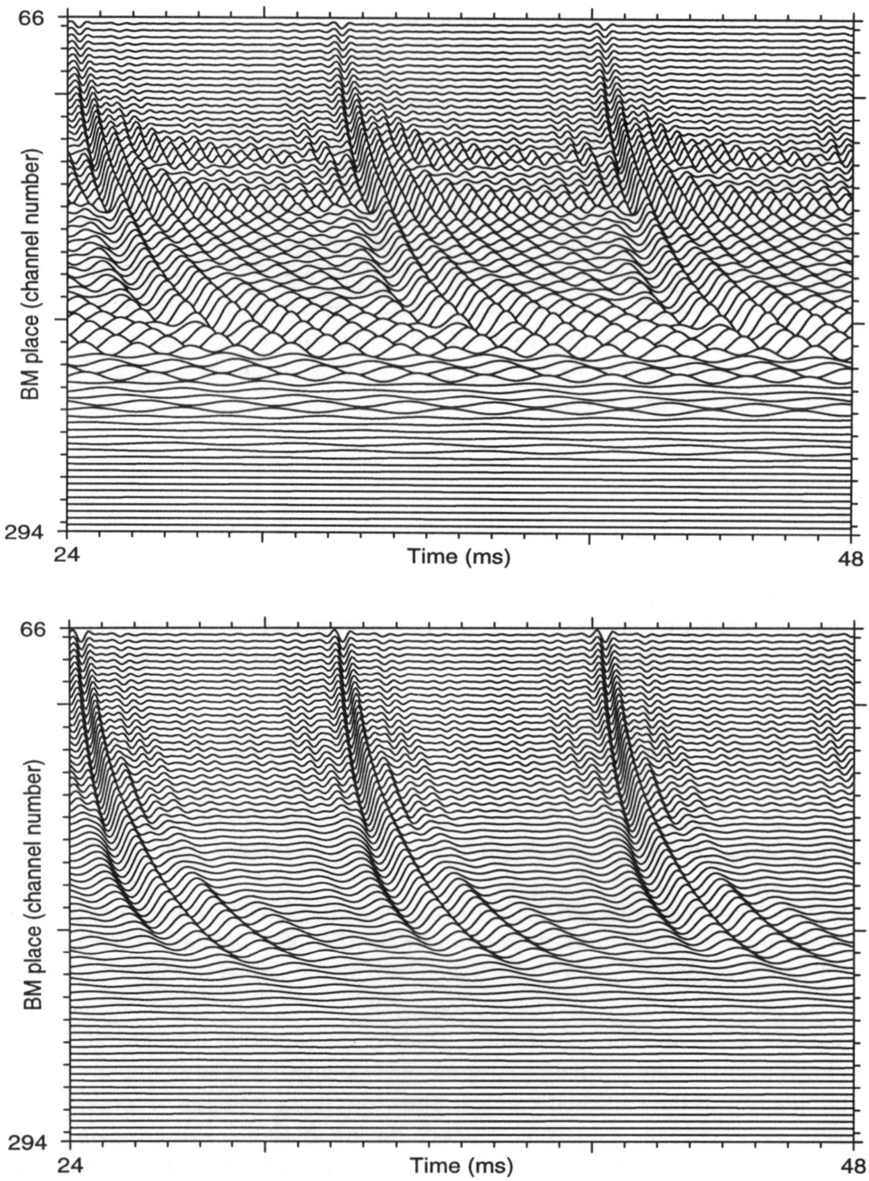

Figure 5. Temporal response of the BM to the vowel /ae/ in the normal cochlea (upper panel) and the damaged cochlea (50% OHC loss) after BM-average compensation (lower panel).

4 Discussion

The effects of OHC loss on the representation of sounds on the BM have been examined with a computational model of the auditory periphery. These are:

- a basalward shift of the place of maximum vibration,
- a decrease gain near the characteristic place,
- a broadening of the tuning curves,
- an altered summation of activity across frequency components, and
- an altered temporal waveform.

These effects depend on level, being most pronounced at low levels and gradually vanishing at high levels. While there is considerable uncertainty about the exact details of the model predictions, owing to the absence of comparative physiological data on the human cochlea and the difficulty of modeling the relevant physiological data from animal experiments, this work has general implications on the theoretical limits of hearing aids.

The effects of OHC loss on BM vibration depend on both *frequency* and *place*. From a hearing aid point of view, only the frequency dimension is accessible, and not the place dimension. Thus, the task of a hearing aid is to try to compensate the (nonlinear) transfer function of a 2-D system while having access to only one of the two dimensions. The simulations presented here illustrate that this task is fundamentally impossible, even when the characteristics of the damaged cochlea are completely available. In particular, BM vibration cannot be restored exactly for a large range of sounds including pure tones, complex tones, vowels, and transient sounds. Only for continuous broadband sounds without sharp spectral peaks that restoration may be possible over a substantial extent of the BM. In general, with only the frequency dimension accessible, it is only possible to restore activity over a very narrow segment of the BM at a given time.

The above discussion does not argue against the use of signal compression techniques to compensate for loudness recruitment in the hearing impaired. Indeed, the primary goal of a hearing aid is to restore the audibility of soft signals while maintaining loud signals at a comfortable sensation level. However, beyond this goal, there are reasons to believe that there is little scope at attempting to compensate for the exact hearing damage. In terms of speech intelligibility the effects of signal compression are marginal [1]. Instead, efforts should be targeted at reducing the extraneous environmental factors like noise and reverberation at the front-end of a hearing aid using directional microphones, microphone-array techniques and/or binaural processing (see examples in [7,11]).

compensate for the exact hearing damage. In terms of speech intelligibility the effects of signal compression are marginal [1]. Instead, efforts should be targeted at reducing the extraneous environmental factors like noise and reverberation at the front-end of a hearing aid using directional microphones, microphone-array techniques and/or binaural processing (see examples in [7,11]).

References

1. Drullman R. and Smoorenburg G.F., Audio-visual perception of compressed speech by profoundly hearing-impaired subjects. *Audiology* **36** (1997) pp. 165-177.
2. Giguère C. and Woodland P.C., A Computational model of the auditory periphery for speech and hearing research. I. Ascending path. *J. Acoust. Soc. Am.* **95** (1994) pp. 331-342.
3. Giguère C., Smoorenburg G.F. and Kunov H., The generation of psychoacoustic combination tones in relation to two-tone supression effects in a computational model. *J. Acoust. Soc. Am.* **102** (1997) pp. 2821-2830.
4. Giguère C., Bosman A.J. and Smoorenburg G.F., Automatic speech recognition experiments with a model of normal and impaired peripheral hearing. *Acustica/Acta Acustica* **83** (1997) pp. 1065-1076.
5. Holube I. and Kollmeier B., Speech intelligibility prediction in hearing-impaired listeners based on a psychoacoustically motivated perception model. *J. Acoust. Soc. Am.* **100** (1996) pp. 1703-1716.
6. Johnstone B.M., Pattuzzi R. and Yates G.K., Basilar membrane measurements and traveling wave. *Hear. Res.* **22** (1986) pp. 147-153.
7. Kollmeier B. (ed.), *Psychoacoustics, Speech and Hearing Aids* (World Scientific, Singapore, 1996).
8. Rabiner L. R. and Gold B., *Theory and Application of Digital signal Processing* (Prentice-Hall, New Jersey, 1975).
9. Rhode W. S. and Cooper N.P., Nonlinear mechanics in the apical turn of chinchilla cochlea in vivo. *Auditory Neuroscience* **3** (1996) pp. 101-121.
10. Ruggero M.A., Responses to sound of the basilar membrane of the mammalian cochlea. *Curr. Opin. Neurobiol.* **2** (1992) pp. 449-456.
11. Smoorenburg G.F. and Bosman A. J. (eds.), Hearing impairment and signal-processing hearing aids II. *Scand. Audiol.* **22** Suppl. 38 (1993).
12. Young E.D. and Sachs M.B., Representation of steady-state vowels in the temporal aspects of discharge patterns of populations of auditory nerve fibers. *J. Acoust. Soc. Am.* **66** (1979) pp. 1381-1403.

MODELING INTENSITY DISCRIMINATION AND DETECTION IN NOISE FOR STIMULI WITH DIFFERENT SPECTRO-TEMPORAL SHAPES

NICOLLE H. VAN SCHIJNDEL, TAMMO HOUTGAST AND JOOST M. FESTEN

Dept. of Otolaryngology, University Hospital VU, P.O. Box 7057, 1007 MB Amsterdam, The Netherlands

E-mail: jm.festen@azvu.nl

A model combining signal information from multiple frequency-time (f-t) windows is developed to address detection and intensity discrimination performance for stimuli with different spectro-temporal shapes. The model is based on the concept of elementary auditory f-t windows for stimulus analysis in the auditory system. Experiments have shown that this f-t window is a critical band wide and has a duration of about 4 ms at 1 kHz (1 ms at 4 kHz). The spectro-temporal shape of the stimulus determines the number of auditory f-t windows excited in the auditory system. The internal noise (coding acuity) in each f-t window is proportional to the stimulus energy (in accordance with Weber's Law). The information of the different f-t windows involved in the task is combined. Simulations with this model show that, for stimuli with different spectro-temporal shapes and equal total energy, the masked detection threshold is independent of the number of f-t windows, whereas the intensity discrimination threshold decreases if the number of f-t windows increases. These trends match our experimental data.

1 Introduction

In this study an attempt is made to model some aspects of the combined spectro-temporal processing of the auditory system. The basis of the model is the multiple window idea, i.e., the concept that the auditory system analyzes sounds with a grid of elementary frequency-time (f-t) windows (see Figure 1). Experiments have shown that a reasonable choice for the auditory f-t window is a Gaussian-shaped window with the width of a critical band and a duration of about 4 ms at 1 kHz (1 ms at 4 kHz) (Van Schijndel *et al.* (submitted)). Incoming signals cover a specific number of auditory f-t windows depending on their spectro-temporal shape. In Figure 1 a few examples are given: the *tone* and the *click* cover a large number of f-t windows, whereas the *plop* covers only one or a few f-t windows. The information of the multiple f-t windows excited by the stimulus can be combined.

In this study, both detection and intensity discrimination will be modeled by means of this multiple-window idea. The results of model simulations will be compared with experimental data on detection and discrimination for stimuli with different spectro-temporal shape.

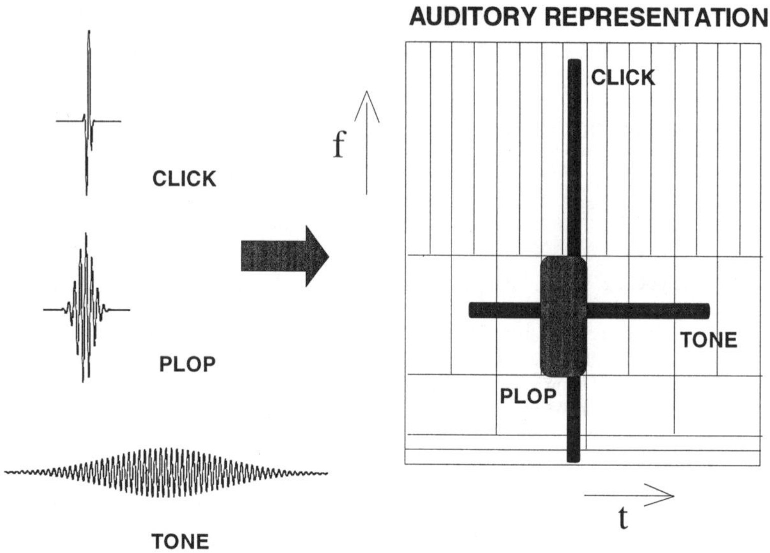

Figure 1. Multiple-window concept. The auditory system analyzes sounds in elementary f-t windows. The number of f-t windows excited by a stimulus depends on its spectro-temporal shape as illustrated here for three types of stimuli.

2 Detection and discrimination model

The input of the detection and discrimination model is a signal embedded in background noise. The external or background noise is Gaussian (random phase and Rayleigh-distributed modulus). The signal covers a particular number (n) of auditory f-t windows. The total energy E of the signal is a constant. Each f-t window gets E/n part of the energy. In the model an 'actual' experiment, i.e., a 3-IFC adaptive procedure is mimicked. Thus, the results of the simulations can be compared directly to the results of experiments.

A scheme of the detection model is presented in Figure 2a. To model the listener's inability to focus exactly on the n f-t windows covered by the signal, for each interval the energy of 50 f-t windows is summed (in the signal interval, these windows include the n windows of the signal). For simplicity, the internal or 'coding' noise is not considered: the observation of extra f-t windows already accounts for internal detection uncertainty. The interval with the highest total energy is decided to be the interval containing the signal.

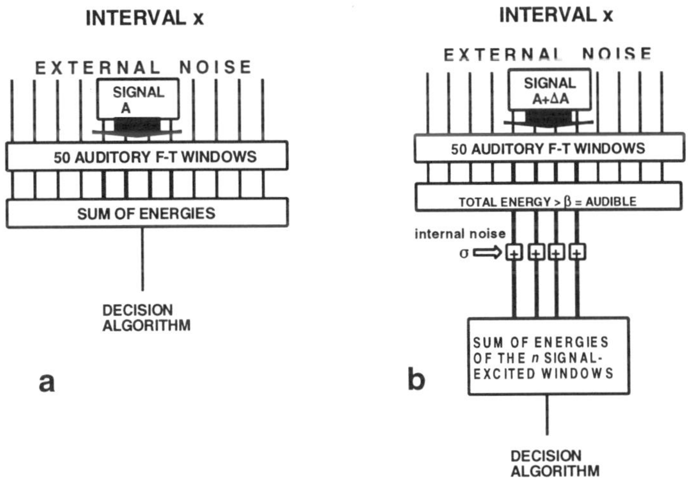

Figure 2. A schematic representation of the detection (a) and discrimination (b) model.

A scheme of the discrimination model is presented in Figure 2b. A two-stage algorithm is used: a detection stage followed by a discrimination stage. In the detection stage a signal is audible if the total energy of 50 f-t windows is higher than β. The constant β is chosen such that the probability that noise alone will have a total energy higher than β is 1 ‰. If the signal is audible, the detection stage is followed by a discrimination stage. In the discrimination stage internal 'coding' noise proportional to the signal energy (in accordance with Weber's Law) is added to the internal representation. In this stage where the signal is always audible, it is assumed that the listener can focus exactly on the n f-t windows covered by the signal. Among the audible signals the signal with the highest energy sum (n f-t windows) is chosen. If none of the signals is audible, the answer is random.

3 Results and discussion

The results of two simulations (circles) are shown in Figure 3a (masked detection threshold) and 3b (intensity discrimination threshold at 10 dB SL). In the figures the corresponding observed data (triangles) are also plotted. Figure 3a shows that both the simulated and observed masked detection thresholds are independent of the number of f-t windows covered by the signal. This agrees with the classical energy detector models, in which detectability depends on the total energy of the signal.

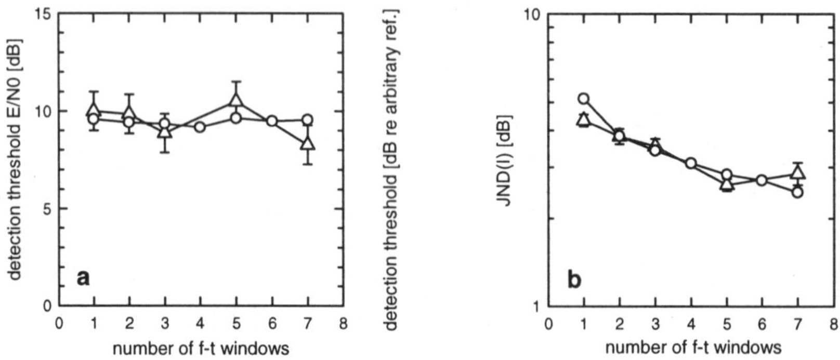

Figure 3. Masked detection threshold (a) and just-noticeable difference in intensity JND(I) (b) as a function of the number of f-t windows covered by the signal. Circles and triangles represent simulated and observed thresholds, respectively. In panel (a), the right axis corresponds to the observed detection threshold. The error bars show the standard error of the mean.

Figure 3b shows that at 10 dB SL both the simulated and observed intensity discrimination thresholds decrease when the number of f-t windows covered by the signal increases. In this model, the coding noise is proportional to the signal energy in each f-t window, and independent between different f-t windows. Therefore, the sum of the energies of the different f-t windows is less influenced by the coding noise, when more f-t windows are summed. As a result, discrimination thresholds will be lower when more f-t windows are covered by the signal.

The simulated trends match the observed data. This indicates that the multiple-window concept is a useful way of addressing the spectro-temporal processing of the auditory system.

4 Acknowledgments

This research was supported by the Netherlands Organization for Scientific Research (NWO).

References

1. Van Schijndel, N. H., Houtgast, T., and Festen, J. M., Intensity discrimination of Gaussian-windows tones: indications for the shape of the auditory frequency-time window, *J. Acoust. Soc. Am.* (submitted).

NEURAL COMPUTATIONS IN BINAURAL HEARING

HERMANN WAGNER
Institut für Biologie II, Lehrstuhl für Zoologie/Tierphysiologie,
RWTH Aachen, Kopernikusstrasse 16, D-52074 Aachen
E-mail: wagner@tyto.bio2.rwth-aachen.de

Binaural hearing helps humans and animals to localize and unmask sounds. Here, binaural computations in the barn owl's auditory system are discussed. Barn owls use the interaural time difference (ITD) for azimuthal sound localization, and they use the interaural level difference (ILD) for elevational sound localization. ITD and ILD and their precursors are processed in separate neural pathways, the time pathway and the intensity pathway, respectively. Representation of ITD involves four main computational steps, while the representation of ILD is accomplished in three steps. In the discussion neural processing in the owl's auditory system is compared with neural computations present in mammals.

1 Introduction

Patients with unilateral deafness or with a cochlear implant connected to only one ear have difficulties in communication in noisy environments and they have also difficulties in sound localization. These observations clearly demonstrate the advantages of binaural hearing. In this article I shall discuss neural computations resulting in the binaural representation of auditory space. I shall mainly concentrate on results described from studies in barn owls, but refer to mammalian results in the discussion. More extensive accounts on this theme may be found in [2, 18, 34].

Binaural hearing takes place in three-dimensional space. In a spherical coordinate system the three spatial parameters are 1) azimuth, the horizontal deviation from the midsagittal plane, 2) elevation, the vertical deviation from the horizontal plane, and 3) the distance between the listener and the sound source. Information about these three parameters has to be extracted from the acoustic signal. A sound may be described by the variation of amplitude and phase with frequency. In the binaural case, further parameters arise from the comparison of the basic signal parameters at the two ears. Thus, we arrive at interaural phase differences (IPDs) and interaural level difference (ILDs) and their variation with frequency. What has to be explained then is how through binaural computations the signal parameters are mapped onto the space parameters.

2 Behavioral correlates

Experiments with earphone stimulation clearly demonstrated that binaural cues are very important for sound localization. Monaural processes in general have much worse spatial resolution than binaural processes [2]. Therefore, one advantage of having two ears is to compare the signals reaching the two ears for sound localization. In the barn owl the interaural time difference (ITD) is the main cue for azimuthal sound localization [2, 20], while the interaural level difference (ILD) is the main cue for elevational sound localization. The use of ILD for elevational sound localization becomes possible in the barn owl, because the axis along which ILD changes is inclined away from the horizontal axis. This inclination reaches 70 deg for 9 kHz [15]. It is unclear of what importance spectral notches and other parameters are for localization precision.

A second advantage of binaural hearing is that it helps in unmasking sounds as measured by binaural masking level differences (BMLD). If a signal and a masker noise are presented at different phases, the signal can be detected more easily than in monaural conditions. The masking level differences depend on the exact interaural conditions and may be as high as some 15 dB [reviewed 2].

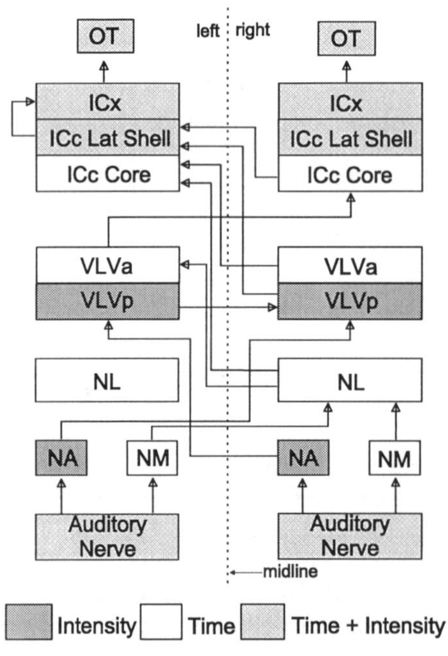

Figure 1. The auditory pathway of the barn owl. There are two parallel sub-pathways, the time pathway and the intensity pathway, that separate at the level of the cochlear nuclei (NA: Nucleus angularis, NM: nucleus magnocellularis), and start to converge again at the level of the lateral shell of the central nucleus of the inferior colliculus (ICc Lat Shell). Note that in most cases only connections originating on one brain side are shown. Further abbreviations: NL: nucleus laminaris, VLVp: posterior division of ventral nucleus of the lateral lemniscus; VLVa: anterior division of ventral nucleus of the lateral lemniscus, ICc Core: core region of central nucleus of inferior colliculus, ICx: external nucleus of inferior colliculus, OT: optic tectum.

3 Auditory pathways

In the barn owl, a clear separation of two functional pathways is evident from physiological recordings (21, 28, Fig. 1). The time pathway processes ITD and its precursors, while the intensity pathway processes level difference and its precursors.

The functional separation is achieved by morphological specializations in the cochlear nuclei [18]. Binaural convergence starts after the separation of the pathways. Thus, the neural computations underlying the mapping of the two spatial coordinates azimuth and elevation takes place independently. This seems reasonable, since two orthogonal coordinates have to be represented, and influence of one parameter on the other should be kept minimal. After the stages of the first binaural convergences almost all neurons exhibit binaural properties. The pathways start to converge again in the lateral shell of the central nucleus of the inferior colliculus (ICc Lat Shell), a substructure of the inferior colliculus that has not been described in mammals. The convergence is completed in the external nucleus of the inferior colliculus (ICx). From there the information is transmitted to integration centers like the optic tectum. Several stations of the inferior colliculus send their information to the forebrain (not shown).

4 Physiological correlates

The most basic test in studying binaural interaction is to audio-visually detect whether a cell responds to stimulation of either ear. Thus, the response types EE, EI, E0, 00-F, etc. were established [8] with E meaning excitatory, I meaning inhibitory, 0 meaning no response, and F meaning facilitation. The first letter in the pairs always refers to contralateral stimulation. EE and EI types are found most frequently, with EE dominating in the time pathway, and EI dominating in the intensity pathway. IE and E0 are rare in the owl. Note that a classification of E0 does not mean that the neuron is monaural, because shunting or silent inhibition needs not be obvious with monaural stimulation. The type 00-F is found only at the higher stations, in the owl especially the ICx and the OT, where many neurons are obligatory binaural.

4.1 Representation of interaural time difference

4.1.1 Coincidence detection

Computation of ITD starts with the process of phase locking in the hair cells. The function of phase locking in representing ITD becomes obvious after the separation of the pathways, and the binaural interaction of magnocellular axons with laminaris somata. In this interaction, the laminaris neurons act as coincidence detectors, while the magnocellular axons act as delay lines as envisioned some 50 years ago by Jeffress [13]. It is remarkable how precise Jeffress's idea is realized in the auditory system of the barn owl [5, 6, 18].

To actually achieve coincidence detection, some regulatory mechanism has to bind together inputs with similar delays. Such a regulatory mechanism to establish the correct wiring of the system is necessary, because the times that have to be measured are in the range of microseconds, while the total delay from the ear drum to the coincidence detector lies in the range of 2-4 milliseconds. Gerstner *et al.* [9] proposed that Hebbian mechanisms might allow the cells to select inputs with the correct delay. This mechanisms is probably active in the soma.

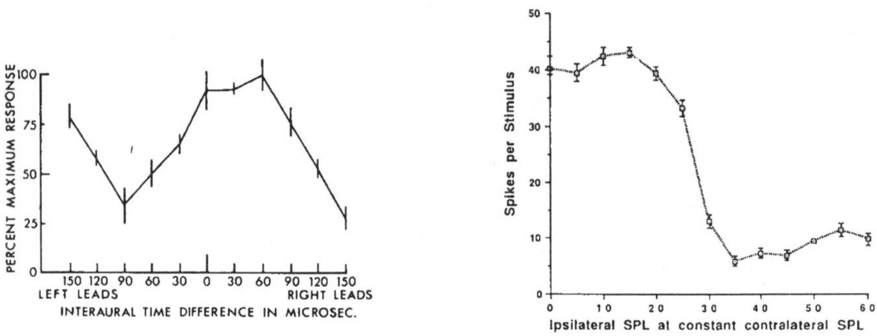

Figure 2. Results of primary binaural computations. left) Output of a coincidence-detecting neuron in nucleus laminaris. Note that the response varies in a periodic manner with ITD. right) Response of an EI-type neuron from VLVp. A) after [18], b) after [1].

While, in principle, binding by a Hebbian mechanism may also work via axo-dendritic synapses and passive or active conductances in a complex dendritic tree, in cells of nucleus laminaris a correlation between frequency tuning and dendritic arborisation exists. Cells tuned to frequencies above some 4 kHz do not have dendritic trees at all, but the afferent axon terminals synapse on the soma [5]. Thus, the reduction of the dendritic tree is a further specialization that underlies temporal sensitivity in the microsecond range.

The output of laminaris neurons upon stimulation with tones or noise of varying ITD is a periodic function exhibiting several maxima of similar response height [Fig. 2 left; Ref. 6]. The response curve is periodic, because laminaris neurons are narrowly tuned to frequency, and thus cannot make use of the information contained in a broad-band signal. The temporal difference of the response peaks in the ITD-curve corresponds to one period of the stimulus tone. Since these narrow-band neurons respond maximally to more than one ITD, they cannot signal ITD unambiguously but only the phase equivalent of the ITD. However, there is usually one characteristic ITD that may be derived by measuring the temporal differences of the signals in the delay lines reaching the coincidence detector or by stimulating with different frequencies. This characteristic ITD or characteristic delay [25] represents the difference in traveling time of the signals from the ear to the coincidence detector. One specialty in the owl is that ITD tuning is found for frequencies up to 9 kHz or about one octave higher than in other animals tested.

In addition to the periodicity of the tuning curves, the tuning to ITD is still weak. An improvement of ITD- tuning is observed, however, in the next station, the ICc core. The underlying computations are not understood. In principle the information could be extracted by a population code and averaging over many weakly tuned neurons. However, the time pathway allows for another possibility: cross-inhibition may take place between the inputs of nucleus laminaris and VLVa in the ICc core.

4.1.2 Across-frequency integration

The computation mentioned in the last paragraph improves tuning, but does not remove the periodicity in the tuning curves. Thus, ICc core neurons do not represent ITD unambiguously. There are two ways to solve this problem. The first one is to restrict analysis of behaviorally relevant ITDs to a range of frequencies at which the ambiguity problem does not occur. If we take the range of physiological ITDs into account, the upper limit would be at some 750 Hz in humans, and at some 3000 Hz in the barn owl. The barn owl cannot make use of this solution, because it has to analyze ITDs in a frequency range at which its head creates measurable ILDs. Such ILDs occur only in the frequency range above some 5 kHz. Therefore, the barn owl needs to employ a different solution to the ambiguity problem. It resolves the ambiguity occurring in the upper frequency range by integrating over several frequencies. Integration across frequencies starts in the lateral shell of the ICc and is completed in the external nucleus of the inferior colliculus, at least three synapses after the computation of coincidence detection (Fig. 1).
Theory suggests how across-frequency integration may take place: if the responses of cells tuned to different frequencies and different interaural phase differences but having the same characteristic delay are combined, responses at the characteristic delay are boosted while the responses at the other "virtual" phase delays are not. Exactly such an integration has been observed (Fig. 3, see also Ref. 27, 31). Response curves from 7 recordings obtained at different depths (and therefore frequency) within the lateral shell of the central nucleus of the inferior colliculus are shown in Fig. 3A. There are multiple peaks in all recordings where a wide enough ITD range has been tested. This array of neurons possesses one ITD at which the relative response is stimulus-independent as shown by statistical analysis (Fig. 3B). This array of neurons projects to one neuron in the next higher station of the auditory pathway, the ICx. By this across-frequency integration a new property - unambiguous representation of ITD - emerges in the ICx neuron [31]. The commonly shared ITD in the array is equivalent to the characteristic delay of the ICx neuron that receives input from the array [27].

Thus, the neural algorithm for the transformation of an ambiguous code to an invariant code in the sound-localization pathway of the owl is across-frequency integration. This algorithm follows the rule that cells tuned to different frequencies but having a constant relation of phase and frequency project onto one cell in the next station. By this integration, and the addition of a tuning to ILD, the cell in a hierarchically higher station becomes not only broadly tuned to frequency, but can also unambiguously represent one ITD or one location in auditory space.

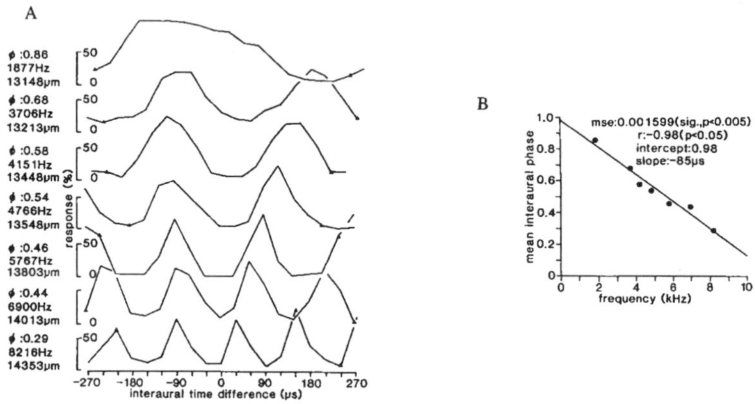

Figure 3. Across-frequency integration. A) Responses of seven cells or cell clusters from the ICc Lat Shell. The neurons were recorded at different depths (denoted in μm), were tuned to different frequencies (Hz) and represented different interaural phase differences (ϕ). B) Interaural phase differences plotted as a function of best frequencies. Note the linear relation (quantified by the mean square error (mse) and the regression coefficient r). Note also the slope of -85 μs, which is close to the -90 μs in A at which all curves share one response peak. The straight line intercepts the y-axis at .98.

The data in Fig. 3 suggest that neurons representing one ITD are combined. This was motivated by the good correlation between theory and observations made with stimuli presented via earphones. Indeed, if the phase-frequency relationship is measured for sources at loci in frontal auditory space, linear relationships result [3, 15]. For lateral auditory space, however, the relation is no longer linear [3]. Therefore, a linear relationship would not represent one locus in space, but many of them in a frequency-dependent way. Thus, the linear theory is a good approximation for frontal space only. For other locations in auditory space, the theory must be extended. The wiring is most likely achieved by behavioral feedback, i.e. those phase-frequency combinations are bound together that lead to the best localization or hunting success.

4.2 *Representation of interaural level difference*

In the computation of interaural level difference (ILD), the sensitivity of neurons to sound pressure level is important. This becomes again obvious in the cochlear nucleus. Neurons in the nucleus angularis that are at the outset of the intensity pathway exhibit a high level sensitivity. The first stage of binaural convergence in the level pathway takes place in VLVp (Fig. 1). In this nucleus neurons that exhibit EI-type responses are found (Fig. 2B). The computation is not a simple inhibition of the contralateral input by the ipsilateral input, however, because, there are dorso-ventral gradients of inhibitions that are due to mutual inhibition between the two VLVps [26]. The EI curves show a sensitivity to ILD, but their response depends also on other stimulus parameters, especially also on stimulus level. Therefore, neurons in VLVp cannot represent one ILD independent of stimulus level. This dependence is largely removed by a further computation taking place in the ICc Lat Shell. Here an excitatory signal from an unknown source is inhibited by EI-type inputs. Adolphs [1] concluded from his experiments that two EI-curves interact with the excitatory signal: one EI-curve having a positive slope, i.e. with low responses at negative ILDs; and one EI-curve with a negative slope. In this way, a peaked ILD curve is created. The location of the peak is robust against changes in sound level and, therefore, allows for an unambiguous representation of ILD.

5 Discussion

Representation of the binaural parameters ITD and ILD is achieved in separate networks. It takes four main computational steps to represent ITD: phase locking, coincidence detection, (possibly) cross-inhibition and across-frequency integration. ILD representation involves three steps: creation and maintenance of level sensitivity, subtraction of the ipsilateral responses from the contralateral response in the EI curves and interaction of EI-type responses in creating a peaked ILD curve. After these computations, neurons represent one combination of ITD and ILD. In the following I shall relate the computational principles underlying the extraction of the binaural signal parameters in the barn owl to computations found in the auditory system of mammals.

Computation of ITD in large mammals seems to involve similar steps as found in the barn owl. The first station of binaural interaction of phase locked signals takes place in the medial superior olive (MSO). MSO neurons are coincidence detectors in large animals [10, 33], but not in small mammals like bats [11]. ITD tuning in MSO neurons is weak like in laminaris neurons of owls. An increase of ITD tuning is also seen in the neurons of the mammalian ascending auditory pathway, and similar mechanisms as discussed in the owl may be active [7]. A recent model of IC neurons, for example, contains inhibitory interactions between the inputs from the two MSOs [4]. Across-frequency integration has not been reported in electrophysiological studies of mammals, although psychoacoustic data from humans suggest that such a computation may take place [29]. One important aspect of representing time in the mammalian auditory system is that there the detection of time difference in the envelope of the signal seems to play an important role [14, see also Kohlrausch, this volume]. In mammals, envelope ITDs can be measured in high frequency neurons, while carrier or ongoing ITDs can only be measured for low frequencies. Representation of ITDs of the signal envelope probably occurs also in the owl, but has not been investigated.

In mammals, ILD changes with azimuth and not with elevation. Thus, ILD is a cue for azimuth in mammals. The initial steps of ILD computation seem to be similar in mammals and owls. Cells that are especially sensitive to stimulus level are found in the ventral cochlear nucleus [24, see also Winter, this volume], and there are IE-type cells, that are first observed in the lateral superior olive [14]. However, peaked ILD curves have not been reported from mammalian studies. Since mammals are able to represent space by ILD, other mechanisms like a scaling of an ILD curve by monaurally detected level in cells with large dynamic ranges like onset choppers (see Winter, this volume) may be active (see also Kohlrausch, this volume).

Since ITD and ILD vary along the same dimension, a really strict separation of the pathways computing ITD and ILD is not necessary. Nevertheless, some separation of two pathways is observed at the level of the cochlear nuclei. On the other hand, the phenomenon of time-intensity trading suggests that more interaction between ITD and ILD computation may exist in mammals than in the owl. This statement should be taken with caution, however, because there are different trading ratios in the time and intensity pathways [for a discussion see 12], and it may be that also in mammals the low-frequency, EE-type time processing pathways exhibits little time intensity trading as it does in the owl [30], and that most of the time-intensity trading happens in the EI-type pathway where only envelopes are detected. Interaction of ITD-sensitive neurons and ILD-sensitive neurons to built 2-D space-specific neurons is not possible in mammals, because ITD and ILD vary along the same spatial dimension.

There are many more binaural phenomena than could be discussed here. The use of binaural information for unmasking is an important behavioral advantage of having two ears. Neural correlates of masking level differences have been described in mammals [22]. Likewise, forebrain pathways also contain many binaural neurons, but new computational principles have not been observed. However, we are just at the beginning of understanding forebrain computations [23]. The most interesting suggestion comes from a study in cats that suggested a time code rather than a rate code being important for representation of binaural space in the auditory cortex [19].

References

1. Adolphs R., Bilateral inhibition generates neuronal responses tuned to interaural level differences in the auditory brainstem of the barn owl. *J. Neurosci.* **13** (1993) pp. 3647-3668.
2. Blauert, J. Spatial Hearing (MIT Press, Cambridge, Massachusetts, 1997).
3. Brainard M.S., Knudsen E.I. and Esterly S.D., Neural derivation of sound source location: resolution of spatial ambiguities in binaural cues. *J. Acoust. Soc. Am.* **91** (1992) pp. 1015-1027.

4. Cai H., Carney L.H., Colburn H.S., A model for binaural response properties of inferior colliculus neurons. I. A model with interaural time difference-sensitive excitatory and inhibitory inputs. *J. Acoust. Soc. Am.* **103** (1998) pp. 475-493
5. Carr C.E. and Boudreau R.E., Organization of the nucleus magnocellularis and the nucleus laminaris in the barn owl: encoding and measuring interaural time differences. *J. Comp. Neurol.* **335** (1993) pp. 337-355.
6. Carr C.E. and Konishi M., A circuit for detection of interaural time differences in the brainstem of the barn owl. *J. Neurosci.* **10** (1990) pp. 3227-3246.
7. Fitzpatrick D.C., Batra R., Stanford T.R. and Kuwada S., A neuronal population code for sound localization. *Nature* **388** (1997) pp. 871-874.
8. Fuzessery Z.M. and Pollak G.D., Determinants of sound location selectivity in bat inferior colliculus: a combined dichotic and free-field stimulation study. *J. Neurophysiol.* **54** (1985) pp. 757.
9. Gerstner W., Kempter R., van Hemmen J.L. and Wagner H., A neuronal learning rule for submillisecond temporal learning. *Nature* **383** (1996) pp. 76-78.
10. Goldberg J.M. and Brown P.B., Response of binaural neurons of dog superior olivary complex to dichotic tonal stimuli: some physiological mechanisms of sound localization. *J. Neurophysiol.* **32** (1969) pp. 613-636.
11. Grothe B. and Park T.J., Sensitivity to interaural time differences in the medial superior olive of a small mammal, the mexican free-tailed bat. *J. Neurosci.* **15** (1998) pp. 6608-6622.
12. Hafter, E.R. Spatial hearing and the duplex theory: how viable is the model? In *Dynamic aspects of neocortical function*, ed. by G. M. Edelman, W. E. Gall and W.M. Cowan (Wiley, New York, 1984) pp. 425-448.
13. Jeffress L.A., A place theory of sound localization. *J. Comp. Physiol. Psychol.* **41** (1948) pp. 35-39.
14. Joris P.X. and Yin T.C., Envelope coding in the lateral superior olive. I. Sensitivity to interaural time differences. *J. Neurophysiol.* **73** (1995) pp. 1043-1062.
15. Keller C.H., Hartung K. and Takahashi T.T., Head-related transfer functions of the barn owl: measurement and neural responses. *Hear. Res.* **118** (1998) pp. 13-34.
16. Knudsen E.I. and Konishi M., Mechanisms of sound localization in the barn owl (Tyto alba). *J. Comp. Physiol. A* **133** (1979) pp. 13-21.
17. Konishi M., Listening with two ears. *Sc. Am.* **268 (4)** (1993) pp. 66-73.
18. Konishi, M., Takahashi, T.T., Wagner, H., Sullivan, W.E. and Carr, C.E. Neurophysiological and anatomical substrates of sound localization in the owl. In *Auditory Function - Neurobiological bases of hearing*, . by G. M. Edelman, W. E. Gall and W.M. Cowan (Wiley, New York, 1988) pp. 721-745.
19. Middlebrooks J.C. Clock A.E., Xu L. and Green D.M., A panoramic code for sound localization by cortical neurons, *Science* **264** (1994) pp. 842-844.

20. Moiseff A. and Konishi M., Neuronal and behavioral sensitivity to binaural time differences in the owl. *J. Neurosci.* **1** (1981) pp. 40-48.
21. Moiseff A. and Konishi M., Binaural characteristics of units in the owl's brainstem auditory pathway: precursors of restricted spatial receptive fields. *J. Neurosci.* **3** (1983) pp. 2553-2562.
22. Palmer A.R., Rees A. and Caird D., Binaural masking and sensitivity to interaural delay in the inferior colliculus. *Phil. Trans. R. Soc. Lond. B* **336** (1992) pp. 415-422.
23. Rauschecker, J.P., Cortical processing of complex sounds. *Curr. Opin. Neurobiol.* **8** (1998). pp. 516-521.
24. Rhode W.S., Encoding timing and intensity in the ventral cochlear nucleus of the cat. *J. Neurophysiol.* **56** (1986) pp. 261-286.
25. Rose J.E., Gross N.G., Geisler C.D. and Hind J.E., Some neural mechanisms in the inferior colliculus of the cat which may be relevant to the localization of a sound source. *J .Neurophysiol.* **29** (1966) pp. 288-314.
26. Takahashi T.T. and Keller C.H., Commissural connections mediate inhibition for the computation of interaural level difference in the barn owl. *J. Comp. Physiol. A* **170** (1992) pp. 161-169.
27. Takahashi T. and Konishi M., Selectivity for interaural time difference in the owl's midbrain. *J. Neurosci.* **6** (1986) pp. 3413-3422.
28. Takahashi T., Moiseff A. and Konishi M., Time and intensity cues are processed independently in the auditory system of the owl. *J. Neurosci.* **4** (1984) pp. 1781-1786.
29. Trahiotis C. and Stern R.M., Across-frequency interaction in lateralization of complex binaural stimuli. *J. Acoust. Soc. Am.* **96** (1996) pp. 3804-3806.
30. Viete S., Pena J.L. and Konishi M., Effects of interaural intensity differences on the processing of interaural time differences in the owl's nucleus laminaris. *J. Neurosci.* **17** (1997) pp. 1815-1824.
31. Wagner H., Takahashi T.T. and Konishi M., Representation of interaural time difference in the central nucleus of the barn owl's inferior colliculus. *J. Neurosci.* **7** (1987) pp. 3105-3116.
32. Yin T.C.T. and Chan J.C.K., Interaural time sensitivity in medial superior olive of cat. *J. Neurophysiol.* **64** (1990) pp. 465-488.
33. Yin T.C.T., Chan J.C.K. and Carney L.H., Effects of interaural time delays of noise stimuli on low-frequency cells in the cat's inferior colliculus. III. Evidence for cross correlation. *J. Neurophysiol.* **58** (1987) pp. 562-583.
34. Yin, T.C.T. and Kuwada, S. Neuronal mechanisms of binaural interaction. In: *Dynamic aspects of neocortical function*, ed. by G. M. Edelman, W. E. Gall and W.M. Cowan (Wiley, New York, 1984) pp. 263-313.

A PHYSIOLOGY-BASED COMPUTATIONAL MODEL FOR SOUND LOCALIZATION

K. HARTUNG

Institute of Communication Acoustics, Ruhr University Bochum, 44780 Bochum, Germany
e-mail: hartung@ika.ruhr-uni-bochum.de

S. J. STERBING

Department of Zoology and Neurobiology, Ruhr University Bochum, 44780 Bochum, Germany
e-mail: sterbing@neurobiologie.ruhr-uni-bochum.de

Electrophysiological investigations in the midbrain of the guinea pig (ICc) revealed, that more than 90% of the neurons were spatially tuned when stimulated with broadband virtual sound sources. Most of the neurons preferred lateral positions, but tuning to frontal or rear directions and to different elevations was also observed. Based on the individual head-related transfer functions of each animal the interaural level differences (ILD), interaural time differences (ITD) and the monaural directivity were calculated in 1/3 octave bands for the upper hemisphere. It was assumed, that the neurons received input from ILD and ITD processors and from monaural pathways. The relative weights of these 72 inputs were estimated by a least square approximation of the neuronal response. The modeled responses were in good agreement with the measured responses. The weights were different for each of the tested neurons. High weights were found for ILD cues around the characteristic frequency of the neurons. Based on this single neuron model a localization model using a population of neurons which were tuned to different directions was tested in a localization task. The model allowed a robust estimation of the direction of the sound source.

1 Introduction

Single unit recordings in the midbrain (central nucleus of the inferior colliculus, ICc) showed, that the vast majority of neurons (approx. 90%) was responding selectively for certain sound directions [1]. One of the main findings was, that front-rear discrimination could be shown on single neuron level. The ICc is an obligatory auditory pathway station and receives among others inputs from the medial (MSO) and the lateral (LSO) part of the superior olivary complex. It is assumed, that the MSO computes the interaural time differences and the LSO mainly the interaural level differences. Furthermore, the ICc receives additional monaural inputs from the cochlear nuclei. It is investigated, whether the spatial tuning of the ICc neurons can be explained by a superposition of physiological plausible representations of the these inputs.

2 Methods

The individual head-related transfer functions (HRTF) of 19 guinea pigs were measured in an anechoic room with miniature microphones (Knowles 3046) placed at the entrance of the ear canal(Details see [1]). Broadband virtual spatial sounds (200 Hz - 16 kHz, 50 ms duration, 5 ms ramp, 80dB SPL) from 122 directions of the entire upper hemisphere were generated off-line and presented via individually calibrated earphones to the anesthetized (ketamine/thiazine) animal in a pseudo-random order. Each position was presented five times. The single unit activity in the ICc was recorded with glass microelectrodes (Impedance: 3-10 MΩ, filled with 3M KCl). The number of spikes for each direction was tested for statistical significant differences against the neighbored directions (Kruskal-Wallis-Test). For each neuron CF was determined with pure tones.

3 Results of the physiological recordings

Virtual sound source signals (VSS) of different bandwidth (white noise, one octave, 1/3 octave) were used to stimulate 46 single neurons. The center frequency of the one-octave and third-octave bands was chosen according to the characteristic frequency of each neuron. The characteristic frequencies were measured using pure tone stimulation of the contralateral ear. The majority (74%) of neurons, which was tuned under the white-noise condition, was also tuned during stimulation with one-octave VSS. The size of the receptive fields (RF) was constant with decreasing bandwidth for 51% of the neurons, increased for 35%, decreased for 7%. Another 7% of the units showed a change of RF position. For stimulation with third-octave bands only 31% of the neurons responded spatial selective. Usually, spike rate increased for narrow-band stimuli.

4 Model structure

Fig. 1 shows the structure of the model. The signals are spatially filtered by the HRTF of the left and right ear. Within 1/3 octave bands the ILD (interaural level difference) and ITD (interaural time difference) of each frequency band are calculated. It is assumed, that within a frequency band different neurons are tuned to different ITD values. A neuron responds maximally if the presented ITD (τ_Δ) matches the preferred ITD ($\tau_{\Delta,opt}$) of that neuron.

This response is modeled with

$$A(\tau_\Delta, f_i) = e^{-\left(\frac{\tau_{\Delta,opt}(f_i)-\tau_\Delta(f_i)}{\sigma_{\tau_\Delta}(f_i)}\right)^2}$$

$\sigma_{\tau\Delta}$ determines the tuning sharpness. The values are between 0.6 and 1.5 ms. The ILD-tuning is modeled by a similar equation using $\sigma_{L\Delta}$values between 5 dB and 10 dB.

$$A(L_\Delta, f_i) = e^{-\left(\frac{L_{\Delta,opt}(f_i) - L_\Delta(f_i)}{\sigma_{L_\Delta}(f_i)}\right)^2}$$

Additionally to this binaural inputs monaural inputs are also contributing to the activity. The activity of each of the monaural inputs is proportional to the power in each 1/3 octave bands at the left and right ear. The values are scaled in such way, that for a constant level of the sound source the values are between 0 and 1 for all frequency bands and directions. The activity of the model neuron is the weighted linear superposition of the activities of the above mentioned inputs. The model neuron can receive maximally one ITD, one ILD and one monaural input from the left and the right ear for each band and integrates these inputs over the whole frequency range.

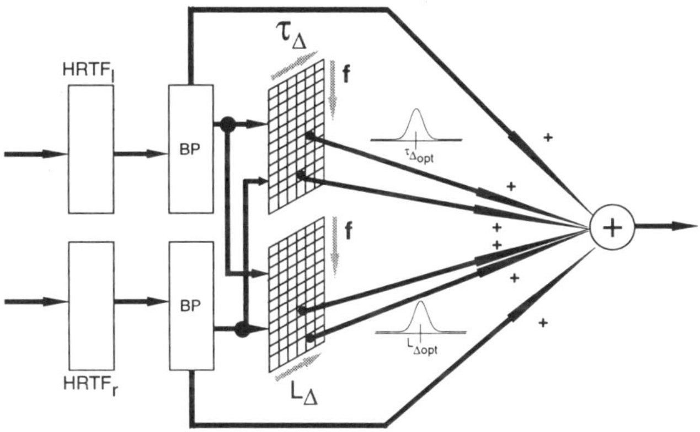

Figure 1. Structure of the model. $HRTF_{l,r}$:head-related transfer functions, BP: Band-pass filter bank, τ_Δ : time difference processors. L_Δ level difference processors.

Fig.2 shows an example of the directional activity of a time and level difference processor. These activities reflect the shape of the ILD and ITD contours of the HRTF. The ILD activity shows prominent asymmetries between sound incidence from the frontal and rear hemisphere, while the ITD activity is symmetric around the interaural axes.

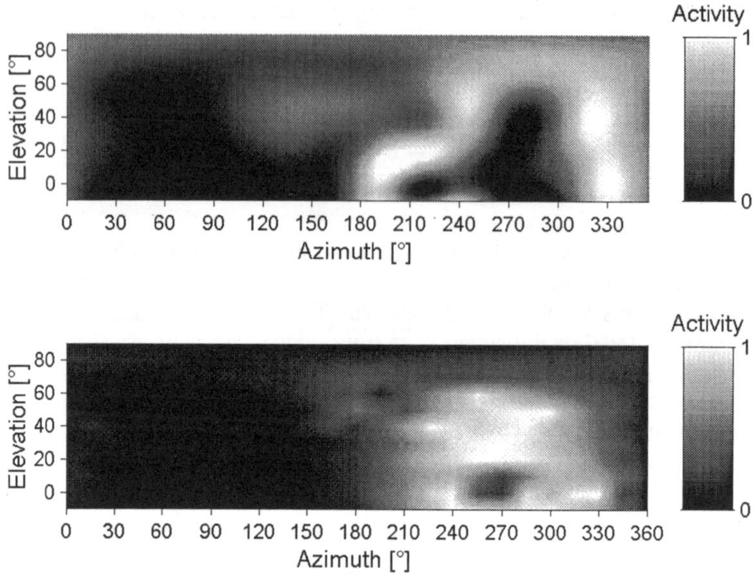

Figure 2. Examples of directional activity of one ILD (upper panel) and one ITD (lower panel).input. The inputs are optimally tuned to sound incidence from 315° azimuth and 0°elevation. The center frequency is 8.2 kHz for the ILD input and 10.3 kHz for the ITD input. The gray scale bar on the right side of each diagram indicates normalized activity. White areas mark directions with maximum activity.

The weight of the different inputs are adjusted in order to minimize the squared error between response of the model and the measured neuronal response over all directions. This least square problem is constrained by the condition, that the weights should not take any negative values. The optimal ITD ($\tau_{\Delta,opt}$) and ILD ($L_{\Delta,opt}$) corresponds to the values at the preferred direction of the neuron, which is modeled.

5 Results

The spatial activity of the model neurons fit the measured response very well. Fig.3 shows an example of the measured and modeled spatial response. The front-rear discrimination and the selectivity for high elevation can be modeled. The distribution of the input weights is different for all tested neurons. The modeled neurons have high weights for the ILD inputs close to the characteristic frequency (CF). The monaural inputs and ITD inputs seem to be less important, but cannot be neglected. In many cases it is observed, that in addition to the inputs close to CF further inputs from distant frequency bands have a significant contribution to the output activity. These results are in accordance with the results of the physiological experiments with narrow band sound sources (1/3 octave, octave). It can be assumed, that the spatial tuning of ICc neurons requires the integration of monaural and interaural cues over a wide frequency range.

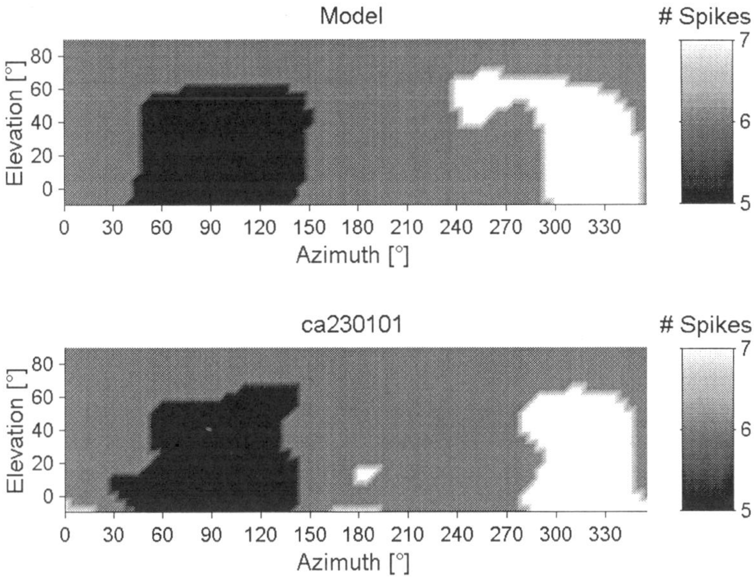

Figure 3. Spatial activity of the model neuron (upper panel) and measured response (lower panel). The gray scale bar on the right side of each diagram indicates the median spike count. White areas mark directions with maximum activity.

6 Localization model

In order to test the efficiency of the representation, a localization model based on a population of many single neurons was created. The localization model consists of great number of model neurons, which are each tuned for different directions (resolution 5°). The preprocessing and the representation of the ILDs and ITDs is the same as in the single cell model. The robustness of the representation of the interaural parameters was tested by a superposition of internal noise on the presented ILD and ITD patterns.

A localization test in the horizontal plane using interaural level and time difference cues derived from a catalogue of human HRTF gave the following results: for all directions, except directly for the front and rear, the estimated direction matched the presented direction. The model responses contained more front-back than back-front-confusions.

7 Discussion

The results of the narrow-band experiments and the model simulations give evidence for the assumption that the spatial tuning of a single neuron in the ICc requires the integration over non adjacent bands in a relatively large frequency range. This is different to the findings of Esterly and coworkers[2]. They modeled the spatial tuning of the neurons in the optic tectum of the barn owl and found that the shape of the spatial receptive fields could be explained by the superposition of the activity of ILD and ITD processor of adjacent frequency band. They used slightly different activation functions for the interaural cues and did not consider monaural cues. Front-back discrimination was not tested, because the presented directions were restricted to the frontal hemisphere. Our results suggest, that the processing in the ICc of the guinea pig is different to the processing in the optic tectum of the barn owl. It seems that the in ICc the different cues are integrated in a very specific manner, which might be more robust against distractors and can explain spatial front-back discrimination. The distribution of the estimated input weights is in agreement with the results of the narrow-band experiments and anatomical studies [3]. It should be remarked, that this input weights are only estimates of the real inputs. Further experiments will test the estimates during the electrophysiological experiment.

8 Acknowledgements

This work was supported by DFG grant Ho 450/23-1-347.

References

1. Hartung K. and Sterbing S. J., Generation of virtual sound sources for the electrophysiological characterization of auditory spatial tuning in the guinea pig, In: J. SYKA (ed.) Acoustical signal processing in the central auditory system , Plenum Press, New York, 1997, pp. 408–412.
2. Brainard M., Knudsen E.and Esterly, S., Neural derivation of sound source location: Resolution of spatial ambiguities in binaural cues. *J. Acoust. Soc. Am.*, **91**(2) (1992), pp. 1015-1027.
3. Malmierca M. S., Rees A., LeBeau F. E., Bjaalie J.G., Laminar organization of frequency-defined local axons within and between the inferior colliculi of the guinea pig. *J. Comp. Neurol.* **357**, (1995), pp. 124-144.

QUALITY OF COINCIDENCE DETECTION AND ITD TUNING: A THEORETICAL FRAMEWORK

RICHARD KEMPTER[†], WULFRAM GERSTNER[‡], HERMANN WAGNER[*], AND J. LEO VAN HEMMEN[†]

[†]*Physik Department der TU München, Theoretische Physik, D-85747 Garching, Germany*
E-mail: kempter,lvh@physik.tu-muenchen.de

[‡]*Swiss Federal Institute of Technology, Center of Neuromimetic Systems, EPFL-DI, CH-1015 Lausanne, Switzerland*
E-mail: Wulfram.Gerstner@di.epfl.ch

[*]*RWTH Aachen, Institut für Biologie II, D-52074 Aachen, Germany*
E-mail: wagner@tyto.bio2.rwth-aachen.de

Coincidence detector neurons increase their firing rate significantly if the input changes from random to coherent. Similarly, neurons in the avian nucleus laminaris vary their firing rate as a function of the interaural time difference (ITD). In both cases, neurons transform temporally coded input into a rate-coded output. To characterize the quality of this transformation we define a new measure, which explicitly takes noisy spike output of neurons into account. As an application, we investigate the coincidence detection properties of an integrate-and-fire (I&F) neuron in dependence on internal parameters and input statistics. We show that there is an optimal threshold and, furthermore, that there is a broad range of near-optimal threshold values. The theoretical results are applied to ITD-tuning of neurons in the laminar nucleus of the barn owl.

1 Introduction

Neurons have two basic operation modes: they can function either as integrators or as coincidence detectors. In the integration mode, the output rate changes as a function of the *mean* input rate and is independent of the temporal fine structure of input spike trains. Neurons act as coincidence detectors, if their mean output firing rate does depend on the temporal structure of input spike trains. Coincidence detection implies that the output firing rate is higher if spikes arrive 'coincidently' as compared to a random spike arrival.

There is, however, no sharp boundary that separates the two basic operation modes. It is known that the membrane time constant [10], the spike threshold and the number of input synapses [1] are relevant parameters. In this contribution we develop a framework that allows to assess how these parameters influence the quality of coincidence detection. We consider stochastic arrival of presynaptic spikes and ask the following question: How well can a neuron detect the degree of synchrony in the input? Can it convert a time-coded signal into a rate-coded one? For I&F neurons we show explicitly the dependence of the precision of neuronal coincidence detection upon, e.g., the membrane time constant, the input statistics, and the voltage threshold at which an action potential is generated. The theoretical results are applied to neurons in the laminar nucleus of the barn owl. For a more comprehensive and mathematically detailed account the reader is referred to Kempter *et al.* [7,8]. For a different model approach see [2].

2 Temporal Coding of the Input

Input spikes are generated stochastically and arrive at each of the $N \gg 1$ synapses of a single neuron with a time-dependent rate $\lambda^{\text{in}}(t) \geq 0$. The probability of having a spike in the interval $[t, t + \Delta t[$ is $\lambda^{\text{in}}(t)\Delta t$ as $\Delta t \to 0$. In this way we obtain a *nonstationary* or *inhomogenous* Poisson process. [8]

As a specific example we consider T-periodic rates, $\lambda^{\text{in}}(t) = \lambda^{\text{in}}(t+T)$, where input spikes are more or less phase-locked to a periodic stimulus. Phase locked spikes are an example of a temporal code. The average number of spikes that arrive during one period at a synapse will be called p. The time-averaged mean input rate is $\bar{\lambda}^{\text{in}} := 1/T \int_{t_0}^{t_0+T} \mathrm{d}t' \, \lambda^{\text{in}}(t')$ and equals p/T. To parameterize the input, we take the function

$$\lambda^{\text{in}}(t) := p \sum\nolimits_{m=-\infty}^{\infty} G_\sigma(t - mT) \tag{1}$$

where $G_\sigma(.)$ denotes a normalized Gaussian distribution with zero mean and standard deviation $\sigma > 0$. At each input terminal spikes arrive with a probability density given by Eq. (1). Note that spikes arriving at different synapses are generated independently.

The standard deviation $\sigma \in [0, \infty[$ of the Gaussian distribution in Eq. (1) can be used to characterize the degree of synchrony of the input. In the case $\sigma = 0$, the input spikes arrive perfectly phase-locked and occur only at times $t_m = mT$ with integer m, and the number of spikes arriving at time t_m has a Poissonian distribution with parameter p. In the case $\sigma > 0$, there is also a temporal jitter of input spikes times around the t_m.

Instead of σ, another measure of synchrony, the so-called *vector strength* r^{in} may be used [5]. This measure can be defined as the ratio of the amplitude of the first Fourier component of the periodic rate $\lambda^{\text{in}}(t)$ and that of the Fourier component of order zero. For the input given in Eq. (1) we find $r^{\text{in}} = \exp[-(2\pi/T)^2 \, \sigma^2/2]$. By construction, we have $0 \leq r^{\text{in}} \leq 1$.

For example, neurons in the avian laminar nucleus are sensitive to the degree of coherence in the input. If a neuron in an *in vitro* preparation is stimulated by synchronized current pulses at times $T, 2T, 3T, \ldots$ (that means $r^{\text{in}} = 1$), then the neuron fires at a high rate. If the same amount of current is delivered by incoherent pulses or constant input current, the neuron is much less active [11]. In vivo, input spikes are not perfectly phase locked but have some jitter around a preferred phase. The maximum input vector strength to a neuron in the laminar nucleus of the barn owl is $r^{\text{in}} \approx 0.5$ for a 5 kHz sound wave. It is achieved, if input from the left and the right ear arrives at the optimal ITD. [3] If left and right ear stimulus are out of phase, the effective input vector strength is $r^{\text{in}} \approx 0$. The dependence of the output firing rate of a laminar neuron upon the relative phase of the stimulus is called the ITD-tuning curve of the neuron. We note that for different positions on the ITD-tuning curve the *mean* input is always the same – only the timing and, hence, the effective vector strength of the input changes.

3 Classification of Coincidence Detectors

In a standard firing rate description a neuron would be characterized by its gain function, which describes the dependence of the output firing *rate* upon the mean input *rate*. Here we take a different approach. We vary the temporal structure of the input under the constraint of a *constant* mean input rate, $\bar{\lambda}^{\text{in}} = \text{const}$. The example of ITD tuning shows that even with a constant mean input rate, a neuron's output rate $\bar{\lambda}^{\text{out}}$ can vary as a function of the temporal structure of the input. In other words, a neuron can 'read out' a temporal code. In order to quantify this effect, we define two measures: the *coherence gain* and the *quality factor*.

3.1 Coherence Gain and Modulation Depth of ITD Tuning

Let us suppose we change the vector strength r^{in} of the input while keeping the mean constant. For each value of r^{in} we measure the output rate $\bar{\lambda}^{\text{out}}(r^{\text{in}})$. We want to compare the output rate for a given vector strength, e.g., $r^{\text{in}} = 0.5$ with the rate for $r^{\text{in}} = 0$. To do so we define the *coherence gain*

$$E(r^{\text{in}}) := \bar{\lambda}^{\text{out}}(r^{\text{in}})/\bar{\lambda}^{\text{out}}(0) \tag{2}$$

Similarly, $E(1)$ would be the ratio between the mean firing rate for fully coherent input and the rate for completely incoherent input. $E(1) \gg 1$ is commonly interpreted as signaling a good coincidence detector. A coherence gain $E(r^{\text{in}} = 1) = E(r^{\text{in}} = 0.5) = 1$ means that a neuron is an integrator.

The notion of coherence gain can be directly interpreted as the modulation depth of an ITD tuning curve. To see how this works, let us return to the example of a 5 kHz neuron in the nucleus laminaris of the barn owl. As we have seen at the end of the previous Section, optimal ITD (that is the maximum of the ITD tuning curve) corresponds to stimulation with vector strength $r^{\text{in}} = 0.5$. The minimum of the ITD tuning curve corresponds to nearly incoherent stimulation and may hence be identified with $r^{\text{in}} = 0$. The modulation depth M of the ITD tuning curve may be defined as the maximum rate divided by the minimum rate, hence $M = E(r^{\text{in}} = 0.5)$. For an example of ITD tuning in a model neuron, see Fig. 2.

3.2 Quality Factor of Coincidence Detection

The coherence gain E, which is a ratio of firing rates, is in general not sufficient for characterizing coincidence detectors. The reason is that in biological systems the mean output firing rate of a neuron cannot be determined exactly within a finite amount of time. The output rate of a neuron is defined as the number of spikes occuring in a time interval divided by the length I of that interval. Since spikes occur often highly irregularly, seemingly randomly [12], the number of spikes in an interval shows up variability. That is to say, the estimated output rate is subject to statistical errors — the shorter I and the smaller the firing rate, the larger the error. This has to be taken into account in characterizing a coincidence detector. To illustrate the limitations of E when quantizing coincidence detection, we consider a neuron that is silent for incoherent input and has a rate of 0.1 Hz for coherent input. Even though E is infinite in this case, the neuron's output could not be used to detect coherent input in a time interval of, say, $I = 100\,\text{ms}$ since most likely not a single spike will occur.

To overcome the shortcomings in the definition of E we introduce a more powerful notion of characterizing a coincidence detector. The output of a coincidence detector must convey a signal concerning the nature of a noisy input (coherent or incoherent) in a *finite* amount of time I. We assume that the number n of output spikes in I is a random quantity, which has some probability density $P(n)$. The mean value $\overline{n} = \sum_{n=1}^{\infty} n\,P(n)$ is $\overline{n} = \bar{\lambda}^{\text{out}}\,I$, where $\bar{\lambda}^{\text{out}}$ is the mean output rate. For two different input vector strengths, e.g., $r^{\text{in}} = 0$ and $r^{\text{in}} = 0.5$, we have two different distributions P with mean values $\overline{n}(r^{\text{in}} = 0)$ and $\overline{n}(r^{\text{in}} = 0.5)$, respectively. The task is to distinguish the two cases based on a single measurement of the number n of output spikes in I. This can be done by means of a number n' which acts as a decision boundary. If n is larger than or equal to the criterion n', then we classify the input to have $r^{\text{in}} = 0.5$; if n is smaller than n', then we classify $r^{\text{in}} = 0$.

Now the key idea is that we have to segregate ***probability distributions.***

If the two distributions have a large overlap, a reliable decision is impossible, whatever n'. If the two distributions are well separated, then there is a high chance of making the right choice. The width of each distribution is characterized by the standard deviation ς. In order to quantify the discriminability of the two distributions we consider their normalized 'distance':

$$\gamma := \frac{\bar{n}(r^{\text{in}}) - \bar{n}(r^{\text{in}})}{\varsigma(r^{\text{in}}) + \varsigma(r^{\text{in}})} = \sqrt{\bar{n}(r^{\text{in}})} - \sqrt{\bar{n}(0)} \ . \tag{3}$$

For the second equality sign we have used that we have $\varsigma^2 = \bar{n}$ for Poisson distributed numbers. A Poisson distribution is a good approximation if spiking is driven by the variance of the neuron's input [13]. We call γ the *quality factor* for coincidence detection [a]. Using the definition of the coherence gain E [cf. Eq. (2)] we obtain from Eq. (3)

$$\gamma = \sqrt{I\,\bar{\lambda}^{\text{out}}(0)}\left(\sqrt{E(r^{\text{in}})} - 1\right) \ . \tag{4}$$

Equation (4) shows how the γ increases with increasing length of the time interval I, the mean output rate $\bar{\lambda}^{\text{out}}(0)$, and the coherence gain $E(r^{\text{in}})$. It is important to realize that $\bar{\lambda}^{\text{out}}(0)$ and $E(r^{\text{in}})$ are *not* independent variables.

4 Integrate-and-Fire Neurons

As a specific example we study the coincidence detection properties of an integrate-and-fire (I&F) neuron that receives stochastic input spike trains through N independent channels as specified in Sec. 2. We show that there is an optimal threshold for coincidence detection and that there is a broad range of near optimal threshold values. Fine tuning is not necessary. Finally, we show how the quality factor depends on neuronal parameters such as the width of the EPSP, the number of synapses N, and the mean input rate $p/T = \bar{\lambda}^{\text{in}}$. Based on an analytical expression for the quality factor we discuss coincidence detection as performed by laminar neurons.

4.1 Neuron Model

The I&F unit fires, if its membrane potential u approaches the threshold ϑ from below. This defines a firing time t^n with integer n. After an output spike, which need not be described explicitly, the membrane potential is reset

[a] Multiplying the quality factor γ by a factor 2 we obtain a quantity similar to the index d' commonly used in psychophysics, where d' is a measure of the discriminability of a signal in noise [6].

to 0. Between two firing events the membrane voltage changes according to the linear differential equation

$$\frac{\mathrm{d}}{\mathrm{d}t}u(t) = -\frac{1}{\tau_m}u(t) + i(t) \ , \tag{5}$$

where i is the total input current and $\tau_m > 0$ is the membrane time constant. The input is due to presynaptic activity. The spike arrival times at a given synapse j, with $1 \leq j \leq N$, are labeled by t_j^f where $f = 1, 2, \ldots$ is a running spike index.

Each presynaptic spike evokes a small postsynaptic current (PSC) which decays exponentially with time constant $\tau_s > 0$. All synapses are equal in the sense that the incoming spikes evoke PSCs of identical shape and amplitude. The total input of the neuron is then taken to be $i(t) = \tau_s^{-1} \sum_{j=1}^{N} \sum_f \exp[-(t - t_j^f)/\tau_s]\theta(t - t_j^f)$ where $\theta(.)$ denotes the Heaviside step function with $\theta(s) = 1$ for $s > 0$ and $\theta(s) = 0$ for $s \leq 0$. It appears because of causality. We substitute the total input into Eq. (5) and integrate. This yields the membrane potential at the hillock,

$$u(t) = \sum\nolimits_{j=1}^{N} \left[\sum\nolimits_f \epsilon(t - t_j^f)\right] + \sum\nolimits_n \eta(t - t^n) \ . \tag{6}$$

The first term on the right in Eq. (6),

$$\epsilon(s) = \tau_m/(\tau_m - \tau_s) \ [\exp(-s/\tau_m) - \exp(-s/\tau_s)] \ \theta(s) \ , \tag{7}$$

describes the time course of an excitatory postsynaptic potential (EPSP). If $\tau_s = \tau_m$, we have instead of Eq. (7) the so-called alpha function, $\epsilon(s) = (s/\tau_m)\exp(-s/\tau_m)\,\theta(s)$. The second contribution to Eq. (6), $\eta(s) = -\vartheta \exp(-s/\tau_m)\,\theta(s)$, accounts for the reset of the membrane potential after each output spike and incorporates neuronal refractoriness.

4.2 The Threshold is not a Critical Parameter for Coincidence Detection

In a series of simulations, we have driven neurons by stochastic spike arrival generated at the time-dependent rate as given in Eq. (1). Figure 1 shows the quality factor γ as a function of the threshold ϑ. For various combinations of neuronal parameters (*a–f* in Fig. 1) there is always an *optimal* choice for ϑ that maximizes γ. Furthermore, γ exhibits a broad maximum and therefore depends only weakly on ϑ. In other words, the exact value of ϑ is *not* a critical parameter for coincidence detection. In contrast to that, γ varies strongly, if we change, e.g., the neuronal time constants τ_m and τ_s, the number N of synapses, or the number p of inputs spikes per period T, where $p/T = \bar{\lambda}^{\text{in}}$. We note that the mean output rate also depends strongly on ϑ. Along each

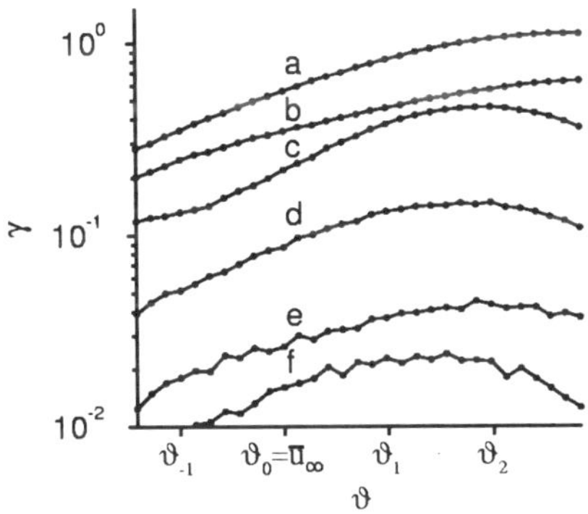

Figure 1. Coincidence detection properties depend on the value of the spiking threshold for different scenarios (a)–(f) of simulation parameters. The quality factor γ [in units of $T^{-1/2}I^{1/2}$, logarithmic scale; cf. Eq. (4)] shows a maximum always above the mean voltage $\bar{u}_\infty = \vartheta_0$, where $\bar{u}_\infty = N\,\bar{\lambda}^{\text{in}}\,\tau_m = 200$. In practical situations it is immaterial, though, whether we take, e.g., $\vartheta_1 = 207$ or $\vartheta_2 = 213$. That is to say, the choice of the threshold is not critical because the maximum is broad. Simulation parameters are: (a) $\tau_m = \tau_s = T/4$, $Np = 800$; (b) $\tau_m = \tau_s = T/2$, $Np = 400$; (c) $\tau_m = T/2$, $\tau_s = T$, $Np = 400$; (d) $\tau_m = \tau_s = T$, $Np = 200$; (e) $\tau_m = 2T, \tau_s = T$, $Np = 100$; (f) $\tau_m = \tau_s = 2T$, $Np = 100$. To simplify comparison, parameters have been chosen so that the mean input current and, hence, $\bar{u}_\infty$ does not change between one simulation and the next. Each point corresponds to 10^4 output spikes (taken from Ref. 8).

graph in Fig. 1 the output rate $\bar{\lambda}^{\text{out}}(0)$ varies by orders of magnitudes (data not shown).

Why is there an optimal threshold? If the threshold is low, the neuron basically acts as an integrator and is not sensitive to the temporal structure of the input. Hence, the coherence gain E is unity and the right-hand side of Eq. (4) vanishes. On the other hand, for very high values of ϑ the firing rate $\bar{\lambda}^{\text{out}}$ approaches zero and so does the right-hand side of Eq. (4). Hence there must be an optimal value of ϑ in between. The idea dangling in the background is that of stochastic resonance [14]. Because of the stochastic spike arrival the membrane potential will always fluctuate around its asymptotic mean $\bar{u}$. The optimal threshold is around $\vartheta \approx \bar{u} + \Delta u$ where Δu is the standard deviation of the fluctuations [9].

In contrast to the quality measure γ, the coherence gain E increases monotonically with ϑ. [7,8] Using the coherence gain E as a measure for coincidence detection, we would therefore have totally missed the fact that there is an *optimal* threshold.

4.3 Analytical Expression for the Quality Factor

We are going to discuss and simplify a mathematical expression [8] for the quality factor of an I&F neuron at its optimal threshold. For $N\,\bar{\lambda}^{\text{in}}\,\tau_m \gg 1$

an approximation to the maximum value of the quality factor is

$$\gamma_{\rm opt} \approx r^{\rm in}\,\sqrt{2\,N\,\bar{\lambda}^{\rm in}\,\tau}\;\frac{|\tilde{\epsilon}(\omega)|}{A\,\tau}\cdot\sqrt{\frac{I}{\tau^{\rm dec}+\tau^{\rm ref}}\,\frac{4}{\sqrt{54\pi}}}\;. \tag{8}$$

The term $r^{\rm in}\sqrt{2\,N\,\bar{\lambda}^{\rm in}\,\tau}\,|\tilde{\epsilon}(\omega)|/(A\,\tau)$ on the right in Eq. (8) is the signal-to-noise ratio of the membrane potential of a neuron that receives spike input with vector strength $r^{\rm in}$ (cf. Sec.2) and rate $\bar{\lambda}^{\rm in}$ through N synapses. The symbol τ denotes the width of the integration window of a neuron (that can be approximated by the full width at half maximum of the EPSP), $|\tilde{\epsilon}(\omega)|$ is the absolute value of the Fourier transform of the EPSP at frequency $\omega = 2\pi/T$, A is the amplitude of the EPSP, I is the length of a time interval, $\tau^{\rm dec}$ is the time necessary for a decay of any fluctuation in the membrane voltage of the neuron (which is of the same order of magnitude as τ_m), and $\tau^{\rm ref}$ is the refractory period of the neuron. Equation (8) remains valid as long as the signal-to-noise ratio is smaller than or at least of order of 1.

We now simplify Eq. (8). Using $\tau_m = \tau_s$ the EPSP in Eq. (7) reduces to an alpha function. The absolute value of the Fourier transform then is $|\tilde{\epsilon}(\omega)| = \tau_m/\left(1+\omega^2\tau_m^2\right)$. If τ_m exceeds T, we obtain $|\tilde{\epsilon}(\omega)| \approx 1/(\omega^2\,\tau_m)$. Following Kempter *et al.* [8] we take $\tau^{\rm dec} = 3/2\,\tau_m$, $\tau^{\rm ref} = 2\tau_m$, $A\sqrt{\tau} = \sqrt{\tau_m/2}$ for I&F neurons, and are left with

$$\gamma_{\rm opt} \approx C\,r^{\rm in}\,\sqrt{N\,\bar{\lambda}^{\rm in}\,I}\;(T/\tau_m)^2 \tag{9}$$

where $C = 8.3\cdot 10^{-3}$. The membrane time constant τ_m and the number N of synapses are the relevant *neuronal* parameters for coincidence detection. Furthermore, the statistics of the input ($r^{\rm in}$, $\bar{\lambda}^{\rm in}$, and T) as well as the width of the time interval I determine a neuron's ability to read out a temporal code. The dimensionless fraction of time scales T/τ_m enters quadratically.

4.4 Application to Laminar Neurons

The membrane time constant (=decay time of the EPSP) of neurons in the nucleus laminaris is extremely short whenever the neuron is depolarized above resting potential. This is due to an outward rectifying current [11]. The data of Reyes *et al.* [11] on chicken – not an auditory specialist – suggest a time constant in the range of 0.2–0.5 ms, but for barn owls it may be even shorter. We therefore take $\tau_m = 0.1$ ms. [4] We consider a neuron in the 5 kHz channel, hence $T = 0.2$ ms. As for the other parameters, $N = 200$, $\bar{\lambda}^{\rm in} = 500$ Hz, and $r^{\rm in} = 0.5$ are reasonable [3]. We suppose an interval or 'waiting time' of $I = 100$ ms. From Eq. (9) we get an estimate $\gamma_{\rm opt} = 1.7$. In simulations with optimal threshold we obtain $\gamma = 0.9$ (Fig. 2), which is off by a factor of two.

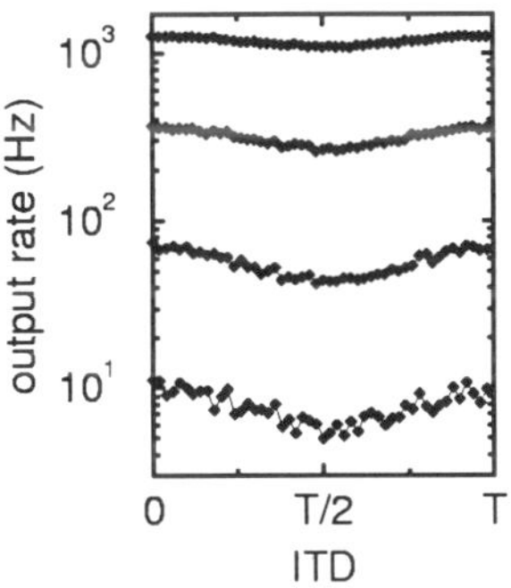

Figure 2. Numerical simulation of an I&F unit with parameters as given in Sec. 4.4 for a laminar neuron. We have plotted the output rate as a function of the ITD for four different threshold values ($\vartheta = 8, 7, 6, 5$ from bottom to top). To obtain ITD tuning curves, we have divided the synapses into two groups, viz., left and right, and 'delayed' one with respect to the other by the ITD as plotted on the horizontal axis. The minimum output rate is always at ITD $= T/2$, where input spikes arrive almost randomly with an effective input vector strength $r^{\text{in}} \approx 0$ [$\bar{\lambda}^{\text{out}} = 5.7, 43, 270, 1100\,\text{Hz}$]. The maximum output rate for coherent input ($r^{\text{in}} = 0.5$) occurs at ITD $= m\,T$ with integer m [$\bar{\lambda}^{\text{out}} = 10, 70, 370, 1260\,\text{Hz}$]. Though for different threshold values we obtain different coherence gains and output rates, which vary by orders of magnitude, the quality factor does not change that much ($\gamma = 0.2, 0.6, 0.9, 0.7$). Data points have been obtained by measuring the number of output spikes in a time interval of 10 s.

What does a quality factor of $\gamma \approx 1$ mean? We remind the reader of the definition of γ in Eq. (3). The quality factor measures the distance between the spike count distributions for random and coherent input. For $\gamma = 1$ the two distributions are just distinguishable and for $\gamma \gg 1$ they are well separated [6].

Even for values of γ smaller than 1, the system as a whole could still function as a coincidence detector, if there is a *pool* of neurons operating in the same pathway and receiving the same type of input. A system can either use the output spike count of a single neuron in an interval I or count the number of spikes of L statistically independent, identical neurons operating in the same pathway during a period of time I/L. [4]

Equation (9) is rather general and could be applied just as well to cortical neurons. With a realistic set of parameters, we [8] find that cortical neurons are as good in reading out a 40 Hz signal as laminar neurons are in reading out a 5 kHz signal, a result that may stimulate further discussions on the role of cortical neurons (integrator or coincidence detector) [1,10].

Acknowledgments

RK has been supported by the Deutsche Forschungsgemeinschaft (DFG) under grant number KL 608/10-1.

References

1. M. Abeles. Role of the cortical neuron: Integrator or coincidence detector? *Isr. J. Med. Sci.*, 18:83–92, 1982.
2. H. Agmon-Snir, C.E. Carr, and J. Rinzel. The role of dendrites in auditory coincidence detection. *Nature*, 393:268–72, 1998.
3. C.E. Carr and M. Konishi. A circuit for detection of interaural time differences in the brain stem of the barn owl. *J. Neurosci.*, 10:3227–46, 1990.
4. W. Gerstner, R. Kempter, J.L. van Hemmen, and H. Wagner. A neuronal learning rule for sub-millisecond temporal coding. *Nature*, 383:76–8, 1996.
5. J.M. Goldberg and P.B. Brown. Response of binaural neurons of dog superior olivary complex to dichotic tonal stimuli: Some physiological mechanisms of sound localization. *J. Neurophysiol.*, 32:613–36, 1969.
6. D.M. Green and J.A. Swets. *Signal Detection Theory and Psychophysics.* Krieger, Huntington, 1974.
7. R. Kempter. *Hebbsches Lernen zeitlicher Codierung: Theorie der Schallortung im Hörsystem der Schleiereule.* Naturwissenschaftliche Reihe, Vol. 17. DDD, Darmstadt, 1997. TU München, Dissertation.
8. R. Kempter, W. Gerstner, J.L. van Hemmen, and H. Wagner. Extracting oscillations: Neuronal coincidence detection with noisy periodic spike input. *Neural Comput.*, 10:1987–2017, 1998.
9. R. Kempter, W. Gerstner, and J.L. van Hemmen. How the threshold of a neuron determines its capacity for coincidence detection. *BioSystems*, 48: 105–112, 1998.
10. P. König, A.K. Engel, and W. Singer. Integrator or coincidence detector? The role of the cortical neuron revisited. *Trends Neurosci.*, 19:130-7, 1996.
11. A.D. Reyes, E.W. Rubel, and W.J. Spain. In vitro analysis of optimal stimuli for phase-locking and time-delayed modulation of firing in avian nucleus laminaris neurons. *J. Neurosci.*, 16:993–1007, 1996.
12. C.F. Stevens and A.M. Zador. Input synchrony and the irregular firing of cortical neurons. *Nature Neurosci.*, 1:210–7, 1998.
13. T.W. Troyer and K.D. Miller. Physiological gain leads to high ISI variability in a simple model of a cortical regular spiking cell. *Neural Comput.*, 9:971–83, 1997.
14. K. Wiesenfeld and F. Moss. Stochastic resonance and the benefits of noise: From ice ages to crayfish and SQUIDs. *Nature*, 373:33–6, 1995.

A NEW BINAURAL DETECTION MODEL BASED ON CONTRALATERAL INHIBITION

JEROEN BREEBAART (1) AND ARMIN KOHLRAUSCH (1,2)

(1) IPO-Center for Research on User-System Interaction, P.O. Box 513, NL-5600 MB Eindhoven, The Netherlands

(2) Philips Research Laboratories Eindhoven, Prof. Holstlaan 4, NL-5656 AA Eindhoven, The Netherlands

Binaural models attempt to explain binaural phenomena in terms of neural mechanisms that extract binaural information from acoustic stimuli. In this paper, a model setup is presented that can be used to simulate binaural detection tasks. In contrast to the most often used cross correlation between the right and left channel, this model is based on contralateral inhibition. The presented model is applied to a wide range of binaural detection experiments. It shows a good fit for changes in masker bandwidth or masker correlation, static and dynamic cues and level and frequency dependencies.

1 Introduction

One of the challenges of binaural modeling lies in the design of a system both for the extraction of static interaural differences (i.e., as in localization tasks) and the detection of dynamically varying interaural differences (i.e., as in detection tasks). Most binaural models heavily rely on some form of a crosscorrelation function of the signals from both ears (cf. Albeck and Konishi, 1995), based on the coincidence mechanisms proposed by Jeffress (1948). Some drawbacks of these models can be summarized as follows.

1. Most correlation models do not integrate the processing of static interaural differences and dynamically varying interaural differences, which results in the difficulty to link detection and localization phenomena.
2. Some models are not able to detect static interaural intensity differences at all.
3. Most models cannot explain the influence of masker bandwidth in a binaural detection task.

In this chapter, we present a new binaural model which bases its predictions on a process of contralateral inhibition. The various functional elements of this particular model, as shown in Fig. 1, are discussed below.

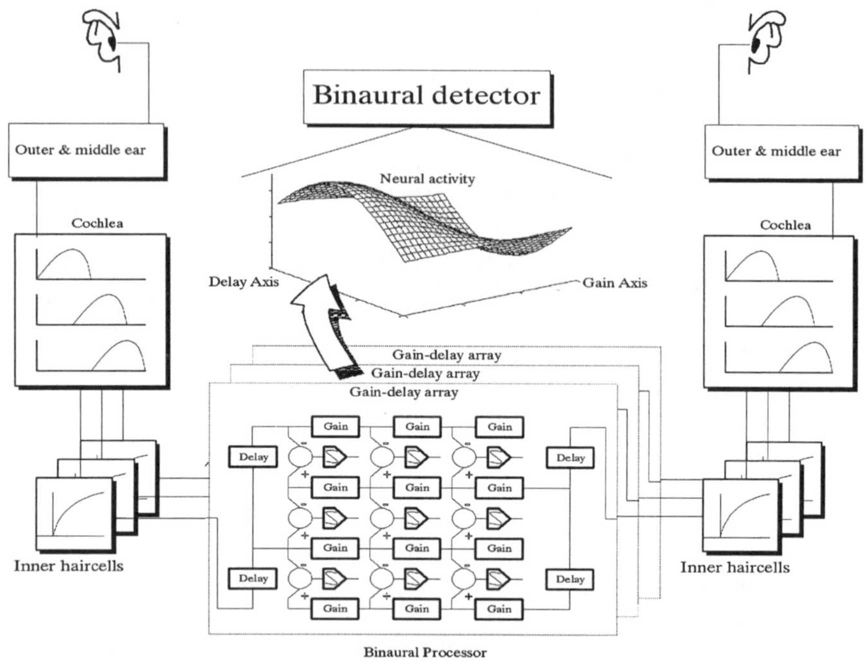

Figure 1 - Structure of the binaural model. Signals presented at both ears enter the outer and middle ear block, followed by the cochlea model, the inner hair-cell model, the binaural processor and finally the binaural detector. The EI-type cells, denoted by circles in the gain-delay array of the binaural processor (see text) record the difference in signals from left and right ear. Finally, a temporal integrator computes the energy of the EI-type cell output. This leads to a neural activity pattern that is analyzed by a binaural detector.

1.1 Outer and middle ear

The first processing block simulates the general spectral shaping of the outer and middle ear, without taking into account the direction specific spectral shaping found in head-related transfer functions. The combined outer and middle-ear response is simulated by a frequency-transfer function with a rolloff of 6 dB/Oct below 1 kHz and -6 dB/Oct above 4 kHz.

1.2 Cochlea and basilar membrane

The second block, the cochlea module, simulates a running spectral analysis of the incoming signals. In the present model, this part is modeled by means of a fourth-order Gammatone filterbank with filters of equivalent rectangular bandwidth. The spacing of the adjacent filters is two per equivalent rectangular bandwidth. The signal levels of this model are expressed in terms of model units (MU), whereby one MU corresponds to a sound pressure of 20 μPa. In order to introduce an absolute threshold, an interaurally-independent noise is added to the filtered signals from the left and right ear with a level of 10 dB re 1 MU.

1.3 Inner hair cells

The third block simulates the effective signal processing of the inner hair cells. Important processing features are:

1. First, there is only response for the positive phases of the temporal waveform. This effect is modeled by a half-wave rectifier.
2. The loss of fine structure is modeled by a 5th-order low-pass filter with a cutoff frequency of 650 Hz.
3. Third, compression. Both physiological and psychophysical findings indicate the presence of nonlinearities in the peripheral auditory system. The model assumes that it is possible to group these nonlinearities together and describe them in terms of a simple time-invariant power law. Therefore, the instantaneous values of the waveforms are raised to the power 0.4.

1.4 Binaural processor

The binaural processor simulates the binaural interaction found in the auditory system. Neurons in the ascending auditory pathway exhibiting binaural interaction are found in the superior olivary complex (SOC) and the inferior colliculus (IC). In both nuclei, two different types of binaural neurons are found: EE-type cells (i.e., coincidence detectors) and EI-type. The EI-type cells are excited by one ear and show *inhibitory* effects when the other ear is stimulated (Excitation-Inhibition).

The activity of EI-type cells has been measured in phsyiological experiments a function of several parameters of the acoustic stimuli (Goldberg and Brown, 1969; Joris and Yin, 1995; Joris, 1996). To account for these neurophysiological results, for each auditory filter, a gain-delay array including EI-type cells is used, as shown in Fig. 1. Each EI-type cell receives the neural activity from the inner hair cell model corresponding to one auditory filter, while each position in the array corresponds to a specific interaural delay and interaural gain. Each cell is characterized by three parameters (filter number, characteristic delay, balance between excitatory and inhibitory influence) and computes a normalized difference-intensity (E') from the incoming signals as described below:

$$E'(\tau,\alpha,n) = \frac{\int \left(e^{-\alpha} l_n(t) - e^{\alpha} r_n(t-\tau)\right)^2 dt}{e^{-2\alpha}\int l_n^2 dt + e^{2\alpha}\int r_n^2 dt}. \qquad \text{(Eq. 1-1)}$$

Here, n represents the Gamma-tone filter number, τ the characteristic delay, α the balance between excitatory and inhibitory influence (here defined as a relative gain), l_n and r_n the signals from Gammatone filter n after being processed by the inner hair cell model of the left and right signal, respectively. To include saturative properties of the EI-type neuron, the following input-output is included:

$$E = E' \exp\{cE'\}. \qquad \text{(Eq. 1-2)}$$

Here, E' is the neural activity as described in Eq. 1-1; E represents the saturated activity and c the saturation coefficient (c=-0.625).

As an example, the neural activity (E) according to Eq. 1-2 as a function of the external interaural time difference for a sinusoid is shown in Fig. 2. The solid line represents the neural activity of the model for α=0.03, τ=250 μs, the circles represent neurophysiological data from Kuwada et al. (1984).

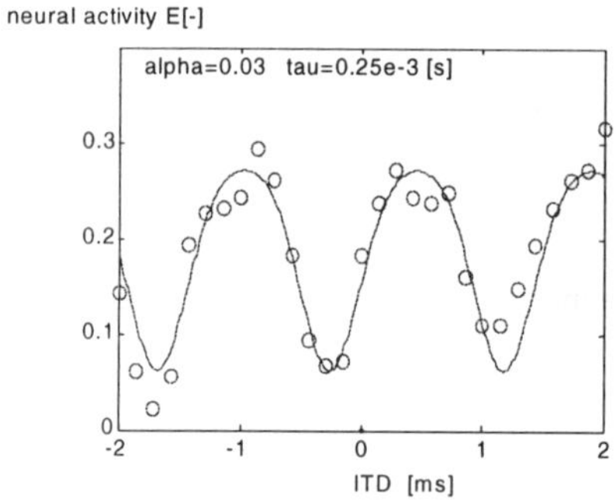

Figure 2. EI-cell activity as a function of the external interaural delay for a sinusoid of 700 Hz center frequency. Solid line: model predictions according to Eq. 4, for α=0.03, τ=250 μs, on-frequency filter. Circles: data adapted from Kuwada et al. (1984), multiplied by a constant factor to convert the spike rate to a normalized neural activity.

1.5 Binaural detection mechanism

The binaural detector extracts binaural information from the EI-activity pattern. The analysis of the pattern resulting from a stimulus is done by means of an optimal detector that weights the information for each individual auditory filter. In this model, a reference is formed by the mean excitation pattern for a masker alone. This reference is called a template, and can be obtained easily by averaging the excitation patterns for several maskers alone intervals. In a 3IFC experiment, where the subject (i.e., the model) has to detect the stimulus containing the test signal, the detection mechanism computes the excitation pattern for each stimulus, and compares it to the template, by integrating the difference between excitation pattern and template for each filter. Here, it is assumed that cells with smaller characteristic delays are more frequent than cells with larger characteristic delays. Therefore, a weighting function $p(\tau)$, as described by Colburn (1977) is applied in the integration process:

$$p(\tau)=\begin{cases} 1 & for \quad |\tau|\le 0.15ms \\ \exp\left(\dfrac{-|\tau|+0.15}{0.6}\right) & for \quad 0.15\le|\tau|\le 2.2ms \\ 0.033\exp\left(\dfrac{-|\tau|+2.2}{2.3}\right) & for \quad |\tau|>2.2ms \end{cases} \qquad \text{(Eq. 1-3)}$$

The internal errors (or internal noise) are modeled by adding a Gaussian-distributed random variable with mean zero and fixed variance to all outputs.

2 Simulation Results

2.1 *Procedure and stimuli*

We compared model predictions with the results from several experimental studies with human observers which were adapted from literature. The procedure used in determining the model predictions was the same as the procedure that was used in the psychoacoustical studies.

2.2 *Effect of masker bandwidth*

To study the dependence of binaural detection thresholds on masker bandwidth, binaural thresholds were simulated for several binaural masking conditions for different center frequencies, as a function of masker bandwidth. These experiments were adapted from Breebaart et al. (1998a). Fig. 3 shows signal-to-noise ratios of the model (filled blocks) in comparison with experimental data (open symbols). The upper-left panel shows the thresholds for an NoSπ condition at 125-Hz center frequency, while the upper-right panel shows the same condition at 500-Hz center frequency. We see that the model predictions fit the experimental data quite well. For bandwidths beyond 100 Hz, the thresholds increase with 5 dB at 125 Hz and 3 dB at 500-Hz center frequency. For larger bandwidths, the thresholds decrease again. Interestingly, this decrease starts at approximately 2 to 3 times the critical bandwidth.

The middle panels show thresholds for the detection of an in-phase sinusoid added to a time-delayed noise masker (NτSo, blocks, τ equals half the period of the center frequency) and a phase-reverted noise masker (NπSo, triangles) at 125-Hz center frequency (left panel) and 500-Hz center frequency (right panel). The open blocks represent experimental NτSo data from Breebaart et al. (1998a). The experimental data for the NπSo conditions are not shown, since Breebaart et al. (1998a) showed that the NτSo and NπSo conditions have equal thresholds. We see that the predictions and the experimental data are very similar, and that the NπSo and NτSo conditions show very similar thresholds, perfectly in line with the experimental observations.

The bottom panels show thresholds for the detection of an out-of-phase multiplied noise added to an in-phase sinusoidal masker as a function of the bandwidth of the noise at 500-Hz center frequency. The left panel shows data for an IID condition ($\theta=0$), while the right panel shows data for an ITD condition ($\theta=\pi/2$). The open symbols represent the experimental data from Breebaart et al. (1998b), the filled symbols represent the model predictions.

For bandwidths greater than 100 Hz, both experimental and predicted thresholds decrease in a very similar way. However, the ITD condition shows slightly higher predicted thresholds for bandwidths between 40 and 160 Hz.

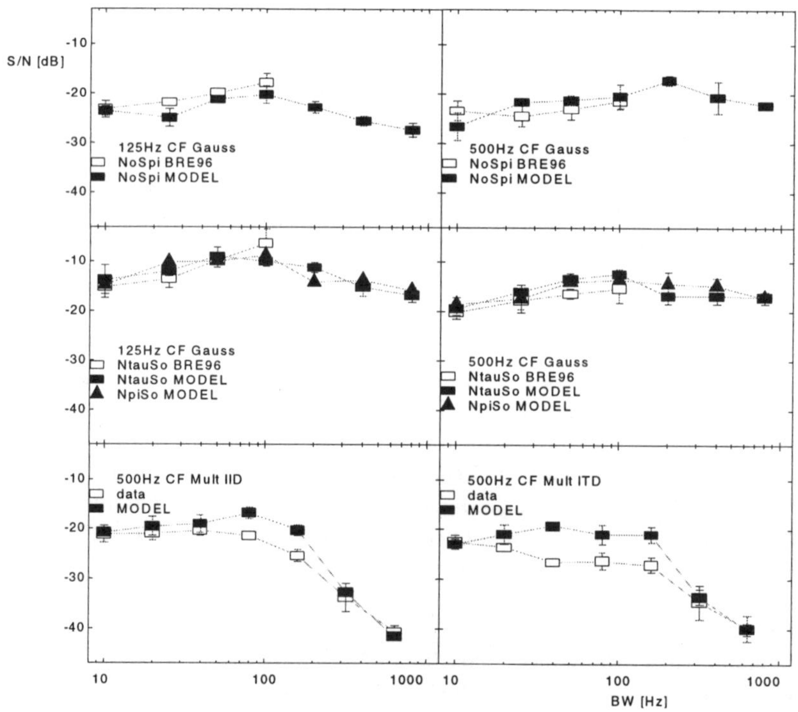

Figure 3 - Bandwidth dependence of the model compared to experimental data. Filled symbols: model predictions. Open symbols: experimental data. Upper panels: NoSπ condition at 125 Hz (left) and 500 Hz (right) center frequency. Middle panels: NπSo and NτSo conditions at 125 Hz (left) and 500 Hz (right) center frequency. Lower panels: NoSπ (sinusoidal masker with out-of-phase multiplied noise test signal for $\theta=0$ (left) and $\theta=\pi/2$ (right)) at 500 Hz center frequency. Errorbars denote the standard deviation of the mean.

2.3 *Frequency dependence of static interaural differences*

The dependence on center frequency of static interaural difference in time (right panel) and intensity (left panel) for sinusoids is shown in Fig. 4. As in the preceding figures, the open symbols represent experimental data adapted from literature, while the filled symbols represent the model predictions. We see that for both the experimental data and the predicted thresholds, the intensity jnd is approximately constant over the whole frequency range. For ITDs, however, the interaural time difference jnd decreases with center frequency for frequencies below 1 kHz, while for frequencies above 1 kHz, the jnds increase rapidly. The increase in sensitivity with increasing center frequency equals a constant phase jnd, while the decrease in sensitivity above 1 kHz is the result of the decrease of phase locking in the inner hair cell model.

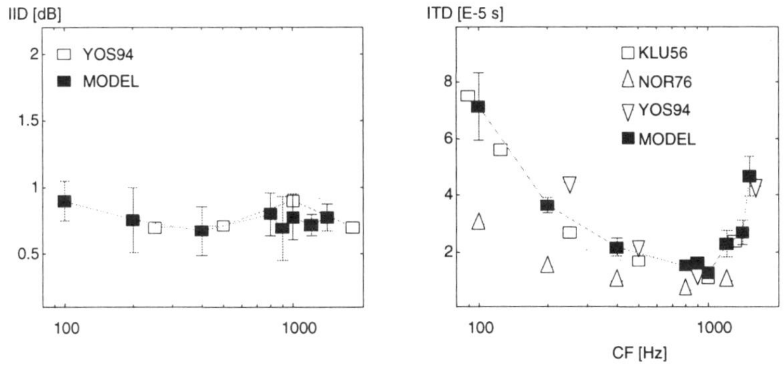

Figure 4 - IID-jnd (left panel) and ITD-jnd (right panel) for sinusoids as a function of center frequency. Open symbols: data adapted from literature. Filled symbols: model predictions. Errorbars denote the standard deviation of the mean. Legend: KLU56=Klumpp and Eady, 1956; NOR76=Nordmark, 1976; YOS94=Yost, 1994.

2.4 *Interaural correlation of the masker*

To study the dependence on interaural correlation of the masking noise, the detection thresholds of the model were determined for several (negative) interaural correlations of a narrowband masking noise (NρSo condition). Thresholds were measured at a center frequency of 125 Hz and a bandwidth of 50 Hz.

The predicted (filled symbols) and experimental (open symbols) thresholds are shown in Fig. 5. The open blocks represent data from Breebaart (1996) for a 50-Hz wide masker, while the open triangles represent data from Robinson and Jeffress (1963) for a broadband noise masker. As the correlation is changed from -1 to -.95, thresholds increase by about 3 dB. As the correlation increases further, thresholds increase with increasing correlation, but with a decreasing slope.

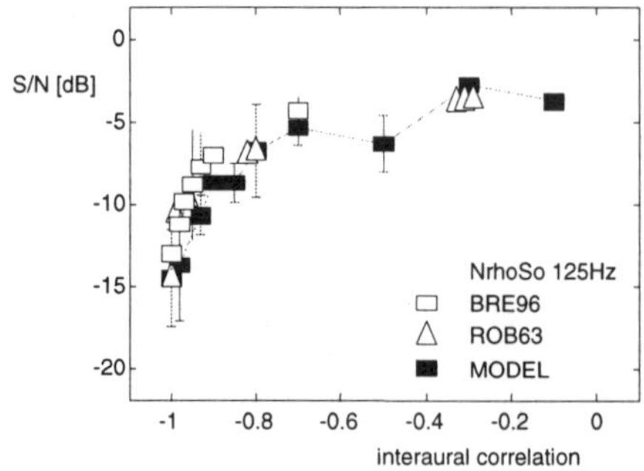

Figure 5 - Detection thresholds as a function of the interaural correlation of the masker at 125Hz center frequency. Errorbars denote the standard deviation of the mean. Filled symbols: model predictions. Legend: BRE96=Breebaart,1996; ROB63=Robinson and Jeffress, 1963.

2.5 *Temporal integration*

During the previous simulations, the changes in the internal activity were computed from the whole stimulus, i.e., 400 ms. Data from Grantham & Wightman (1978) and Kollmeier & Gilkey (1990) show that the binaural system has a limited time resolution. To incorporate a limited time resolution in the model, the following condition from Kollmeier and Gilkey (1990) was used.

Here, the first 375 ms of the masker consist of a diotic noise of 500-Hz bandwidth centered at 500 Hz. The next 375 ms (i.e., from 375 to 750 ms) consist of a phase-inverted noise of the same bandwidth and center frequency. Thus, at the temporal center, the masker switches from No to Nπ. The test signal consists of a 20-ms out-of-phase sinusoid, ramped by a 2-ms Hanning window. Furthermore, the total signal consisting of masker plus test signal is windowed temporally by a

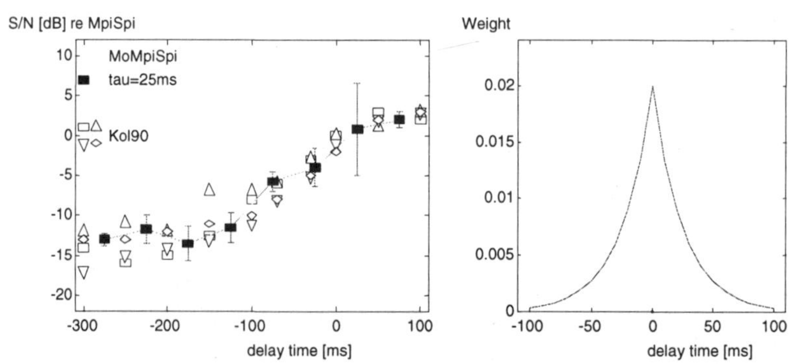

Figure 6 – Signal-to-noise ratios as a function of the delay time of the signal-offset; 0 corresponds to the phase-switch of the masker (left panel) and temporal window (right panel). Note that according to the data from Kollmeier and Gilkey (1990), the thresholds are expressed in comparison to NπSπ thresholds. The errorbars denote the standard deviation of the mean.

symmetric exponential window with a variable offset. By varying the temporal position of the signal relative to the phase transition of the masker, the condition shifts from NoSπ to NπSπ, resulting in a change in the detectability with temporal position.

The position of the temporal window resulting in the lowest detection thresholds is not always given by centering the window on the signal. To determine the optimal position of the temporal window for a given signal offset, the window was shifted from -70 to +70 ms away from the temporal center of the test signal. The lowest signal-to-noise ratio achieved was used as a detection threshold for that signal offset.

The simulation results (filled symbols) and data from Kollmeier and Gilkey (1990) are shown in Fig. 6 (left panel). The right panel shows the temporal window with a constant of 25 ms.

2.6 *Masker-level dependence*

In Fig. 7, the signal level at threshold is shown for a 500-Hz out-of-phase sinusoid added to an in-phase 500-Hz-wide noise as a function of the spectral level of the noise.

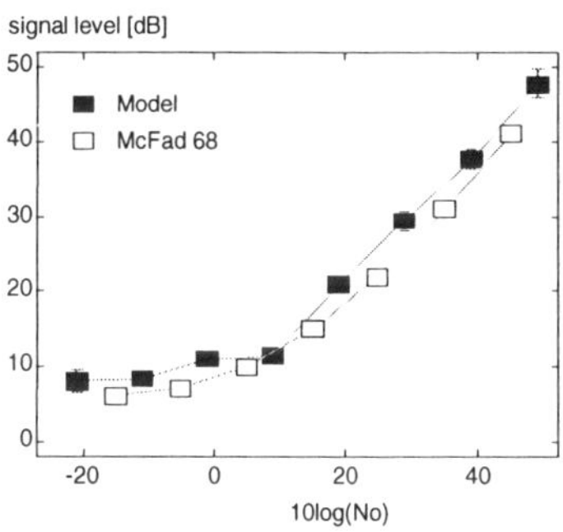

Figure 7 - signal level at threshold for a 500-Hz out-of-phase sinusoid added to a 500-Hz wide masker as a function of the spectral energy density of the noise. Open symbols: data adapted from literature (McFadden, 1968). Filled symbols: model predictions.

2.7 Discussion of the bandwidth dependence of the model

In the model, three effects play an important role in the dependence of detection thresholds on the bandwidth of the masker (see Fig. 3):

1. **The peripheral filtering**. For narrowband maskers, the masker power in the on-frequency filter is (almost) equal to the total power of the masker. With increasing masker bandwidth, the masker power in the on-frequency filter decreases due to the bandpass characteristic of the filter if the total masker power is kept constant. For masker bandwidths that extend the bandwidth of the peripheral filter, a -3 dB/Oct decrease of noise energy is expected, resulting in lower signal-to-noise ratios. This effect plays a minor role for masker bandwidths smaller than the ERB.
2. **The change of information across filters**. For a narrowband diotic masker, the output variable of the on-frequency filter is (almost) equal to the output variable of the adjacent filters. Therefore, the model effectively computes the mean of these output variables and uses it as a decision variable. Since the errors induced by the internal noise are independent in each filter, the model reduces the internal error, resulting in low detection thresholds. With increasing masker bandwidth, the off-frequency filters contain more and more noise power and a relatively decreasing amount of the signal to be detected. Therefore, the output variable of the off-frequency filters contains less information about the presence of an out-of-phase signal, resulting in less cancellation of the internal noise. Therefore, the thresholds increase with increasing masker bandwidth, even for bandwidths up to 3 times the ERB. This explains the phenomenon that for an $NoS\pi$ condition, the effective auditory filter bandwidth is wider than the ERB.
3. **Variance due to decorrelation**. The output of an EI-type cell can be written in terms of the interaural correlation at a delay equal to the internal delay of the EI-type cell (Breebaart et al., 1997). Since the EI-activity is computed as a funtion of the internal delay, the change in activity of the EI-type cell due to the addition of the test-signal can be written in terms of a change of the interaural correlation function. With increasing masker bandwidth, the subsequent periods of the masker become less similar (i.e., damping of the cross-correlation function). Furthermore, the variance of the normalized cross-correlation function increases with increasing masker bandwidth. This increase of the variance results in higher detection thresholds with increasing masker bandwidth. The damping of the cross-correlation function is limited by the peripheral filtering. Therefore, this effect only plays a role for bandwidths smaller than the ERB.

3 Conclusions

The presented model is able to predict the dependence of binaural thresholds on several stimulus parameters. The tested conditions show a good fit for changes in bandwidth or correlation of the masker, signal durations, static and dynamic cues, level and frequency dependencies.

The combination of information across auditory filters seems to work fine if the separate information is summed according to the described optimal criterion. This approach also accounts for the wider effective auditory filters for an $NoS\pi$ condition. However, not all phenomena concerning binaural detection are covered. For example, the phenomenon of binaural interference (Bernstein and Trahiotis, 1995) cannot be accounted for by this model nor do thresholds increase as a result of distortion products at the basilar membrane (Van der Heijden et al., 1997). These effects need a more sophisticated model of the binaural detection mechanism and the basilar membrane, respectively.

Summarizing, a system based on contralateral inhibition can account for a wide range of effects found in binaural detection tasks. One of the main advantages of an EI-type cell based system in comparison with an EE-type (i.e., correlation) based system is the incorporation of static interaural intensity differences.

4 References

1. Albeck, Y., Konishi, M. (1995). Responses of neurons in the auditory pathway of the barn owl to partially correlated binaural signals. *J. Neurophysiol.* 74, 1689-1700.
2. Bernstein, L.R., Trahiotis, C. (1995). Binaural spectral interference in detection and discrimination paradigms. In *Advances in Hearing Research: proceedings of the 10th international symposium on hearing*. Eds. G.A. manley, G.M. Klump, C. Koppl, H. Fastl, H. Oekinghaus.
3. Breebaart, J. (1996). *Modeling the influence of masker bandkwidth on BMLDs*. Institute for Perception Research (IPO), Eindhoven, The Netherlands. Report 1109.
4. Breebaart, J. (1997). *Binaural signal detection: experiments, modeling and applications. Applications for sound source coding*. Institute for Perception Research (IPO), Eindhoven, The Netherlands. Report 1155.
5. Breebaart, J., Van de Par, S., Kohlrausch, A. (1998a). Binaural signal detection with phase-shifted and time-delayed noise maskers. *J. Acoust. Soc. Am.* 103, 2079-2083.
6. Breebaart, J., Van de Par, S., Kohlrausch, A. (1998b). The contribution of static and dynamically varying ITDs and IIDs on binaural detection. *J. Acoust. Soc. Am.* Submitted for publication.

7. Glasberg, B.R., Moore, B.C.J. (1990). Derivation of auditory filter shapes from notched-noise data. *Hearing Research*. 47, 103-138.
8. Goldberg, J.M., Brown, P.B. (1969). Response of binaural neurons of dog superior olivary complex: an anatomical and electrophysiological study. *J. Neurophysiol.* 31, 613-636.
9. Grantham, D.W., Wightman, F.L. (1978). Detectability of varying interaural temporal differences. *J. Acoust. Soc. Am.* 63, 511-523.
10. Jeffress, L.A. (1948). A place theory of sound localization. *Journal of Comparative and Physiological Psychology*. 41, 35-39.
11. Joris, P.X. (1996). Envelope coding in the lateral superior live. II. Characteristic delays and comparison with responses in the medial superior olive. *J. Neurophysiol.* 76, 2137-2156.
12. Joris, P.X., Yin, T.C.T. (1995). Envelope coding in the lateral superior olive. I. Sensitivity to interaural time differences. *J. Neurophysiol.* 73, 1043-1062.
13. Klumpp, R.G., Eady, H.R. (1956). Some measurements of interaural time difference thresholds. *J. Acoust. Soc. Am.* 28, 859-860.
14. Kollmeier, B., Gilkey, R.H. (1990). Binaural forward and backward masking: evidence for sluggishness in binaural detection. *J. Acoust. Soc. Am.* 87, 1709-1719.
15. Kuwada, Y., Yin, T.C.T., Syka, J., Buunen, J.F., Wickesberg, R.E. (1984). Binaural interaction in low-frequency neurons in he inferior colliculus of the cat. IV. Comparison of monaural and binaural response properties. *J. Neurophysiology.* 51, 1306-1325.
16. McFadden, D. (1968). Masking level differences determined with and without interaural disparities in masker intensity. *J. Acoust. Soc. Am.* 44, 212-223.
17. Nordmark, J.O. (1976). Binaural time discrimination. *J. Acoust. Soc. Am.* 60, 870-880.
18. Oxenham, A.J., Moore, B.C.J. (1994). Modeling the additivity of nonsimultaneous masking. *Hearing Research*. 80, 105-118.
19. Van der Heijden, M., Trahiotis, C., Kohlrausch, A., Van de Par, S. (1997). Binaural signal detection with spectrally non-overlapping signals and maskers: Evidence for masking by aural distortion products. Submitted for J. Acoust. Soc. Am.
20. Yost, W.A. (1994). *Fundamentals of hearing. An introduction.* Academic Press, London.

MODELING THE 'EFFECTIVE' BINAURAL SIGNAL PROCESSING IN DETECTION EXPERIMENTS

CARSTEN ZERBS, TORSTEN DAU, BIRGER KOLLMEIER
AG Medizinische Physik, Universität Oldenburg, 26111 Oldenburg, Germany
E-mail: cz@medi.physik.uni-oldenburg.de

1 Introduction

A large part of the performance of the auditory system in tasks like localization, object identification or noise reduction in a noisy environment (the so-called "cocktail party-effect") is achieved due to binaural processing, i. e. making use of information coded in differences between both ear signals. In understanding the underlying mechanisms of such binaural phenomena the development of binaural models can play an important role.

To overcome shortcomings of several proposed models, a binaural processing unit following classical ideas of binaural processing was added to a functional model of monaural processing which is able to explain a large variety of monaural detection data [1]. So not only BMLDs but also masked thresholds are predicted.

2 Model and Method

In the monaural stage the processing starts with a gammatone filterbank to simulate the frequency-place transformation of the basilar membrane. A half-wave rectifier and a 1 kHz lowpass filter (for envelope extraction) and a set of five adaptation loops (that account for effects of adaptation, compression and temporal masking) are following.

The binaural stage is also a simple functional unit with a small set of parameters. At the output of the adaptation circuits the monaural signal in one channel is temporally delayed or, for symmetry, advanced (according to the *equalization* process in the EC-model by Durlach [2]) by an amount determined using the position of the maximum of the interaural cross correlation function, which is evaluated in a restricted temporal range. Both signals are subtracted from each other to reduce the noise masker in the resulting difference signal. Like in the EC-model this (*cancellation*) process is assumed to work not perfectly (time *jitter*). The difference is then full-wave rectified. The signals of the binaural and monaural channels are filtered by an low-pass filter with a

cut-off frequency of 8 Hz.
The decision algorithm is implemented as an optimal detector.

Masked thresholds were simulated and measured using an adaptive 3AFC procedure. The threshold was adjusted by a 1-up 2-down algorithm. In all configurations the task consisted in detecting a tonal signal in a running noise masker.

3 Results

The following figures show the results of binaural detection experiments and corresponding simulations in some exemplary configurations.

Fig. 1 compares the thresholds obtained in measurements [4] and simulations in the three configurations N_0S_0, $N_\pi S_0$ and N_0S_π, as a function of the test-tone frequency. Fig. 2 shows the results for the configuration $N_\tau S_0$ as a function of the interaural masker time delay. The damped periodic course in the threshold pattern is an effect of the relative interaural phase of signal and masker for different delays and decreasing interaural masker correlation with increasing interaural delay.

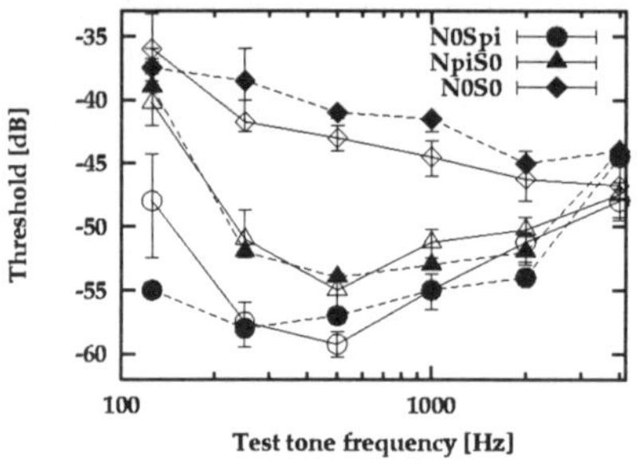

Figure 1. Simulated (filled symbols) and measured (open symbols) masked thresholds as a function of the test tone frequency for the three configurations N_0S_0, $N_\pi S_0$ and N_0S_π.

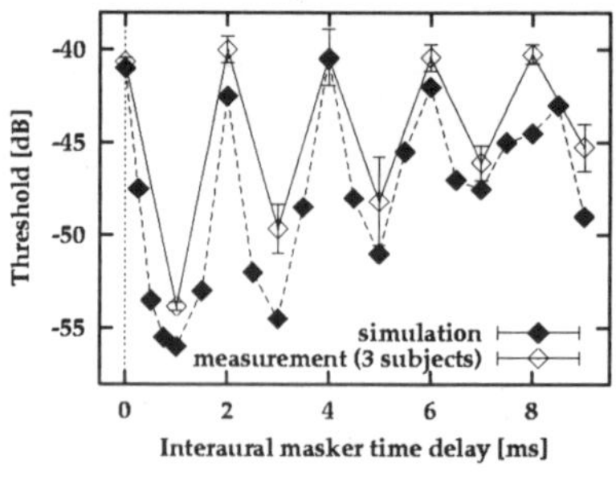

Figure 2. Simulated (filled symbols) and measured (open symbols) masked thresholds as function of the interaural time delay of the masker for the configuration $N_\tau S_0$.

Fig. 3 shows the results of an experiment in which an interaural masker time delay is combined with an interaural masker phase difference. Both factors are chosen in an opposite to direction to cancel out as much as possible. For increasing external delays the spectral range in which the cancellation mechanism of the model works optimally decreases. The BMLD decreases with less effective noise suppression. Fig. 4 shows thresholds for configurations where the interaural masker correlation is "switched" in the frequency range at 500 Hz. Thresholds are measured [5] as a function of the test-tone frequency in the resulting configurations $N_{0\pi}S_\pi$ (correlation switched from 1 to -1) and $N_{\pi 0}S_\pi$ (from -1 to 1). Both are compared to the standard N_0S_0 and N_0S_π configurations.

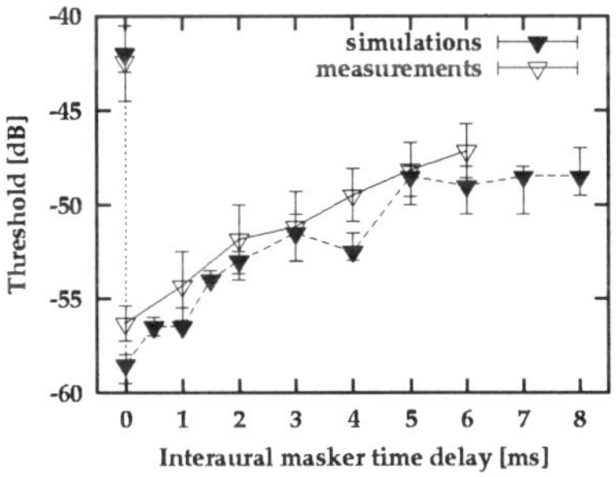

Figure 3. Thresholds for a binaural combination of interaural time and phase differences ($N_{\tau=t_0,\varphi=-\omega_0 t_0}S_\pi$) and for the diotic reference situation. $\omega_0 = 2\pi\cdot$ test tone frequency.

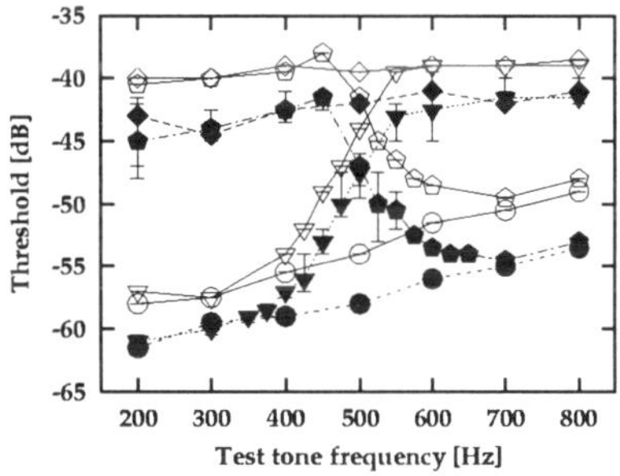

Figure 4. Interaural parameters of the masker changing with frequency. $N_{0\pi}S_\pi$, $N_{\pi 0}S_\pi$, N_0S_0 and N_0S_π. Experimental data from Kohlrausch.

Fig. 5 deals with test-tone integration, i. e. dependence of thresholds of the length of the signal in experiments and simulations.

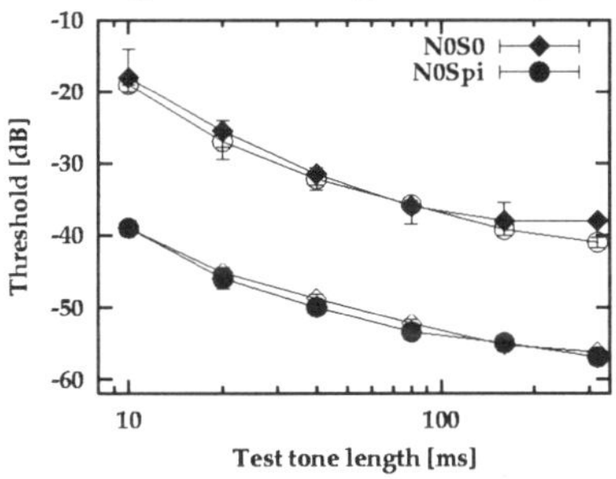

Figure 5. Test tone integration in the configurations N_0S_0 and N_0S_π.

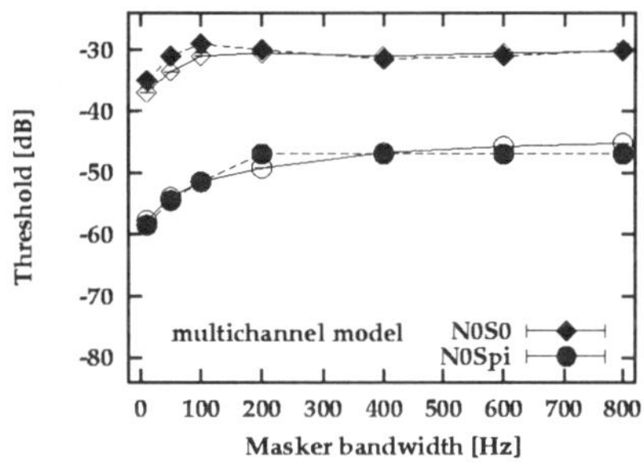

Figure 6. N_0S_0 and N_0S_π thresholds as a function of the masker bandwidth. Noise masker with constant spectral level (50 dB/Hz). Simulation considering 10 frequency channels.

Experiments to estimate critical bandwidths can reveal much broader critical bands in binaural (dichotic) configurations than in monaural (or diotic) configuration (e. g. Hall et al [3], Zurek and Durlach [7]). Two suggested mechanisms which may account for such effects are tested within the framework of the model.

The first is the concept of a binaural sluggishness, i. e. the inability of the binaural auditory system to process fast fluctuations of interaural differences of both ear signal (like proposed by Zurek and Durlach [7].).

The second concept assumes that detection in certain binaural conditions is performed in more than one critical band, as long as only *internal* noise limits the detection. This multichannel detection process would lead to broader critical band *estimates*, but is achieved due to a combinatory process [6].

To test the second hypothesis simulations have been performed where binaural processing takes place in several frequency channels. Fig. 6 compares the Hall [3] data with such simulations. Fig. 7 shows the same situation, but for a 20 dB lower masker level. The N_0S_π threshold as function of the masker bandwidth leads to a smaller estimate of the critical bandwidth.

The results of an experiment performed with notched noise are plotted in Fig. 6. The configurations are again N_0S_0 and N_0S_π, thresholds from experiments and simulations are plotted as a function of the notchwidth of the masker. Again the model produces too large BMLDs at low masker levels.

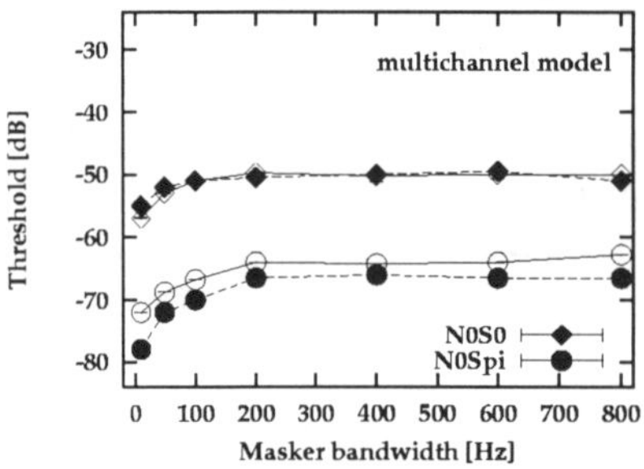

Figure 7. The same like in Fig. 6 but with a noise masker presented with constant spectral level of 30 dB/Hz. Binaural detection performed in 10 channels.

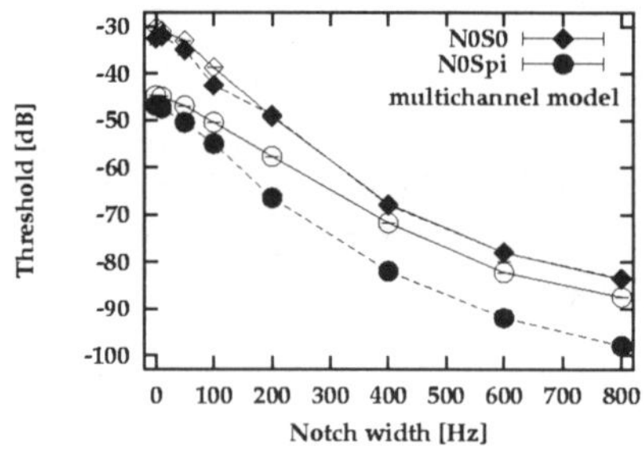

Figure 8. N_0S_0 and N_0S_π thresholds measured with notched noise as a function of the notch width. Spectral level of the masker: 50 dB/Hz.

References

1. Dau et al., *J. Acoust. Soc. Am.* **99**, 3615–3631 (1996).
2. Durlach, N. I., in *Foundations of modern auditory theory.*. N. Y., Acad. Press, (1972).
3. Hall, J. W. et al. *J. Acoust. Soc. Am.* **73**, 894–898 (1983).
4. Holube, I. et al., in *Fortschritte der Akustik – DAGA 95*, Hrsg. W. Arnold und S. Hirsekorn.
5. Kohlrausch, A., *J. Acoust. Soc. Am.* **84**, 573–583 (1988).
6. Kohlrausch, A. et al. *J. Acoust. Soc. Am.* **103**, 3082 (1998).
7. Zurek, P. M.; Durlach, N. I., *J. Acoust. Soc. Am.* **81**, 459–464 (1987).

ON THE FOUR FACTORS INVOLVED IN SENSORINEURAL HEARING LOSS

BIRGER KOLLMEIER

AG Medizinische Physik, Carl-von-Ossietzky Universität Oldenburg, D-26111 Oldenburg Germany

E-mail biko@medi.physik.uni-oldenburg.de

A review is given on past work in our working group that focussed on identifying and modeling factors involved in sensorineural hearing loss. Only the first two factors (i.e., the attenuation component or "linear" portion of the hearing loss as well as compression loss or "nonlinear" portion of the hearing loss) are primarily of cochlear origin. The "binaural loss", i.e., a deterioration in binaural interaction encountered in most hearing-impaired listeners, constitutes the third factor which is a consequence of the distorted transformation from the cochlear output into the "internal representation" of the incoming acoustical signals. Finally, the fourth factor is the "central component" which may be viewed as an increased internal noise at the stage of the "internal representation". This factor may both be due to noise in the transformation process from the acoustical signal to the internal representation and due to increased additive noise in the central processing mechanisms. The experimental data are interpreted in terms of a model of "effective" signal processing in the brain. Based on these findings, conclusions for the design of "intelligent" hearing aids can be drawn.

1 Introduction

It has often been argued that the audiogram (i.e., the frequency-dependent individual pure-tone threshold for air conduction and bone conduction) is sufficient to characterize any given patient's hearing disability. Undoubtedly, the *linear component* of a hearing loss (such as the conductive part and the "attenuation component" of a sensorineural hearing loss as opposed to the „distortion component“, c. f. Plomp [17]) can well be characterized by this component and - in most patients - characterizes the largest component of the hearing loss. However, two patients with exactly the same audiogram may differ significantly in their performance in various auditory tasks. This paper therefore deals with the "extra" factors involved in sensorineural hearing loss besides the linear attenuation component.

2 Cochlear damage: Attenuation loss vs. compression loss

We define the attenuation loss (first component) as the frequency-dependent linear portion of the total hearing loss that can be compensated by, e.g., an appropriate frequency-dependent amplification at intermediate levels. Obviously, this frequency dependence is not directly given by the audiogram but rather by a number close to the half-gain rule known from the prescription of linear hearing aids. In an individual patient, it can be derived from the necessary amplification for each frequency to obtain an intermediate loudness impression (i.e., the difference in L25 between the individual hearing-impaired listener and the average normal listener, see below). In physiological terms, the attenuation loss can be attributed to a decreased sensitivity in inner hair cells and - to a lesser degree - to that of the outer hair cells.

The exclusive loss in outer hair cell activity causes a reduction in compression in the auditory system. This causes the nonlinear level-dependent part of the hearing loss (second component). It corresponds to the well-known recruitment phenomenon in patients with cochlear hearing loss, i.e., a reduced dynamic range between the absolute threshold in quiet and the discomfort level. A direct psychoacoustic method to characterize the recruitment phenomenon is the categorical loudness scaling (Pascoe [15], Heller [6], Hohmann & Kollmeier [7]). In the optimized method described by Kollmeier [10], the subject is requested to denote the perceived loudness of a given sound (in our case: a third-octave noise band with a pseudorandomly selected level) on a categorical scale including the categories "very soft", "soft", "intermediate", "loud", "very loud", and "too loud" and four intermediate categories. In order to obtain reliable and reproducible results within a reasonable amount of time, a minimum number of ten response categories and an adaptive adjustment of the initial level range to be tested appears to be optimal (Brand et al. [2]).

A typical result of categorical loudness scaling (CS) with a normal listener and a sensorineural hearing-impaired listener for a broad-band (open symbols) and a narrow-band noise stimulus (filled symbols) is given in Fig. 1 (left panel). While in normal listeners the effect of loudness summation across frequency is clearly visible (i.e., broad-band sounds are louder than a narrow-band sound with the same physical level), this effect of loudness summation disappears in sensorineurally hearing-impaired listeners with loudness recruitment. Fig. 1(right panel) shows the data with the same two subjects using an absolute magnitude estimation (AME) paradigm. In comparison with CS, the AME method is more context-sensitive, more variable across subjects, but shows the same effects of loudness summation and other physical parameter dependencies. Different attempts have been made to model loudness perception and to connect the reduced loudness summation to the increased slope of the loudness function in hearing-impaired listeners [5], [14], [12]. With the two-component loudness models first introduced by Launer et al. [12] and later (in a different version) adopted by Moore et al. [14], the compression loss is assumed to be a second component independent of the attenuation loss. Using these assumptions and modifications of the original loudness model for normal listeners, the altered loudness perception in hearing-impaired listeners can be modeled in a satisfactory way. This even extends to categorical loudness data which can be modeled if an appropriate transformation is adapted between the sone-scale and the categorical scale (Blum et al. [1]).

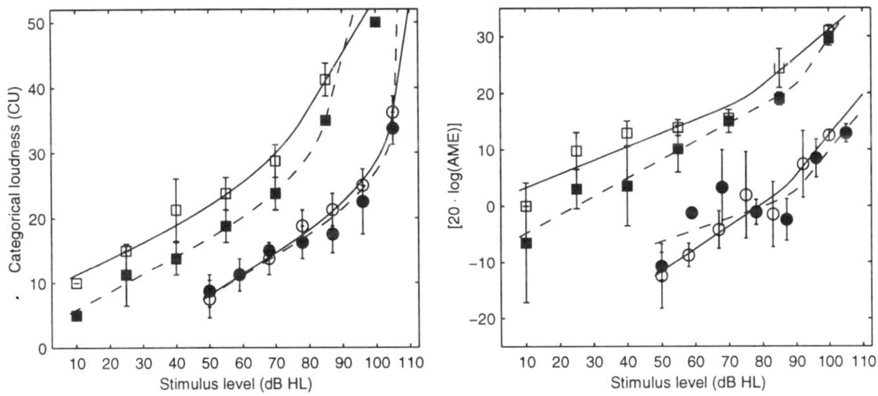

Fig. 1: Categorical loudness scaling (left panel) and absolute magnitude estimation (right panel) for a narrow-band noise stimulus with one bark bandwidth (stars and dashed lines) and a broad-band stimulus of six bark bandwidth at 4 kHz (open circles and solid lines). The respective two upper curves in each panel show the averaged data and fitted model function for a normal listener, while the respective two lower functions show the data for a hearing-impaired listener with a sensorineural hearing loss of 40 dB. The data points give the average value and intraindividual standard deviation from four scaling trials. Note that the AME-scale in the right panel is based on an individual scale without any normalization.

3 Central loss: Binaural processing component vs. central „detector noise“ component

Besides the recruitment phenomenon discussed above, hearing-impaired listeners often complain about the impaired "cocktail-party effect", i.e., their reduced ability to understand speech in noisy environments. A certain part of this deterioration can be explained in terms of the reduced audibility (first factor) and the reduced compression (second factor) which is also connected to a reduced spectral resolution and a reduced apparent temporal resolution. Yet another significant component in this deterioration is the loss in binaural noise reduction abilities in these patients (third factor): While normal listeners can use a central comparison and interaction between the signals from both ears to suppress parts of the ambient noise, hearing-impaired listeners show a deficit in this binaural interaction.

The three factors discussed so far primarily relate to *processing* deficits in the auditory system, i.e., an altered transformation from the acoustic waveform to some internal representation. They have to be separated from the central *„detector noise“* deficits (fourth factor) that relate to the deficiency of the central nervous system in interpreting these patterns (such as e.g., detection or discrimination of sounds, interpreting aspects of the internal representation of certain speech elements, or other tasks). These deficits occur both for monaural and binaural tasks even if the cochlear deficits are accounted for. One consequence of „detector noise" is a general increase in detection thresholds for hearing-impaired listeners while the dependence on temporal and spectral parameters remains unchanged. For example, such an effect was observed for modulation detection as a function of modulation frequency (Derleth et al. [4]).

Many studies have been performed to study the deterioration in binaural interaction in relation to the hearing loss and to the deterioration of various other psychoacoustically defined parameters of monaural and binaural temporal and spectral performance [9], [8]. If the average hearing loss and the age-related hearing loss of the subjects is partialed out (i.e., by evaluating partial rank correlations or by performing a factor analysis), only very few and mostly unsystematic relations remain between the different parameters that characterize monaural and binaural hearing. This finding reflects the fact that patients with a similar audiogram and similar monaural functions were found in this study to perform the binaural tasks in a very different way so that a prediction from the other parameters is not possible and no single „binaural component" of the hearing loss can be found. This supports the existence of an additional (fourth) component that is independent of monaural and binaural processing components.

Another way to directly demonstrate the binaural processing deficits in hearing-impaired listeners with respect to cocktail-party situations was demonstrated by Peissig & Kollmeier [16]: They measured the intelligibility level difference in a (simulated) spatial situation with a target speaker in front and an increasing number of interfering speakers being fixed at +135 degrees or +255 degrees. An additional interfering speaker („jammer") was varied in apparent direction of incidence from 0 to 360 degrees. The speech reception threshold (SRT) was obtained as a function of the angle of incidence. By comparing the results with continuous noise as jammers and the results from a continuous talker as jammer, the following conclusions were drawn:

- Subjects can benefit from the spatial separation between target and jammer by a reduction in SRT of up to 10 dB.
- When more than one jammer is present, the benefit still holds if both jammers come from approximately the same direction. The release from masking breaks down if the two jammers come from very different directions (i.e., from different hemispheres).
- If speech pauses are present in the jammer (interfering talker as jammer), normal listeners utilize these pauses to switch their respective suppression characteristic to suppress an additional noise source emanating from a different direction.

These findings support the notion that normal listeners can tune their two ears to act as an "optimum" two-sensor system that is capable of suppressing one sound source (i.e., one incidence direction) at a given instance of time, but can quickly switch this direction of suppression (e.g., in speech pauses). Hearing-impaired listeners were found to differ in their ambient noise suppression capabilities in a complex way that cannot be predicted from the audiogram or other audiometric features: While some hearing-impaired listeners perform nearly normal, most listeners appeared to be impaired in two ways, i.e.,

- a reduced maximum suppression of the ambient noise source (reduced maximum effect of the ILD). Within the current framwork, this relates to the third component (reduced binaural processing capability).
- a reduced ability to utilize speech pauses in one jammer to suppress the other jammer (i.e., reduced difference between continuous noise as jammers and interfering talkers as jammers). This relates both to increased time constants involved in central binaural processing and to the fourth factor (i. e., reduced central resolution).

4 Model approach

An approach to incorporate the compression loss into the „Oldenburg Perception Model" was described by Kollmeier et al. [10] who tried to model different temporal effects in hearing-impaired listeners by modifying the model of the "effective" auditory signal processing developed by Dau et al. [3]. The adaptive compressive nonlinearity of the model had to be divided into a peripheral compressive part and a more central part that correspond to the central auditory system subsequent to the auditory nerve. By appropriately changing the peripheral nonlinearity and not changing any other parameters (i.e., no change of the time constants involved), they could simultaneously model the following phenomena encountered in hearing-impaired listeners:

- Recruitment: An altered activity pattern as a function of input level that corresponds to an increase in the slope of the loudness function. To achieve a match with the individual's loudness function, the compression characteristic has to be changed in an appropriate way.
- Forward masking: The shallower slopes in forward masking curves for hearing-impaired listeners were shown to be a consequence of the altered setting of the threshold and the compressive nonlinearity, but not due to any change in "effective" time constants [10].
- Temporal integration: The smaller difference in the detection threshold in noise between a short probe tone and a long probe tone for hearing-impaired listeners than in normal listeners could also be attributed to the changes in the compressive properties of the (rather than changes in time constants) model.

- Modulation detection: The same dependence of the modulation detection threshold on modulation frequency is normally found for normal listeners and hearing-impaired listeners (Derleth et al. [4]). This is also predicted by the modified auditory model. However, the absolute value of the thresholds are generally higher in hearing-impaired listeners than in normal listeners. Since this effect is not produced by any altered time constant or other processing parameter of the peripheral transformation stage, it has to be modelled as an increased "internal" noise encountered in the hearing-impaired listeners. It is therefore attributed to the fourth component in the model concept.

So far, the first component of the hearing loss (audibility) has been incorporated by using the appropriate peripheral noise to model the frequency-dependent "effective" linear hearing loss at intermediate levels, while the second component (compression loss) has been modeled by modifying the peripheral compressive properties of the model and keeping some adaptation loops in the more central part. While the fourth component (i.e., the central loss) can to a first approximation be modeled by adding more noise at the stage of the "internal representation", the third component (i.e., binaural loss) has not yet been modeled in a satisfactory way. In fact, first approaches to model aspects of binaural interaction within the framework of the current model for normal listeners has only been performed quite recently (Zerbs et al. [18]). They introduced a binaural interaction at the level subsequent to the adaptation loops and used interaural time delays with a subsequent summation of both binaural channels. A modification of this scheme to be applicable for hearing-impaired listeners is still subject to further research.

5 Discussion

The review of some of the work with hearing-impaired listeners from the author's group given above can only summarise parts of the evidence that exists in the literature for the different factors involved in sensorineural hearing loss. For futural work, however, it is necessary to structure this evidence by appropriate auditory models that help to differentiate between known effects (i.e. effects, that can be explained on the basis of our current knowledge and model assumptions) and „unexplained“ effects. Certainly, the model described above is only one candidate for such a guideline in futural work. Ideally, an optimised model can be adjusted by the diagnosticall parameters obtained for each individual patient within a minimum amount of measurement time at a given level of measurement accuracy. Based on such an adjusted model, an optimised treatment of the hearing disorder and rehabilitation with appropriately selected and adjusted hearing aids might become realistic in the future.

6 Acknowledgement

Supported by BMBF/PT-AUG. Thanks to Karin Bramstedt and Angelika Sievers for typing the manuscript and to all members of the AG Medinzinische Physik for their support.

References

1. Blum, R., *et al.*, Modellierung von Kategorial- und Verhältnislautheit. *Zeitschrift für Audiologie* (1999): (submitted).
2. Brand, T., V. Hohmann, and B. Kollmeier, Adaptive categorical loudness scaling, in: A. Schick and M. Klatte (ed.) BIS, Universität Oldenburg: Oldenburg.1997 pp. 603-610.
3. Dau, T., B. Kollmeier, and A. Kohlrausch, Modeling auditory processing of amplitude modulation: I. Detection and masking with narrow-band carriers. *J. Acoustical Soc. Am.* **102**(5) (1997): pp. 2892-2905.
4. Derleth, R.P., T. Dau, and B. Kollmeier, Perception of amplitude-modulated narrowband noise by sensorineural hearing-impaired listeners, in: B. Kollmeier (ed.): *Psychoacoustics, Speech and Hearing Aids,* World Scientific: Singapore.(1996): pp. 39-44.
5. Florentine, M. and E. Zwicker, A model of loudness summation applied to noise induced hearing loss. *Hearing Research* **1** (1979): pp. 121--132.
6. Heller, O., Hörfeldaudiometrie mit dem Verfahren der Kategorienunterteilung (KU). *Psycholog. Beiträge* **27** (1985): pp. 478--493.
7. Hohmann, V. and B. Kollmeier, Weiterentwicklung und klinischer Einsatz der Hörfeldskalierung. *Audiologische Akustik* **34(2)** (1995): pp. 48-64.
8. Holube, I., Experimente und Modellvorstellungen zur Psychoakustik und zum Sprachverstehen bei Normal- und Schwerhöringen', Ph.D. Thesis, Universität Göttingen.1993.
9. Kinkel, M. and B. Kollmeier, Binaurales Hören bei Normal- und Schwerhörigen II: Analyse der Ergebnisse. *Audiologische Akustik* **31** (1992): pp. 22-33.
10. Kollmeier, B., Hörflächenskalierung - Grundlagen und Anwendung der kategorialen Lautheitsskalierung fuer Hördiagnostik und Hörgeräteversorgung. Heidelberg: median Verlag (1997)
11. Kollmeier, B., R.-P. Derleth, and T. Dau, Modeling the "effective" auditory signal processing for hearing-impaired listeners, in A.R. Palmer, *et al.* (ed.): *Psychophysical and Physiological Advances in Hearing,* Whurr Publisher: London. (1998): pp. 482-490.
12. Launer, S., V. Hohmann, and B. Kollmeier, Categorical loudness summation-experiments and models, in: A. Schick (ed.): *Contributions to psychological acoustics,* BIS Oldenburg: Oldenburg. 1993 pp. 233-250.
13. Launer, S., V. Hohmann, and B. Kollmeier, Modeling loudness growth and loudness summation in hearing-impaired listeners, in W. Jesteadt (ed.): *Modeling sensorineural hearing loss.*,. Lawrence Erlbaum Assoc.: Mahwah, NJ. (1997): pp. 175-185.
14. Moore, B.C.J., B.R. Glasberg, and D.A. Vickers, Factors influencing loudness perception in people with cochlear hearing loss, in B. Kollmeier, (ed.): *Psychoacoustics, speech and hearing aids*, World Scientific: Singapore. (1996): pp. 7-18.

15. Pascoe, D.P., An approach to hearing aid selection. *Hear. Instrum.* **29** (1978): pp. 12--16.
16. Peissig, J. and B. Kollmeier, Directivity of binaural noise reduction in spatial multiple noise-source arrangements for normal and impaired listeners. *Journal of the Acoustical Society of America* **101**(3) (1997): pp. 1660-1670.
17. Plomp, R., A signal--to--noise ratio model for the speech perception threshold of the hearing impaired. *J. Acoust. Soc. Am.* **79** (1986): pp. 772--780.
18. Zerbs, C., T. Dau, and B. Kollmeier, Modeling the `effective' binaural signal processing in detection experiments, in: (ed.)Singapore.1999 pp. .

TEMPORAL INTEGRATION AND MANDATORY PROCESSING IN AUDITORY MODELS

ROY D. PATTERSON, STEFAN UPPENKAMP

CNBH, Dept. of Physiology, University of Cambridge, U.K.
E-mail: roy.patterson@mrc-cbu.cam.ac.uk

In Cambridge, Oldenburg and Eindhoven, new efforts are being made to construct reasonably comprehensive, multi-channel, auditory models based on physiological knowledge. These are time-domain models that take sound waves as input rather than their power spectra, and they are intended to explain a substantial range of perceptual data associated with temporal auditory processing. In the first part of this paper, we review the spectral analysis and neural transduction stages of these models, basically to note the high degree of similarity and the implied common purpose. The output of these time-domain, multi-channel, cochlea simulations is a Neural Activity Pattern, or NAP, that preserves phase-locked time intervals up to 3-5 kHz. In the second part of the paper, we ask whether auditory modules proposed to extract perceptual properties like pitch, loudness, and modulation, all operate independently and in parallel on the NAP; or whether there is a mandatory stage of temporal integration prior to the distribution of the information to the independent, parallel modules.

Introduction

The primary issue in this paper is a comparison between the Auditory Image Model (AIM), where mandatory 'strobed' temporal integration converts the NAP into a stabilised image and all perceptual modules apply their processing to the image rather than the NAP (Patterson *et al.*, 1992), and the Model of the Effective Signal Processing (MESP) developed originally in Göttingen and now used in Eindhoven and Oldenburg (Dau *et al.*, 1996; Dau *et al.*, 1997a). This model applies a multi-channel modulation filterbank to the NAP to produce a representation which will be referred to as a 'modulation image'. The dimensions of this image are channel centre frequency and modulation rate which is very similar to the channel and time-interval dimensions of the auditory image. The questions of interest, then, are a) whether the modulation processing stage in MESP is considered mandatory as strobed temporal integration is in AIM, b) whether the two forms of image are essentially the same, and c) whether the models might profitably be merged to produce a more general auditory model. Unless otherwise specified, the version of AIM is that defined in Patterson *et al.* (1995), and the version of MESP is that defined in Dau *et al.* (1997a).

1 Simulation of the Neural Activity Pattern in the Auditory nerve.

The processing of a wave by the Auditory Image Model (AIM) begins with a band-pass filter designed to represent the middle ear. The filter is adapted from one suggested by Lutman and Martin (1979). In MESP, the middle-ear transfer function can be included either by a cascade of several low-pass filters or with a resonance filter simulating an equal loudness level contour at medium level.

Spectral Analysis: The next stages simulates the spectral analysis performed in the cochlea. In both AIM and MESP, the spectral analysis is simulated by a bank of auditory filters which converts the output of the middle ear into an array of filtered waves, one for each channel of the filterbank. The set of waves is the model's representation of basilar membrane motion (BMM). In AIM as well as in MESP, the filters are distributed linearly along a frequency scale measured in Equivalent Rectangular Bandwidths (ERBs) as proposed by Glasberg and Moore (1990). With suitable alteration to the two constants of the ERB function, it is also possible to produce a reasonable Bark-scale filterbank. The auditory filter in both AIM and MESP is a fourth-order, linear, gammatone filter (Patterson *et al.*, 1992). The impulse response of the gammatone filter provides an excellent fit to the impulse response of primary auditory neurons in cats, and its amplitude characteristic is very similar to that of the 'roex' filter commonly used to represent the human auditory filter (see Patterson, 1994a for details). An example of the BMM produced by the vowel in 'hat' is presented in the top panel of Figure 2 in Patterson *et al.* (1995). AIM also provides an optional, level-dependent auditory filterbank which is a version of Strube's (1985) transmission line approximation to cochlear hydrodynamics, as implemented by Giguere and Woodland (1994). A different version of Strube's (1985) filterbank is sometimes used in MESP. So at the level of the spectral analysis, AIM and MESP are very similar indeed. For brevity, we will restrict discussion to the gammatone filterbank in the paper that follows as it is entirely sufficient to the argument.

Neural Transduction: The next stage of AIM and MESP simulates the mechanical/neural transduction process performed by the inner haircells. It converts the BMM into a neural activity pattern (NAP), which is the representation of the afferent activity in the auditory nerve and/or the anterior ventral cochlear nucleus. In AIM the transduction is simulated by a bank of two-dimensional adaptive-thresholding units (Patterson and Holdsworth, 1996). The process begins with half-wave rectification of the BMM followed by logarithmic compression; then it applies adaptation in time and suppression across frequency. The adaptation and suppression are coupled and they jointly sharpen features like vowel formants in the compressed BMM representation. Briefly, an adaptive threshold value is maintained for each channel and updated at the sampling rate. The new value is the largest of a) the previous value reduced by a fast-acting temporal decay factor, b) the previous value reduced by a longer-term temporal decay factor, c) the adapted level in the channel immediately above, reduced by a frequency spread factor, or d) the adapted level in the channel immediately below, reduced by the same frequency spread factor. The mechanism produces output whenever the input exceeds the adaptive threshold, and the output level is the difference between the input and the adaptive threshold. In MESP the process again begins with half-wave rectification. Then the compression and temporal adaptation are applied by a five-stage adaptation unit; each stage performs square root compression for stationary input with a time constant of adaptation that is longer than the previous one, varying from 5 ms to 500 ms. This results in nearly logarithmic compression for stationary signals and enables MESP to explain both forward masking data and intensity discrimination data (Püschel, 1988; Dau *et al.*, 1996). There is no cross channel interaction in the MESP version of neural transduction like the frequency adaptation in AIM, but this is not an important issue in the current paper. The middle panel in Figure 2 of Patterson *et al.* (1995) shows the NAP obtained with adaptive thresholding in response to the BMM of the vowel in 'hat'. The phase lag in the BMM is preserved in the NAP, the waves are half-wave rectified, and speech features observed in the BMM are sharpened in the NAP. Similar effects are observed in the NAPs produced by MESP.

2 The temporal integration paradox

Sustained notes and vowels produce static auditory images with fixed loudness, fixed pitch, and fixed timbre indicating that some form of fairly lengthy temporal integration is applied to the NAP in the production of our initial perception of a sound – our auditory image. For some reason, however, this integration does not seem to destroy the phase-locking information, when the sound is periodic or quasi-periodic. This temporal integration paradox is illustrated in Figure 1 using a SAM tone with a 2000 Hz carrier and a 120 Hz modulation rate. The top row shows the wave and the second row shows the output of the gammatone filter centred at 2000 Hz. The third row shows the NAP. The loss of phase-locking is implemented as a two-stage lowpass filter with a time constant of 0.133 ms (a frequency cutoff of 1200 Hz), which means that phase-locking information is still evident in this channel. Note, however, that the output oscillates far too fast to explain the stable loudness that this sound produces. The fourth panel shows the NAP when we assume that the limit is the leaky integrator used to explain the detection of amplitude modulation. Following Viemeister (1979), it is implemented as a single stage lowpass filter with a time constant in the range of 2.5 ms. It reduces the phase-locking information substantially but the output still oscillates too fast to explain the stable loudness that we hear. The fifth panel shows the NAP when we assume that the limit is the temporal window used to explain the detection of tones presented between noise bursts. Following Plack and Moore (1991), it

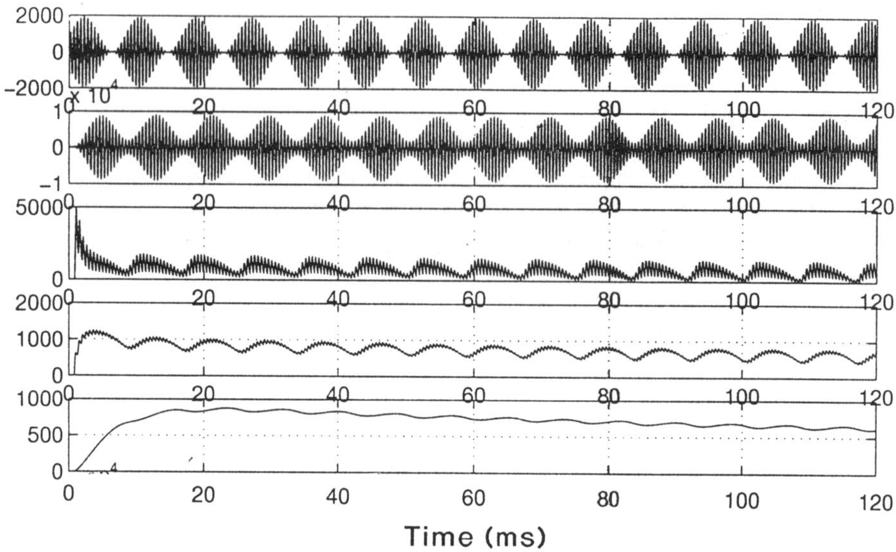

Figure 1. Auditory representations of an amplitude-modulated tone, 2 kHz carrier, 120 Hz modulation rate. Top: waveform, second row: output of gamma tone filter centred at 2 kHz, third row: Neural activity pattern, fourth row: single-stage low-pass filtered NAP, time constant 2.5 ms, fifth row: two-stage low-pass filtered NAP, time constant 5ms.

is implemented as a two-stage lowpass filter with a 5-ms time constant. It produces a rather smooth output that is more like the loudness we hear, but it eliminates the phase-locking information entirely. It also rises to its sustained level in about 10 ms which is too fast to explain the growth of loudness at the onset of a sound. So, integration times long enough to produce the stability we hear, destroy the phase-locking information observed in the auditory nerve – phase locking that is necessary to explain pitch perception (Yost *et al.*, 1996, 1998), sound quality (Patterson, 1994b; Patterson *et al.*, 1996) and phase perception (see Patterson, 1987 for a review).

3 Comparison of time-interval dimension in AIM with modulation-rate dimension in MESP

3.1 Temporal Integration in AIM

In order to solve the temporal integration paradox, AIM employs a form of strobed temporal integration to convert the NAP into an auditory image; the details are presented in Patterson *et al.* (1992) and Patterson (1994b). Briefly, a bank of delay lines is used to form a buffer store for the NAP data as it flows from the cochlea.

There is one delay line per channel and each channel of the buffer is assigned a strobe unit which monitors activity in that channel looking for local maxima in the stream of NAP pulses. The oscillating function in the middle of Figure 2 shows an enlargement of the NAP of a 2-kHz tone modulated at a 120-Hz rate (a section of row three of Figure 1). Along the top of the NAP is plotted the adaptive threshold of the strobe unit and at the bottom of the figure is the strobe points that themechanism identifies. When a strobe point is found, the unit initiates temporal integration in that channel; that is, it transfers a copy of the NAP at that instant to the corresponding channel of an image buffer and adds it point-for-point with whatever is already there. The corresponding channel of the auditory image is presented in Figure 3. The local maximum is mapped to the 0-ms point in the image. The multi-channel version of this STI process produces AIM's representation of our auditory image of a sound. Periodic and quasi-periodic sounds cause regular strobing, which leads to simulated auditory images that are static, or nearly static, and which have the same temporal resolution as the NAP. Dynamic sounds are represented as a sequence of auditory image frames. If the rate of change in a sound is not too rapid, as is diphthongs, features are seen to move smoothly as the sound proceeds, much as characters move smoothly in animated cartoons. AIM has not been successful to date, however, in explaining modulation detection quantitatively (Akeroyd and Patterson, 1997).

3.2 Modulation Perception in MESP

Envelope fluctuations are important in perception – particularly speech perception – and they are the focus in the MESP. The approach is based on the classical temporal modulation transfer functions (TMTFs) measured by Viemeister (1979). However, recognising the limitations of the leaky integrator as a model of temporal integration, MESP provides a more sophisticated method of modulation processing which avoids true integration, at least in the initial stages. The modulation frequencies associated with perception are extracted by calculating a modulation spectrum for each channel of the NAP and assembling the array of modulation spectra into a 'modulation image'. Initially, the modulation spectrum was obtained by calculating the Fourier transform of the Hilbert envelope of the acoustic wave (Kollmeier and Koch, 1994). Recently, a more realistic method has been developed in which the modulation spectrum for each NAP channel is produced with a modulation filterbank. This approach is supported by physiological data on modulation-frequency specificity. The modulation filterbank is the final processing stage before the formation of a template which is the basis of the optimal detector mechanism in MESP (Dau *et al.* 1997a). The form and scaling of the modulation filters has been derived from psychoacoustical experiments on modulation detection and modulation masking (Dau *et al.* 1997a). To date, applications have largely been restricted to single channel versions of MESP. The generalisation to a multi-

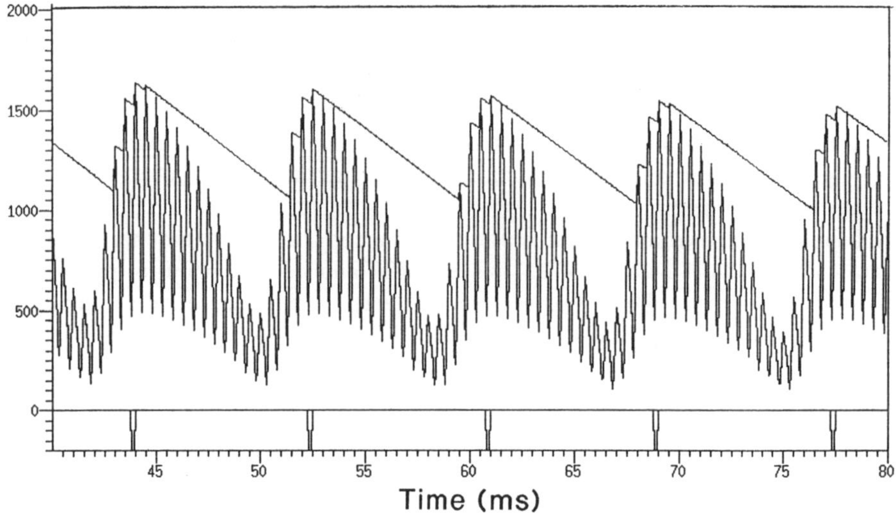

Figure 2. Enlarged section from the third row of Figure 1. Middle trace: NAP, top trace: adaptive threshold of the strobe unit, bottom trace: actual strobe points found by the mechanism.

channel version is straightforward and input to the detection mechanism in the multi-channel version of MESP would be a matrix with dimensions very similar to those of the auditory image; namely, channel centre frequency, modulation rate and time (Dau *et al.*, 1997b). Thus, the general form of the modulation image is very similar to that of the auditory image.

3.3 Comparison of Dimensions in AIM and MESP

The primary differences between the auditory image and the modulation image have to do with the range, scale, and resolution of the time-interval and modulation-rate dimensions, respectively. In AIM, the range of the time-interval dimension is 0-40 ms, the scale is linear, and the resolution is sufficient to represent sub-millisecond fine structure in the NAP throughout its length. The range is determined by the lower limit of pitch. It is assumed that the time-interval dimension is just long enough to accommodate one period of the lowest pitch that humans perceive which is about 35 ms, or 30 Hz. In AIM this is achieved by assuming that activity in the NAP buffer decays 2.5 %/ms as it flows down the buffer so that there is no activity beyond 40 ms. This is the reason the auditory image in Figure 3 slopes downwards from right to left. It is assumed that stimuli with periods longer that 40 ms have no pitch (that is, they do not support musical-interval discrimination); that the resolution in the range beyond 40 ms becomes much coarser, and that the repetition of features that we perceive in such stimuli is processed using summary statistics derived from the auditory image. We will return to this issue

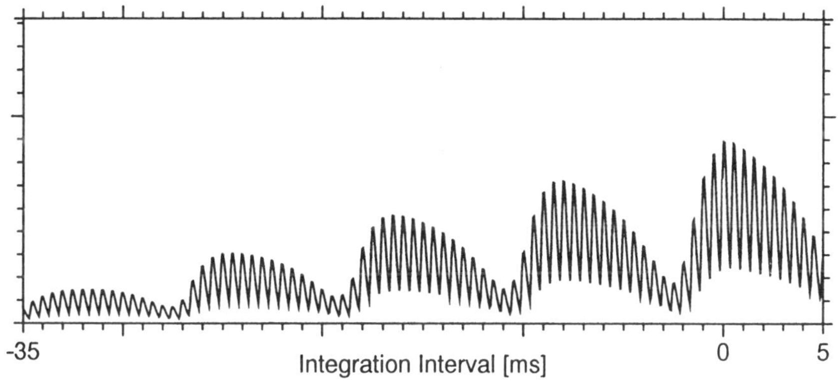

Figure 3. Auditory image constructed from the NAP of the amplitude modulated tone using strobed temporal integration, only the 2kHz channel is shown.

when discussing a combined version of AIM and MESP in the next Section. With regard to resolution within the pitch range, studies with Iterated Rippled Noise (IRN) stimuli indicate that fine-structure with sub-millisecond accuracy is necessary to explain the discrimination of high-pass filtered IRNs with positive and negative gains (Yost *et al.*, 1998). The fine structure is also used to explain timbre and timbre differences; the regular fine structure in Figure 3 indicates that the sound is a tone rather than a noise (Irino and Patterson, 1996). There is no particular reason for using a linear axis for the time-interval dimension; a logarithmic axis would appear to be more appropriate for explaining musical-interval perception (Patterson, 1986).

In MESP, the range of the modulation-rate dimension is about 1-500 Hz; in terms of modulation periods this is 1000-2 ms. It includes the 30-500 Hz range where modulation is perceived as a pitch and the 1-20 Hz range where modulation is perceived as a fluctuation in loudness over time. The scale is logarithmic, and the resolution is proportional to the modulation rate. The resolution, however, is much less than that of the auditory image in the pitch range because it is not the NAP itself but rather the Hilbert envelope of the NAP that is the input to the modulation filterbank. The lack of fine-structure information in the modulation image means that, with regard to pitch, MESP would not support discrimination of IRNs with positive and negative gains, and with respect to timbre, it would be largely insensitive to the dramatic changes we hear in the perception of high-pass filtered IRN as the number of iterations is increased from 0 to 10. The preservation of fine structure in the auditory image enables AIM to segregate the tonal and noisy components of the perception and to assign relative loudness values to the two components (Patterson *et al.*, 1996).

4 Combining AIM and MESP for Modulation Perception

The lack of temporal fine structure in the modulation image means that it is not possible to build a combined serial model in which the modulation-filterbank stage proceeds the construction of the auditory image; the resulting auditory image would not be able to explain pitch perception. It would, of course, be possible to assemble the auditory image and the modulation image in parallel, but at this point it seems more interesting to pursue the serial option with the auditory image before the modulation filterbank. For example, if we could extract modulation information from the auditory image and cast it in the form of a temporal envelope, then these waves could be passed directly to a modulation filterbank to produce a modulation spectrum much like that produced by MESP currently.

At low modulation rates, we hear modulation primarily as a change in loudness over time. As a result MESP preserves modulation phase in this region as well as modulation magnitude. There are no vertical features in the auditory image of such a sound because the period of the modulation is greater than 40 ms, but there is a regular change in level in any active channels, and this level is simply summarised by the value at 0 ms in the channel. So, a continuous record of the level of activity in the image as a function of time would provide a form of envelope with both amplitude and phase information, much like the envelope currently used as input to the low-frequency channels of the modulation filterbank. At higher rates, we hear modulation as flutter, roughness, periodicity pitch, or sinusoidal sidebands, depending on the ratio of the carrier frequency to the modulation frequency. MESP extracts the modulation information using the Hilbert transform; the phase information is discarded because phase changes do not affect modulation detection in this region. AIM stabilises the modulation pattern at frequencies above about 30 Hz (Fig. 3) because it strobes on modulation peaks. In the process, it removes information about modulator phase because the strobe point in the NAP is mapped into the auditory image at the 0 ms point.

The magnitude of the modulation could be extracted as in MESP by passing the Hilbert envelope of the auditory image to the modulation filterbank and discarding the fixed, cosine phase information. In this case, the outputs of the low- and high-frequency channels of the modulation filterbank can be joined to produce a modulation spectrum which could be used to develop a template for the optimal decision process. Alternately, one could use the height of the first peak in the auditory image directly as the estimate of modulation magnitude for high-frequency channels in the modulation spectrum. That is, a temporally compressed version of the auditory image plotted on a logarithmic scale with short periods towards the right (as currently), would be a representation of the modulation spectrum for high modulation rates. This has the advantage that the height of the first peak in the auditory image is also the measure of pitch strength in AIM. This modulation spectrum for frequencies over 30 Hz would then have to be combined with a low-rate modulation spectrum obtained from tracking level changes over time as outlined above. It seems likely, however, that the combined model would be more comprehensive and would not involve much more computation, if any.

The model should probably be tested first on its ability to explain the detection of sinusoidal modulation of sinusoidal and noise carriers to determine whether the long time constant of the auditory image (30 ms) interferes with the detection of modulation at higher rates. It would also be interesting to ascertain whether it can explain the discrimination performance reported by Akeroyd and Patterson (1997) with modulated noises having supra-threshold, half-cosine modulators. The 'down' cosine modulator is more discriminable from SAM noise than the 'up' cosine modulator. If the combined model can explain this discrimination asymmetry as well, then it might be presented with the much greater test of explaining the discrimination of modulation rate. For complex sounds, the percentage change in modulation rate drops rapidly from the range 5-10% to the range 0.5-1% as the modulation rate moves past the lower limit of pitch, 30 Hz (Krumbholz et al., this volume). In the MESP it is not clear how this can be explained because there is no dramatic change in the representation in this region. In the auditory image, a modulation peak with well defined fine structure appears in the image as the period decreases below 35 ms, and this feature provides a basis for a more precise estimate of the modulation rate. At the same time, it explains why the perceptual cues for modulation rate discrimination deteriorate at this point on the modulation axis.

References

1. Akeroyd M.A. and Patterson R.D., A comparison of detection and discrimination of temporally asymmetry in amplitude modulation. J. Acoust. Soc. Am. **101** (1997) pp. 430-439
2. Dau T., Püschel D. and Kohlrausch A., A quantitative model of the 'effective' signal processing in the auditory system. I. Model structure. J. Acoust. Soc. Am. **99** (1996) pp. 3615-3622.
3. Dau T., Kollmeier B. and Kohlrausch A., Modeling auditory processing of amplitude modulation. I. Detection and masking with narrow-band carriers. J. Acoust. Soc. Am. **102** (1997a) pp. 2892-2905.
4. Dau T., Kollmeier B. and Kohlrausch A., Modeling auditory processing of amplitude modulation. II. Spectral and temporal integration. J. Acoust. Soc. Am. **102** (1997b) pp. 2906-2919.
5. Giguere C. and Woodland P.C., A computational model of the auditory periphery for speech and hearing research. I. Ascending path. J. Acoust. Soc. Am. **95** (1994) pp. 331-342.
6. Glasberg B.R. and Moore B.C.J., Derivation of auditory filter shapes from notched-noise data. Hear. Res. **47** (1990) pp. 103-38.

7. Irino T. and Patterson R.D., Temporal asymmetry in the auditory system," J. Acoust. Soc. Am. **99** (1996) pp. 2316-2331.
8. Kollmeier B. and Koch R., Speech enhancement based on physiological and psychoacoustical models of modulation perception and binaural interaction. J. Acoust. Soc. Am. **95** (1994) pp. 1593-1602.
9. Lutman M.E., Martin A.M., Development of an electroacoustic analogue model of the middle ear and acoustic reflex. J. Sound Vib. **64** (1979) pp. 133-157.
10. Patterson R.D., Spiral detection of periodicity and the spiral form of musical scales. Psychology of Music **14** (1986) pp. 44-61.
11. Patterson R.D., A pulse ribbon model of monaural phase perception. J. Acoust. Soc. Am., **82** (1987) pp. 1560-1586.
12. Patterson R.D., The sound of a sinusoid: Spectral models. J. Acoust. Soc. Am. **96** (1994a) pp. 1409-1418.
13. Patterson R.D., The sound of a sinusoid: Time-interval models. J. Acoust. Soc. Am. **96** (1994b) pp. 1419-1428.
14. Patterson R.D. and Holdsworth J., A functional model of neural activity patterns and auditory images. In *Advances in Speech, Hearing and Language Processing*, ed. by W. A. Ainsworth, Vol 3. Part B. (JAI Press, London, 1996) pp. 547-563.
15. Patterson R.D., Robinson K., Holdsworth J., McKeown D., Zhang C. and Allerhand M., Complex sounds and auditory images. In *Auditory physiology and perception*, ed. by Y. Cazals, L. Demany and K. Horner (Pergamon, Oxford, 1992) pp. 429-446.
16. Patterson R.D., Allerhand M. and Giguere C., Time-domain modelling of peripheral auditory processing: A modular architecture and a software platform. J. Acoust. Soc. Am. **98** (1995) pp. 1890-1894.
17. Patterson R.D., Handel S., Yost W.A. and Datta A.J.. The relative strength of the tone and noise components in iterated rippled noise. J. Acoust. Soc. Am. **100** (1996) pp.3286-3294.
18. Plack C.J., and Moore B.C.J., Decrement detection in normal and impaired ears. J. Acoust. Soc. Am. **90** (1991) pp. 3069-3076.
19. Püschel D., Prinzipien der zeitlichen Analyse beim Hören. Dissertation, Universität Göttingen (1988).
20. Strube H.W., A computationally efficient basilar-membrane model. Acustica **58** (1985) pp. 207-214.
21. Viemeister N.F., Temporal modulation transfer functions based upon modulation thresholds. J. Acoust. Soc. Am. **66** (1979) pp. 1364-1380.
22. Yost W.A., Patterson R.D., Sheft S., A time-domain description for the pitch strength of iterated rippled noise. J. Acoust. Soc. Am. **99** (1996) 1066-1078.
23. Yost W.A., Patterson R.D., Sheft, S., The role of the envelope in processing iterated rippled noise. J. Acoust. Soc. Am. (in press, Oct 1998).

MODELING ACROSS-FREQUENCY PROCESSING OF AMPLITUDE MODULATION

TORSTEN DAU AND JESKO L. VERHEY

Arbeitsgruppe Medizinische Physik and Graduiertenkolleg Psychoakustik, Universität Oldenburg, D - 26111 Oldenburg
Germany
E-mail: torsten.dau@medi.physik.uni-oldenburg.de

Recently, a model was presented which is able to predict human performance in a variety of experimental conditions related to modulation detection and modulation masking [2,3]. As a substantial part of the signal processing, a modulation filterbank was introduced to analyze the envelope fluctuations of the stimuli in each peripheral auditory filter. Modulation channels account for effects of frequency selectivity in the modulation frequency domain. However, the model does not cover conditions which require *across-channel* processing. Such across-channel processing of temporally modulated complex stimuli might occur, for example, in conditions of modulation detection interference (MDI), and comodulation masking release (CMR). A new modeling approach is presented that *integrates* envelope fluctuations across frequency while preserving the model's capabilities for describing modulation masking as well as spectral masking in the audio-frequency domain. The model can account for the main features of MDI, but it does not account for CMR.

1 Introduction

To investigate sensitivity to dynamically varying sounds, a number of researchers have employed relatively simple signals such as sinusoidally amplitude modulated tones or noises. Interestingly, modulation-frequency specificity has been observed in different physiological studies of neural responses to amplitude modulated tones. Langner and Schreiner [7] found a highly systematically organized map of best modulation frequencies (BMF) within the inferior colliculus of the cat. Each neuron - tuned to a certain modulation rate - may be considered as a filter in the modulation domain, and the array of neurons may be considered collectively as a modulation filterbank. This modulation filterbank has been suggested by some researchers as an explanation for certain *perceptual* phenomena. For example, Houtgast [6] performed psychoacoustical experiments on modulation masking: masking was most effective when the test modulation fell within the masker-modulation band. Fleischer [5] and Dau et al. [2] examined the ability to detect sinusoidal amplitude modulation of narrowband noise carriers. The data can be accounted for quantitatively in terms of a model based on a modulation filterbank. As a functional realization of a modulation filterbank, Dau et al. [3] proposed a model in which the modulation analysis is performed independently and in parallel in each stimulated peripheral channel.

The model does not contain any combination of amplitude modulation (AM) processing *across* carrier frequencies prior to the decision stage. Therefore, the model must fail in conditions where the auditory system appears to be able to fuse spectral components that are modulated with the same temporal pattern together. One phenomenon associated with across-channel processing is called modulation detection interference (MDI). The ability to detect amplitude modulation of a target tone is severely impaired if another interfering signal is modulated at the same rate as the target. Yost et al. [9] suggested a scheme for grouping spectral components that share a common form of modulation. They postulated modulation channels whereby all units excited by a common modulation provide converging input to a higher neural center. Interestingly, recent physiological studies by Schulze and Langner [8] showed that most of the units in the low-frequency region of the auditory cortex (of the gerbil) responded to AM-stimuli with a spectrum located completely outside the unit's frequency response range. These cortical units seem to integrate information over wide spectral ranges.

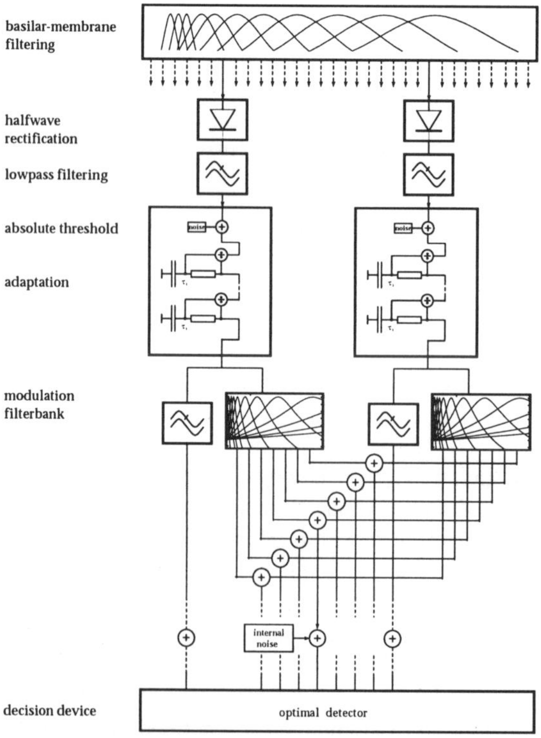

Figure 1. Block diagram of the across-channel modulation-filterbank model. The resulting modulation channels are tuned to modulation frequency but respond to a broad range of carrier frequencies. Low-frequency modulations are processed separately within each peripheral channel. For details, see text.

2 Across-channel modulation-filterbank model

Figure 1 presents a functional realization of the scheme suggested in [9]. The first processing stages of the across-channel modulation filterbank model are the same as in the original model version. Within the new model, it is assumed that the envelope in each peripheral filter is lowpass filtered at a cut-off frequency of 2.5 Hz including the dc-component. This allows energy extraction within each peripheral channel and enables the model to account for classical spectral masking data. Second, the outputs of the several modulation channels tuned to the same modulation frequencies are summed together. Thus, the resulting channels are tuned to modulation rate but respond to a broad range of carrier frequencies. To the output of all these integrated patterns, as well as to the output of the lowpass channels, internal noise is added as a random process. The new modelling approach can not be separated from the original one in modulation detection conditions with random noise carriers. Thus, the across-channel model essentially reproduces the simulations described in [2,3]. In addition, the model allows modulation interference from widely separated spectral regions.

3 Measurements and simulations of MDI

A new experiment was performed. A target tone at 4 kHz was presented and the threshold for an imposed test modulation was measured in the presence of a narrowband noise centered at 1 kHz. Thresholds for 10 Hz and 50 Hz were obtained as a function of the bandwidth of the noise. The noise exhibits inherent modulations which should interfere with the imposed test modulation at the target frequency. Parameter in the experiment was the level difference (ΔL= 0, 20, 30 dB) between the target tone and the interfering noise whereby the level of the target tone was kept constant at 65 dB SPL. An adaptive 3IFC adaptive procedure (1-up 2-down) was used to determine thresholds. Target tone and interfering noise as well as the modulation had a duration of 500 ms including 20 ms Hanning ramps.

Figure 2 shows the experimental data for two subjects. The upper panels show the detection thresholds for 10 Hz modulation. Peaked curves can be observed with most masking at a noise bandwidth of 20 Hz. With increasing level of the noise, threshold increases for each noise bandwidth. This effect is smallest for the sinusoidal interferer (Δf = 0 Hz). The lower panels of Fig. 2 represent the threshold curves for 50 Hz modulation. Thresholds increase up to a bandwidth of 50 Hz and remain constant or slightly decrease at larger bandwidths.

Figure 3 shows corresponding model predictions for 10 Hz (left panel) and 50 Hz (right panel) modulation, respectively. Only the conditions for ΔL = 0 dB (filled circles) and for ΔL = 20 dB (filled boxes) are shown. The main characteristics of the experimentally derived masking patterns can be accounted for by the model. In particular, the shift of the maximum in the patterns with changing signal modulation agrees with the data. However, the model underestimates the level effect on masking observed in the data.

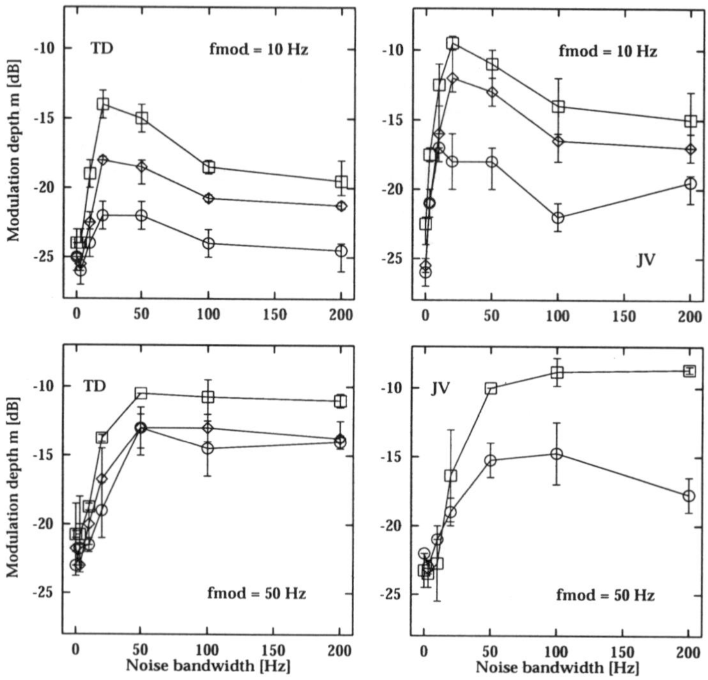

Figure 2. Modulation detection thresholds for 10 Hz (top) and 50 Hz (bottom) as a function of the bandwidth of the interfering noise at 1 kHz. ΔL = 0 dB (O), 20 dB (◇), 30 dB (□). Target-tone frequency: 4 kHz.

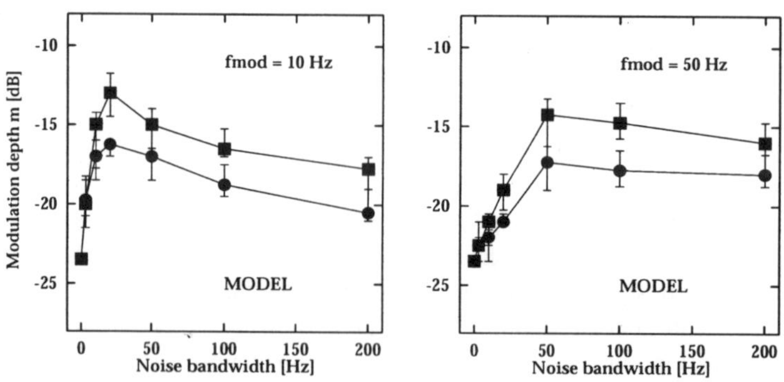

Figure 3. Predicted thresholds for the experimental conditions from Fig.2, obtained with the across-channel model from Fig. 1. ΔL = 0 dB (●), 30 dB (■).

4 Measurements and simulations of CMR

The stimuli used in the present study were similar to those in the study of Cohen [1]. The masker consisted of two 25-Hz wide noise bands, one centred at the signal frequency (on-frequency band) at 1 kHz, and one separated from the on-frequency band, centred at 4 kHz (flanking band). The subject's task was to detect a 1 kHz signal as a function of the masker level for two experimental conditions. In the reference condition, both noise bands were uncorrelated. In the CMR-condition, the two bands had correlated envelopes. The magnitude of CMR is usually defined as the difference in signal threshold for the two conditions. Figure 4 shows the experimental data for two subjects (left two panels). The CMR-effect amounts 7 to 10 dB independent of the masker level. The third panel of Fig. 4 shows model predictions on the basis of the across-channel modulation-filterbank model. The model does *not* lead to a release of masking in the correlated condition of CMR. Thresholds are the same in the reference and the correlated condition.

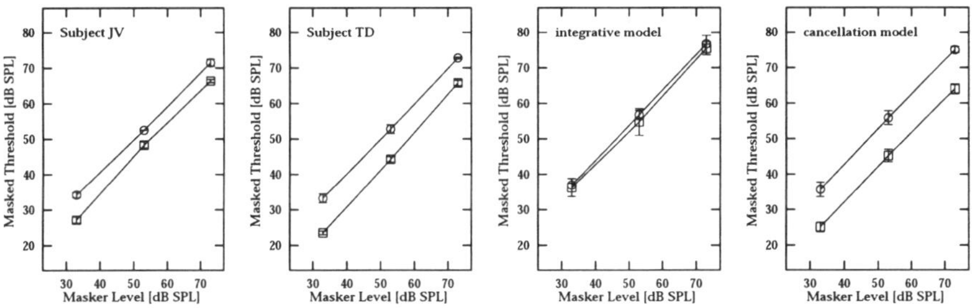

Figure 4. Masked thresholds for the reference condition (O) and the CMR-condition (□) as a function of the masker level. Measured data: Panels 1 and 2. Simulated data: Panels 3 and 4.

In addition, an alternative modelling approach was tested, adapted from the binaural cancellation model by Durlach [4]. Within this model, the temporal patterns at the output of the modulation analysis in the peripheral channels at 1 and 4 kHz are subtracted from one another. This leads to a signal enhancement in the correlated condition whereas it does not in the reference condition. Corresponding simulations are shown in the right panel of the figure. The cancellation model accounts for the release of masking in this experiment. Such an approach, however, would clearly fail in conditions of modulation detection.

5 Summary and conclusions

- The across-channel model suggested in the present study contains a modulation filterbank in which information about modulation is integrated across the whole frequency range prior to the decision stage. To ensure the possibility of predicting spectral masking, a low-frequency part of the modulation spectrum is processed separately within each peripheral channel.

- While the across-channel model preserves the capabilities of the original modulation-filterbank model of describing TMTFs and modulaion masking, it also accounts for the main effects of MDI. The model fails, however, in another condition of across-spectral processing: CMR.
- The proposed cancellation model accounts for CMR in conditions with comodulated waveforms at widely separated frequencies. However, such a model would clearly fail in modulation detection conditions.
- The simulations from the present study suggest that different coding strategies are probably used by the auditory system in the different experimental conditions. Further critical experiments and modeling efforts are needed to better understand the underlying mechanisms.

References

1. Cohen, M.F., Comodulation masking release over a three octave range, *J. Acoust. Soc. Am.* **90** (1991) pp. 1381-1384.
2. Dau T., Kollmeier B. and Kohlrausch A., Modeling auditory processing of amplitude modulation. I. Detection and masking with narrow-band carriers, *J. Acoust. Soc. Am.* **102** (1997a) pp. 2892-2905.
3. Dau T., Kollmeier B. and Kohlrausch A., Modeling auditory processing of amplitude modulation. II. Spectral and temporal integration, *J. Acoust. Soc. Am.* **102** (1997b) pp. 2906-2919.
4. Durlach N.I., Binaural signal detection: Equalization and cancellation theory. In Tobias J.V. (ed) Foundations of Modern Auditory Theory, Vol. 2. (1972) New York: Academic Press, pp. 369-462.
5. Fleischer, H., Modulationsschwellen von Schmalbandrauschen, Acustica **51**, 154-161.
6. Houtgast T., Frequency selectivity in amplitude-modulation detection, *J. Acoust. Soc. Am.* **85** (1989) pp. 1676-1680.
7. Langner G. and Schreiner, C., Periodicity coding in the inferior colliculus of the cat. I. Neuronal mechanism, *J. Neurophysiol.* **60** (1988), pp. 1799-1822.
8. Schulze H. and Langner G., Periodicity coding in the primary auditory cortex of the Mongolian gerbil (Meriones unguiculatus): two different coding strategies for pitch and rhythm?, *J. Comp. Physiol. A* **181** (1997), pp. 651-663.
9. Yost W. A., Sheft S., and Opie J., Modulation interference in detection and discrimination of amplitude modulation, *J. Acoust. Soc. Am.* **86** (1989) pp. 2138-2147.

AUDITORY REPRESENTATIONS OF OPTIMISED CHIRP SIGNALS

STEFAN UPPENKAMP, ROY D. PATTERSON
CNBH, Dept. of Physiology, University of Cambridge, U.K.,
E-mail: stefan.uppenkamp@mrc-cbu.cam.ac.uk

Optimised chirp signals for ABR, recently suggested by Dau *et al.* (1998) have been tested with the Auditory Image Model (AIM, Patterson *et al.*, 1995). AIM is a comprehensive model of the peripheral auditory system, including a stage of strobed temporal integration to produce a stabilised image of repeating neural activity patterns. The chirps are calculated to compensate for spatial dispersion along the cochlea in order to stimulate all auditory nerve fibres at the same moment in time. Outputs of the several stages of the model in response to the chirps illustrate the form that the internal representation of these signals might take. Both a linear gammatone filterbank, and a transmission line model of basilar membrane motion produce responses that are reasonably consistent with what would be expected from the chirp signals, i.e., a distinct alignment of cochlear excitation over time. At the higher stages of auditory processing, some of the intended alignment seems to get degraded. The auditory images of upward and downward sweeps show less asymmetry than would be expected from the excitation along the cochlea.

Introduction

Broad band stimulation of auditory brainstem responses (ABR) is usually initiated by clicks, i.e. very short rectangular pulses about 40 ms in length. While click stimuli actually sound like 'clicks', there is no doubt that some smearing of the temporal concentration takes place in the cochlea due to spatial dispersion along the basilar membrane. This might result in brainstem responses which are dominated by contributions from 'high frequency' places in the cochlea only. One approach to overcoming this problem with click stimulation of ABR was recently suggested by Dau *et al.* (1998) and Wegner *et al.* (1997). They used optimised chirp signals with increasing instantaneous frequency in which $d\omega/dt$ was calculated to compensate for spatial dispersion along the cochlear partition according to the one-dimensional, linear model of the cochlea by de Boer (1980). Dau *et al.* (1998) demonstrated that these chirp signals are useful stimuli for enhancing ABR wave V. Their use also enables band-limited excitation to achieve frequency specific responses. Nevertheless, these signals were calculated to compensate for mechanical properties along the cochlea, whereas ABR represents the summed activity of all nuclei from CN to IC in the brain stem. Thus, it is not clear whether the intended properties are fully preserved in the IC or higher centres.

1 Functional version of AIM

The functional version of AIM includes a linear gammatone filterbank for simulation of the basilar membrane motion (BMM), a stage of two-dimensional adaptive thresholding (Patterson and Holdsworth, 1996) to transform BMM in neural activity patterns (NAP), and finally a stage of strobed temporal integration to transform the NAP into a stabilised auditory image (SAI). The forward chirp was designed to compensate for spatial dispersion along the cochlea, resulting in an alignment of activity over time in BMM and NAP. As expected, there is no alignment in both these processing stages for the backward chirp. On listening to them, one finds that the perceptual difference is actually very small, and in AIM this reflects the additional stage of temporal integration in the auditory system that removes between-channel phase differences. This stage of phase alignment is probably involved in making a true click stimulus sound like a 'click', irrespective of spatial dispersion along the cochlea. In the stabilised auditory images created by strobed temporal integration of the NAPs of both chirp signals, most of the differences between forward and backward chirp are removed, suggesting no perceptual difference for these signals. A similar finding for monotic phase waves appears in Patterson (1987). However, the similarity of the images of both forward and backward chirps might still partially be caused by the processing specific to the functional version of AIM.

2 Physiological version of AIM

The gammatone filterbank is a simple, level-independent representation of basilar membrane motion (BMM) designed to explain psychoacoustical results from spectral masking. A more physiological representation of BMM is the level-dependent implementation of the one-dimensional transmission line model by Strube (1985), including accurate phase characteristics as well as the nonlinear behaviour of BM motion (Giguere and Woodland, 1994). The output of this model in response to chirp trains shows shorter impulse responses at higher levels due to the increasing filter bandwidth. The physiological version of AIM includes the transmission line simulation for BMM, an inner hair-cell model (Meddis, 1988) for generating NAPs, and strobed temporal integration to transform the periodic NAP into a stabilised auditory image. Figure 1 shows the NAP and SAI for a 60 dB SPL sequence of chirps at a repetition period of 12 ms, both for forward chirps (i.e, increasing instantaneous frequency, left) and backward chirps (decreasing instantaneous frequency, right). At low levels, forward and backward chirps result in only slightly different images,

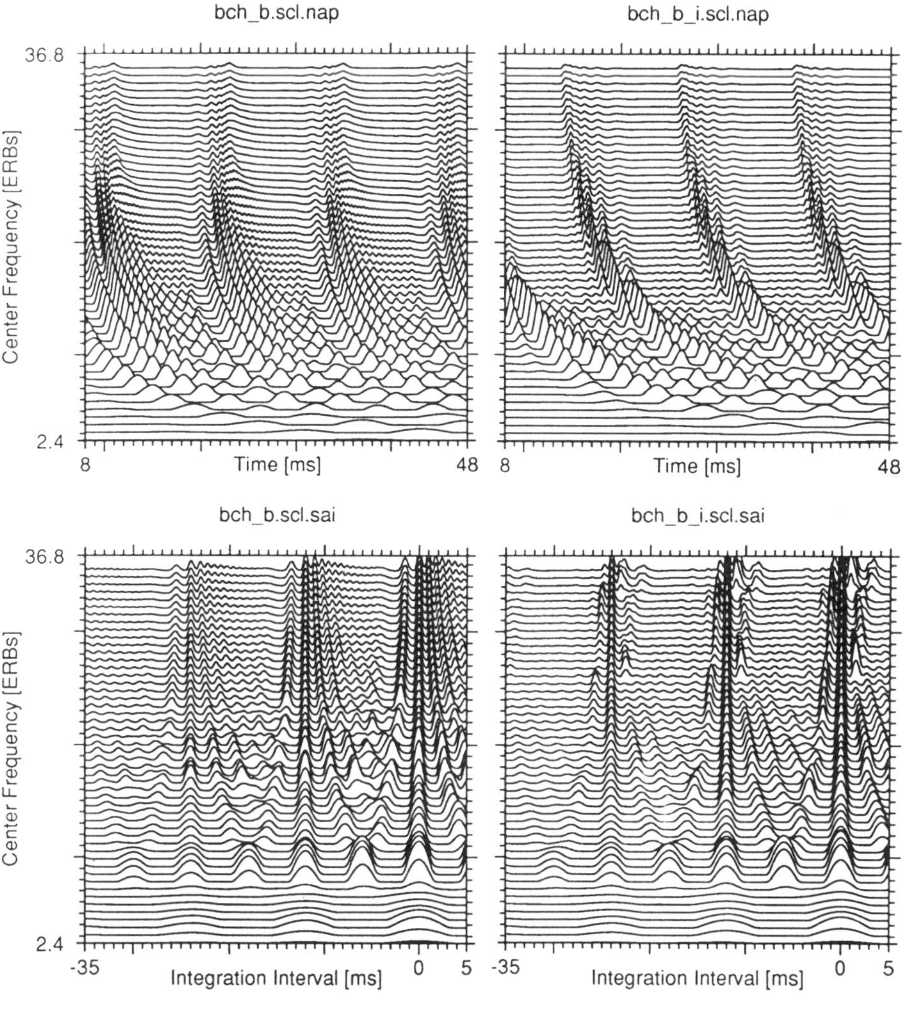

Figure 1: Neural activity patterns (top row) and auditory images (bottom row) for a forward (left column) and backward (right column) chirp signal with instantaneous frequency changing from 100 Hz to 10 kHz. Basilar membrane motion was simulated with a level-dependent transmission line model. The input level was set to 60 dB SPL.

chirps should be strongly dependent on presentation level. Informal listening confirms both that the strength of discrimination increases markedly with level and that it is mainly restricted to frequencies above about 2 kHz.

Acknowledgements

Supported by MRC. We thank Torsten Dau and Oliver Wegner for providing us with their routine to calculate the stimuli.

References

1. de Boer, E., Auditory physics. Physical principles in hearing theory I, Phys. Rep. **62** (1980) pp. 87-174.
2. Dau T., Wegner O., Mellert V., and Kollmeier B, Auditory brainstem responses (ABR) with optimized chirp signals compensating spatial basilar membrane dispersion, in *Abstracts of the 21st ARO midwinter research meeting* (1998), p. 33, abstract No. 130
3. Giguere C., Woodland P., A computational model of the auditory periphery for speech and hearing research. I. Ascending path, *J. Acoust. Soc. Am.* **95** (1994) pp. 331-342.
4. Meddis R., Simulation of auditory-neural transduction: further studies, *J. Acoust. Soc. Am.* **83** (1988) pp. 1056-1063.
5. Patterson R.D., A pulse ribbon model of monaural phase perception, *J. Acoust. Soc. Am.* **82** (1987) pp. 1560-1586.
6. Patterson R.D., Holdsworth J., A functional model of neural activity patterns and auditory images, in *Advances in Speech, Hearing and Language Processing*, Vol. 3, Part B, ed. W.A. Ainsworth, JAI Press, Greenwich, Connecticut, 1996, pp. 547-563
7. Patterson R.D., Allerhand M.H., and Giguere C., Time-domain modelling of peripheral auditory processing: A modular architecture and a software platform, *J. Acoust. Soc. Am.* **98** (1995) pp. 1890-1894.
8. Strube H.W., A computationally efficient basilar-membrane model, *Acustica* **58** (1985) pp. 207-214.
9. Wegner O., Dau T., and Uppenkamp S., Auditory brainstem responses with optimized chirp signals compensating BM dispersion, in *Contributions to Psychological Acoustics - 7th Oldenburg Symposium on Psychoacoustics*, ed. A. Schick, M. Klatte, BIS, Oldenburg, 1997, pp. 83-90

MODELING MASKING PATTERNS FOR SINUSOIDAL AND NARROWBAND NOISE MASKERS

R. P. DERLETH, T. DAU AND B. KOLLMEIER
AG Medizinische Physik, Graduiertenkolleg Psychoakustik
Universität Oldenburg, D-26111Oldenburg
E-mail: peter@medi.physik.uni-oldenburg.de

1 Introduction

In a recent paper by Moore et al. [2] masking patterns for sinusoids and narrowband noise were presented. Data were collected for all four masker-signal combinations that can be generated with these two classes of signals. The different experimental conditions were run using the same experimental procedures in each case with the same subjects. Apart from influences by distortion products, a strong influence of temporal fluctuations on the masking patterns for sinusoidal maskers was found for masker-signal frequency separations up to a few hundred Hz. The present study shows simulations based on the modulation-filterbank model by Dau et al. [1] which is capable to use envelope fluctuations, for example beats, as a detection cue. The model does not include any nonlinearity in its peripheral filtering stage. Thus, the simulations might enable to separate between temporal effects like beating and effects caused by distortion products.

2 Experiments and Simulations

A tone or a 80-Hz wide running Gaussian noise was used as the signal or as the masker. The stimuli were gated synchronously (10 ms raised-cosine rise/fall times) and had a steady-state duration of 200 ms. Simulations were performed for masker levels of 45, 65 and 85 dB. All four masker-signal combinations were tested (**nn:** noise-signal, noise-masker; **nt:** noise-signal, tone-masker; **tn:** tone-signal, noise-masker; **tt:** tone-signal, tone-masker). In the **tt**-condition, a 90 degree phase-shift between signal and masker was chosen, while random phase-shifts were used in all other conditions. Simulations were performed with the modulation-filterbank model [1]. All peripheral channels ranging from half an octave below to one octave above the signal frequency were examined. The modulation analysis was performed separately and independent in each peripheral channel. The center frequency of the highest available modulation filter was restricted to half of the center frequency of the peripheral channel but never exceeded 1 kHz.

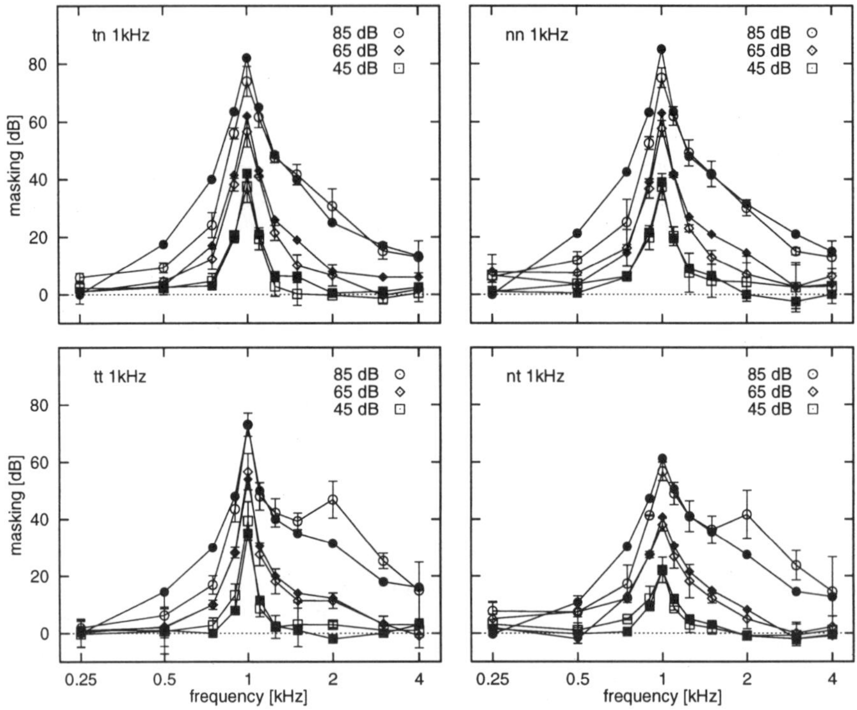

Figure 1: Measured (open symbols) and predicted (filled symbols) masking patterns for all four signal-masker conditions for masker levels of 85 (circles), 65 (diamonds) and 45 dB (squares). The mean experimental data for three subjects are replotted from Moore et al. [2]. Simulations were performed with the modulation-filterbank model using the same 3AFC 1-up 3-down procedure as in the measurements. The masker (center) frequency was 1 kHz and the signal (center) frequency was 0.25, 0.5, 0.75, 0.9, 1.0, 1.1, 1.25, 1.5, 2.0, 3.0 and 4.0 kHz, respectively.

Figure 1 shows mean masking data for three subjects (open symbols, replotted from [2]). The two upper panels show conditions with noise as the masker: The masking patterns with the tonal signal (**tn**) and the noise-signal (**nn**) are very similar in shape and absolute values. This is also true for the patterns with the tone-masker, shown in the two lower panels, except for the on-frequency **tt**-condition (left). In this particular condition, masking is as high as in the conditions with noise-maskers whereas it is reduced by 18 dB in the **nt**-condition (right). The masking patterns obtained with the tonal masker show less masking for signal-masker frequency separations up to about 250 Hz and more masking for frequency separations ≥ 1 kHz when compared to the patterns obtained with the noise masker. In [2] it was suggested that the reduction in masking for small frequency separations (< 250 Hz) is at least partly caused by the detection of temporal cues like beating. The filled symbols in Fig. 1 represent corresponding model predictions obtained with the modulation-filterbank model. For low and medium masker levels, the simulations are in good agreement with the experimental data. In particular, the sharp tuning in the **tt**-condition and the masking release in the on-frequency **nt**-condition are quantitatively accounted for by the model. Some discrepancies between predictions and data occur for the highest masker level. The predicted masking patterns are too shallow on their low-frequency side, independent of the experimental condition. The use of the symmetrical Gammatone-filters [3] leads to an over-estimation of masking of up to 20 dB in this region. A further mismatch can be observed for the tonal masker on the high-frequency side of the masking pattern. For signal frequencies of 2 and 3 kHz, the model underestimates masking by 15 and 9 dB, respectively.

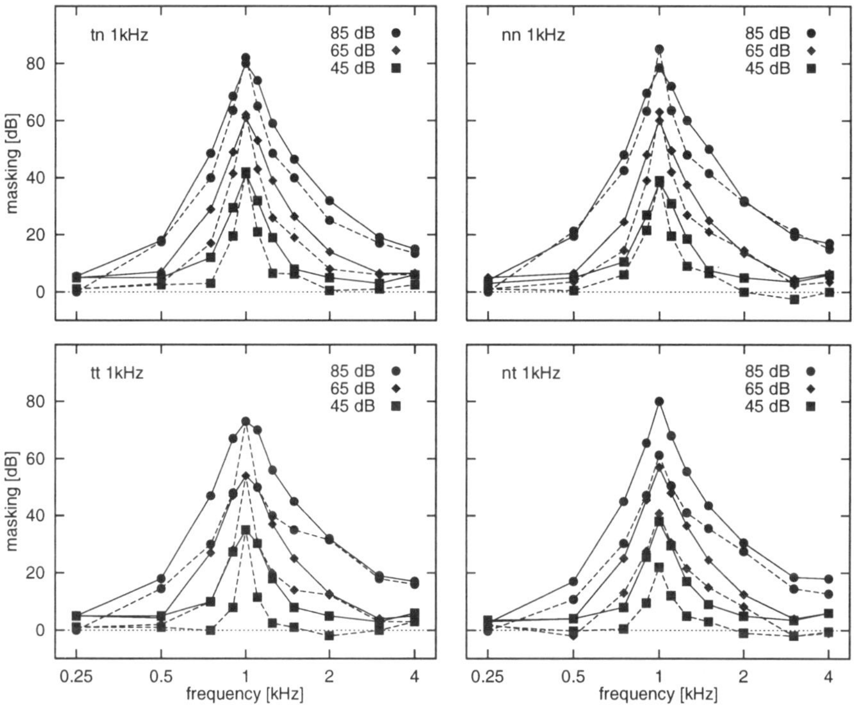

Figure 2: Predicted masking (same experiment as in Fig. 1) obtained with two model approaches. The dashed curves indicate masking obtained with the modulation-filterbank model. The solid curves indicate masking obtained with a model version that basically works as an energy detector.

Figure 2 illustrates how effectively temporal cues are used in the modulation-filterbank model. The solid curves represent simulated thresholds based on a model that includes a lowpass filter with a cut-off frequency of 2.5 Hz instead of the modulation filterbank. This strategy is close to that of an energy-detection model. Such a model predicts nearly identical thresholds across the different signal-masker conditions which apparently is in contrast to the data. The modulation-filterbank model (dashed curves, replotted from Fig. 1) results in up to 20 dB lower thresholds for frequency separations ≤ 250 Hz between signal and masker, depending on the specific condition. Thus, higher-frequency modulations are used within the model for the detection of the signal, especially in the conditions **tt** and **nt**. These temporal cues are less useful in the conditions **tn** and **nn**.

3 Summary and Conclusions

As pointed out in [2], the shape of the masking patterns for signal frequencies above and below the masker frequency is mainly determined by the characteristics of the masker. Masking patterns obtained with a tone masker show sharper tuning for small signal-masker frequency separations and larger masking for large signal-masker frequency separations than masking patterns obtained with a noise-masker. When signal and masker frequency are equal, thresholds are equal for all masker-signal combinations except for the **nt**-condition, where the masking is reduced by about 18 dB. The modulation-filterbank model accounts for the sharper tuning and also for the reduction in masking. In contrast, an energy detection model (Fig. 2) would result in identical masking patterns for all four signal-masker conditions. For low and medium masker levels, the processing of higher-frequency envelope fluctuations within the modulation-filterbank model accounts very well for the differences in the masking patterns. For the highest masker level, an asymmetrical filter with a steeper high-frequency slope must be assumed to account for the data. The reason for the differences on the high-frequency side of the measured patterns obtained with tone and noise maskers is still somewhat unclear. The occurance of harmonic distortion produced by the peripheral nonlinearity most likely plays a role in these conditions.

Acknowledgments

This work was supported by the DFG.

References

1. Dau, T., Kollmeier, B. and Kohlrausch, A. (**1997**). „Modeling auditory processing of amplitude modulation: I. Detection and masking with narrowband carriers," J. Acoust. Soc. Am. **102**, 2892 - 2905.
2. Moore, B. C. J., Alcántara, J. I., Dau, T. (**1998**). „Masking patterns for sinusoidal and narrowband noise maskers," J. Acoust. Soc. Am. **104**, 1023 - 1038.
3. Patterson, R.D., Nimmo-Smith, I., Holdsworth, J. and Rice, P. (**1987**). „An efficient auditory filterbank based on the gammatone function," Paper presented at a meeting of the IOC Speech Group on Auditory Modelling at RSRE, December 14 - 15.

IV. Speech Perception and Applications of Psychoacoustics

SILICON COCHLEA: A DIGITAL VLSI IMPLEMENTATION OF A PSYCHOACOUSTICALLY AND PHYSIOLOGICALLY MOTIVATED SPEECH PREPROCESSOR

A. SCHWARZ AND B. MERTSCHING

IMA Group, Computer Science Department, University of Hamburg Germany
E-mail: schwarz@informatik.uni-hamburg.de

M. BRUCKE AND W. NEBEL

VLSI Group, Computer Science Department, University of Oldenburg Germany
E-mail: Matthias.Brucke@Informatik.Uni-Oldenburg.DE

M. HANSEN AND B. KOLLMEIER

Medical Physics Group, Physics Department, University of Oldenburg Germany
E-mail: biko@medi.physik.uni-oldenburg.de

A digital VLSI implementation of an algorithm modeling the "effective" signal processing of the human auditory system is presented. The model consists of several stages psychoacoustically and physiologically motivated by the signal processing in the human ear and was successfully applied to various speech processing applications. The processing scheme was partitioned for implementation in a set of three chips. It is demonstrated how an application of the model has been used to determine the necessary wordlengths for a transfer of the floating point algorithm into a version suitable for a hardware implementation. The developed synthesizable VHDL-descriptions of this fixed point version are now being evaluated in real time on FPGAs and will be manufactured as ASICs in a later version.

1 1 Introduction

The binaural perception model introduced by [1], describes the "effective" signal processing in the human auditory system and provides an appropriate internal representation of acoustic signals. It's capabilities could be successfully demonstrated as a preprocessing algorithm for speech recognition [2], objective speech quality measurement [3] and digital hearing aids. The algorithm includes

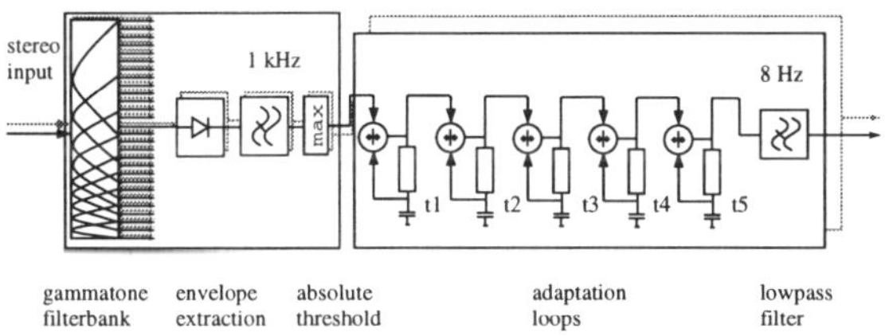

Figure 1. Processing scheme of the binaural perception model introduced by [1].

a gammatone filter bank (30 bandpass filters equidistant distributed on the ERB scale from 73 to 6700 Hz) to model spectral properties of the human ear like spectral masking and frequency-dependent bandwidth of auditory filters. A stage modeling inner hair cell behavior (envelope extraction) is followed by five adaptation loops (with time constants between 5 and 500 ms) to consider dynamical effects as nonlinear adaptive compression and temporal masking (see Fig. 1).

2 2Hardware Design Specifications

Due to the complexity the design was partitioned into three chips (Fig. 2). Besides the serial data interfaces chip 1 contains the 30-channel binaural filter bank, the envelope extraction and a module computing phase differences and magnitude quotients in each stereo output of the channels. A single bandpass filter is multiplexed through all 30 channels and twice again for the left and right side signal. Including a six-stage pipelined multiplier and one adder/subtractor this kernel is realized by a quad cascade of a single stage complex IIR filter. This saves chip area but requires a 50 MHz system clock to operate with a 16276 Hz sampling frequency. RAM units save temporary data and filter constants are read from a ROM.

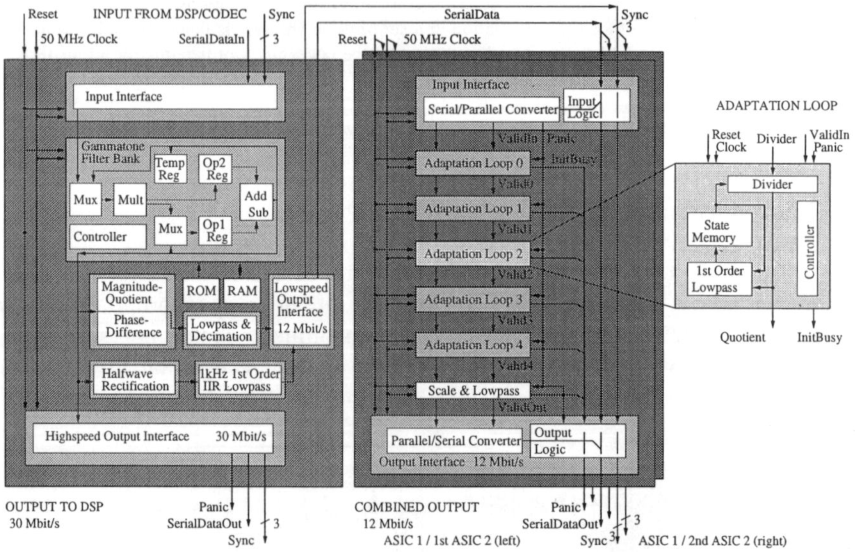

Figure 2. Structure and wiring of the internal components of the three chips. For further processing three system outputs are available. A high speed interface (30 MBit/s) provides real and imaginary parts of the right and left side data for all filter bank channels (chip 1). The adaptive compressed data of the left (1st chip 2) and the right side signal (2nd chip 2) and phase- and amplitude information of the filter bank outputs are combined within a second data stream (12 MBit/s). The 4-wire serial interfaces of the chipset (16 bit data words) support a direct interface to the serial ports of most DSP-devices.

Each of the five adaptation loops contains a divider whose quotient is fed back by a 1st order IIR lowpass providing the divisor. This feedback and the necessary signal dynamic forces large wordlengths. Therefore the dividers are the most area expensive components and need more than 50 % of chip 2 area. Downsampling (4-time) and data serialization in chip 1 allow a single chain of five loops monaurally implemented in two chips (two of chip 2). Each loop contains a ring memory saving the divisors (lowpass outputs) for the 30 serial processed frequency channels.

3 3Algorithm Transformation from Floating Point To Fixed Point

To gain an optimal fixed point implementation different methods are applied to the linear filter bank and the nonlinear adaptation loops respectively. The wordlength of the kernel in the gammatone filter bank could be derived by a signal/noise criterion. Allowing an error of 0.001 a minimal wordlength of 24 bits is necessary for the lowest filter bank channel (Fig. 3). This quantization can be used for all channels.

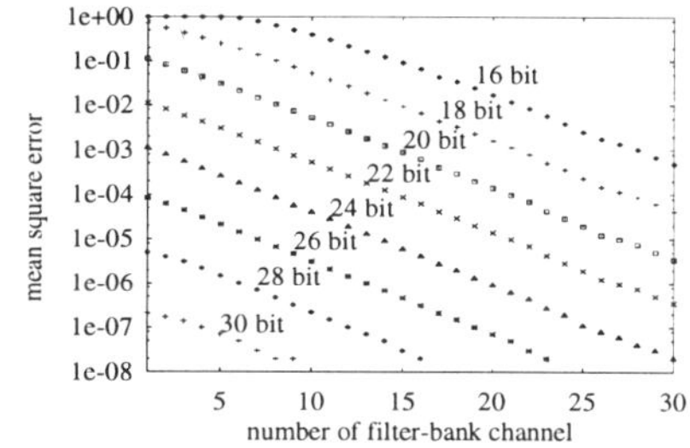

Figure 3. Mean square error (relative error, i.e. noise-to-signal ratio) of fixed point implementations with varying wordlengths in comparison with a floating point implementation by analyzing the impulse response of the gammatone filters. Note that due to the increased analysis bandwidth the error for a given wordlength decreases with increasing center frequency / channel number.

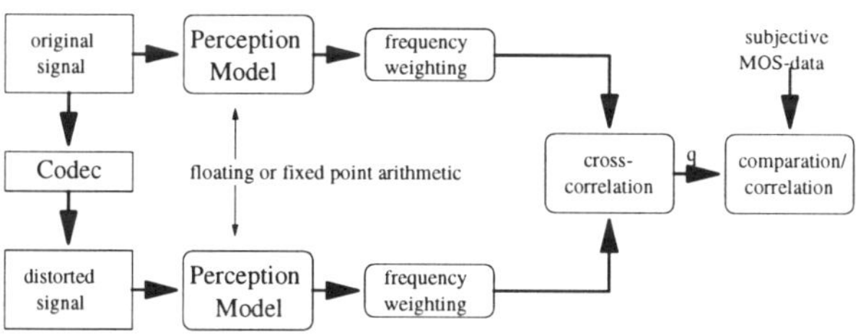

Figure 4. An objective speech quality measure q can be determined with the perception model for signals processed by e.g. a telecommunication codec [3]. To prove this the measure q is correlated with the subjective Mean Opinion Score (MOS) of the test signals. Because this testbench is very sensitive to limited number precision and signal dynamic in the perception model it can be used to evaluate modifications due to fixed point quantization and arithmetic. An optimal quantization of the nonlinear adaptation loops (small wordlengths i.e. small chip area vs. reliable signal processing) was found by empirical wordlength variation. The results were verified varying the input signal levels by up to 50 dB.

The determination of an optimal quantization in the adaptation loops is quite more difficult because of their strong nonlinearity. Figure 4. shows how the speech quality measurement is used as a testbench. Figure 5. demonstrates the results indicating that the dividers in the loops need at least 34 bits fixed point precision.

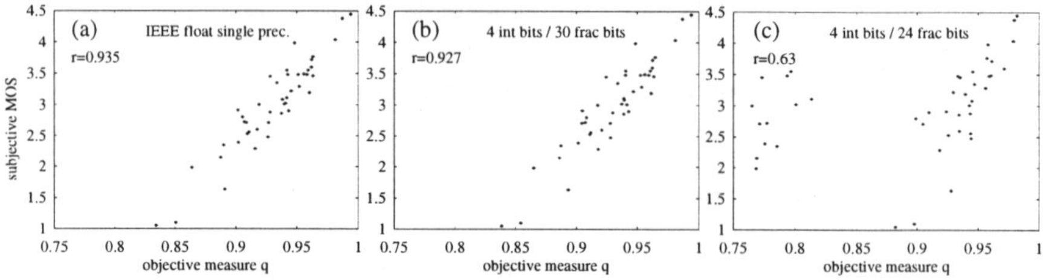

Figure 5. Results for a complete objective speech quality measurement with the ETSI-halfrate selection test data using an IEEE 32 bit floating point implementation (a), a sufficient fixed point implementation (b) and an insufficient quantization (c). Diagram (a) shows that the model works correctly and the objective speech quality measure is well correlated with the subjective MOS (indicated by the linear correlation coefficient r). Using a resolution of 4 integer and 30 fraction bits nearly no losses due to fixed point quantization errors can be found in diagram (b). But enormous losses in the data correlation appear in (c) after reducing the wordlength to 4 integer and 24 fraction bits.

4 4Conclusion

In this paper we presented our work on the digital VLSI implementation of a speech perception model. The hardware design of the algorithm was derived from a floating point C/C++ notation. An application of the model (speech quality measurement) is used to determine optimized wordlengths for a fixed point arithmetic in a dedicated hardware. The development of the perception model as a FPGA/ASIC for a target system, e.g. a PC-card, provides efficient co-processing power and allows real time implementations of complex auditory-based speech processing algorithms.

References

1. Dau, T., Püschel, D. and Kohlrausch, A., A quantitative model of the „effective“ signal processing in the auditory system I., Journal of the Acoustical Society of Amerika (JASA) 99 (6), 1996, pp. 3633-3631.
2. Tchorz, T., Wesselkamp, M. and Kollmeier, B., Gehörgerechte Merkmalsextraktion zur robusten Spracherkennung in Störgeräuschen, Fortschritte der Akustik–DAGA 96, pp. 532-533, DEGA, Oldenburg, Germany.
3. Hansen M. and Kollmeier B., Using a quantitative psychoacoustical signal representation for objective speech quality measurement, Proc. ICASSP-97, Intl. Conf. on Acoustics, Speech and Signal Proc., p. 1387, Munich, Germany.

MODELING THE SPEECH-RECEPTION THRESHOLD FOR AMPLITUDE-COMPRESSED SPEECH

JOOST M. FESTEN AND JANETTE N. VAN DIJKHUIZEN *

Dept. of Otolaryngology, University Hospital VU,
P.O. Box 7057, 1007 MB Amsterdam, The Netherlands
E-mail: jm.festen@azvu.nl

For normal-hearing listeners reduction of modulations in speech transmission yields reduced intelligibility. Modulation transfer is even a good predictor for intelligibility in many listening conditions. For hearing-impaired listeners it is often argued that the limited dynamic range of the ear and the associated loudness recruitment needs to be compensated for by a compressive mapping of the level variations within speech (syllabic compression). However, because the results of numerous experiments on syllabic compression are generally negative, it is interesting to see whether the results for hearing-impaired listeners can be accounted for by a reduced transfer of modulations. Intelligibility scores for 16 normal-hearing listeners and 16 hearing-impaired listeners were obtained for speech after syllabic compression with as parameters the number of processing channels and the compression ratio. Simple modulation transfer accounts only partly for the variability in intelligibility scores. A much better result is obtained with a phase-locked transfer of modulations, using only output modulations in quarter octaves that are in-phase with modulations of the input signal. Both for normal-hearing listeners and for hearing-impaired listeners the data of various compression conditions can be equated in terms of the phase-locked modulation transfer.

1 Introduction

Sensorineural hearing impairment is often accompanied by an abnormally steep increase of loudness with stimulus level, called recruitment. To restore normal loudness perception and to fit speech in the limited dynamic range of the impaired ear many hearing aids apply some form of amplitude compression. Modern hearing aids often use multi-band compression, because both the needs of the ear and the time-varying demands of the signal are a function of frequency. However, the effects of fast-acting, or syllabic, compression on intelligibility as measured in various studies [2] are equivocal, ranging from negative effects to no effect. These confusing results may be a consequence of the strong dependence of compression effectiveness on its time constants and on the bandwidth of the compression channels.

The reduction of level variations with syllabic compression was studied a.o. by Stone and Moore [9]. For sinusoidal modulations they showed that the effective modulation reduction can be considerably less than the target reduction, particularly for modulation rates that are high with respect to the applied time constants. For real speech they showed that, with wideband compression, the overall shape of the

* Present address: ENT Department, Leiden University Hospital, The Netherlands.

wideband envelope was virtually unaffected; only the level variations were reduced. Verschuure *et al.* [10] measured modulation transfer and effective compression for various compression systems. They found that compression in commercial hearing aids usually affects only level variations among words, but is too slow for articulatory level variations among phonemes. Using real speech and wideband compression, they showed that, although wideband modulations were strongly reduced, virtually no reduction was obtained in individual octave bands.

Speech intelligibility after multichannel syllabic compression was studied by Hohmann and Kollmeier [6] for normal-hearing listeners. They used monosyllabic words and applied moderate compression ratios up to three in 23 critical-band-wide channels with 5-ms attack time and 50-ms release time. For low signal-to-noise ratio the results showed a gradual decrease intelligibility with increasing compression ratio. However, the reduction of intelligibility was less than the degradation that was predicted from calculations based on modulation transfer [7].

When compression is fast enough to reduce articulatory level variations in speech, its effect is not limited to the time domain. A reduction of temporal modulations in a number of independent frequency channels leads unavoidably to reduced spectral modulations [4]. Both temporal and spectral modulations of speech are important for intelligibility. Hearing-impaired listeners often have a limited dynamic range and are therefore possibly hypersensitive to temporal modulations, but at the same time their frequency resolution is reduced. While compression may compensate the recruitment, it does not compensate poor frequency selectivity. On the contrary, multichannel compression reduces spectral modulations (for instance of vowels) which determine the spectral variability in speech [4,5,8]. In this study we measured the effect of real syllabic compression on speech intelligibility in quiet for normal-hearing and hearing-impaired listeners. For a number of conditions the results of this experiment are compared with the transfer of modulations in our processing.

2 Speech intelligibility

The intelligibility of amplitude-compressed speech in quiet was measured for 16 normal-hearing listeners and 16 listeners with sensorineural hearing loss. The hearing-impaired listeners had flat to moderately sloping audiograms with an average hearing loss of 55 dB and a residual dynamic range of about 45 dB. Short sentences (250 - 4000 Hz) were amplitude compressed with compression ratio of 1, 2, 4, and ∞ in a single wideband channel or after subdivision in 2, 4, 8, or 16 frequency channels. The compression time constants were very short with an attack time of a few milliseconds and a release time of 20 ms (*syllabic compression*). The compression onset was at a level 30 dB below the average speech level (*full-range compression*). The speech was presented to the normal-hearing listeners at 75 dBA and to the hearing-impaired listeners halfway their individual dynamic range. The average intelligibility scores for both groups are presented below. Apart from a ceiling effect for the normal-hearing group, these results show a gradual decline of intelligibility with increasing compression ratio and with an increasing number of channels.

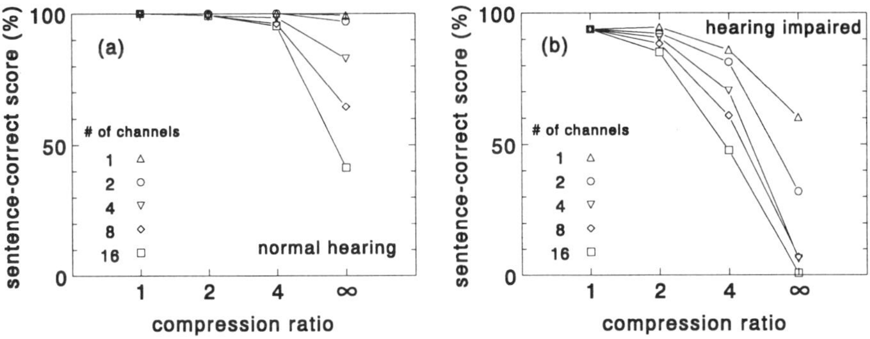

Figure 1. Intelligibility scores for sentences with syllabic compression as a function of compression ratio. The conditions range from single-channel compression (4 octaves) to compression in 16 quarter-octave channels. Panel (a) for normal-hearing listeners and Panel (b) for hearing-impaired listeners.

3 Modulation transfer

For transmission channels with noise and reverberation the quality of speech transmission can be very well expressed by the transfer of modulations [7]. Because noise and reverberation mainly reduce modulations, the integrity of the modulations after transmission can be estimated sufficiently from an evaluation of the attenuation of modulations. With amplitude compression the situation is different: modulations may not only be attenuated, but also new modulations may be added. An example is given in Fig. 2. Panel (a) shows the envelope of a sentence in an octave frequency band before and after compression. In the compressed envelope the modulations are a reduced version of the input modulations. Panel (b) shows envelopes from the same octave band, only now the compressor was fed with the wideband signal. Two effects are apparent: first, the overall modulation reduction in panel (b) is much less than in panel (a), and second, the shape of the envelope has considerably changed. These new modulations are introduced because the compressor control signal contains modulations different from those in the observed frequency band.

Modulations may be important for intelligibility, but it seems reasonable to regard modulations without a relation to the modulations in the input signal as spurious. For speech intelligibility it is assumed that only those modulations contribute that are in phase with input modulations. Therefore, when evaluating syllabic compression in terms of modulation transfer we focus on phase-locked modulations. In the calculations presented below modulations are taken from quarter-octave frequency bands. The phase-locked modulation transfer was calculated for a number of compression conditions from the intelligibility experiment.

The procedure for calculating the phase-locked modulation transfer was a quarter-octave version of the calculations described by Drullman *et al.* [3]. The calculations were performed on 32 concatenated sentences with amplitude compression. The original and compressed signals were split up in 16 quarter octaves and envelopes were determined.

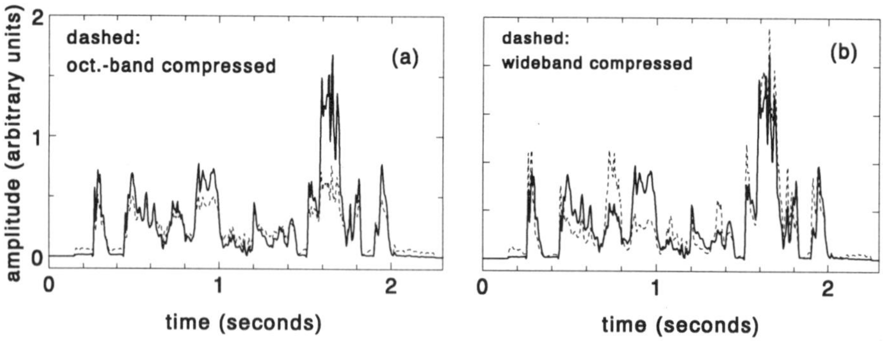

Figure 2. Envelope of a single sentence in an octave band from 2-4 kHz. Panel (a) shows the original envelope (fully drawn) and the envelope after octave-band compression (dashed). Panel (b) shows the original envelope again together with the envelope after wideband compression (dashed).

The phase-locked modulation transfer function is defined as the cross-spectral density of these envelopes relative to the input-envelope auto-spectral density. A single transfer index is obtained by unweighted averaging over the relevant modulation frequencies from 0.5 to 20 Hz and a weighting over frequency bands according to the ANSI Speech Intelligibility Index [1]. In Fig. 3 the intelligibility scores for wideband, octave-band, and ¼-oct.-band compression from the above experiments are plotted as a function of phase-locked modulation transfer. In this figure the different curves from Fig. 1 are, for each group of listeners, reduced to a single curve relating speech intelligibility and temporal modulation transfer.

For both listener groups Fig. 3 shows a gradual decrease of intelligibility with decreasing modulation transfer. The only difference between the two groups of listeners is that, for the same score, the hearing-impaired group needs stronger modulations then the normally hearing group. If positive effects of syllabic amplitude compression were to be expected, it would be for listeners with a narrow dynamic range. Therefore, from the 16 hearing-impaired listeners in Fig. 3 the data of four listeners with the narrowest dynamic range and four with the widest range are plotted separately. As the figure shows, the differences with the average data are extremely small.

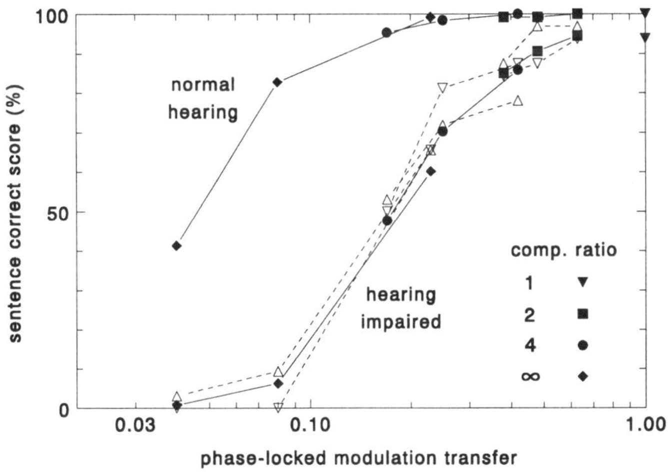

Figure 3. Intelligibility scores as a function of phase-locked modulation transfer for sentences processed with syllabic compression. Different compression ratios are indicated with different symbols. For each compression ratio calculations were performed for the single-band, four-band, and 16-band conditions yielding gradually poorer modulation transfer. Open triangles show average results for subgroups of the hearing-impaired listeners; up-pointing triangles for four listeners with a dynamic range wider than 55dB, down-pointing triangles for four listeners with a dynamic range narrower than 40 dB.

4 Discussion

With syllabic compression in a number of separate frequency channels not only temporal modulations, but also spectral modulations are reduced. This combined reduction of modulations has strong effects on speech intelligibility. Speech intelligibility after syllabic compression, can be very well accounted for by the associated temporal modulation reduction. The good fit already obtained with solely temporal modulation reduction is a consequence of the high correlation between spectral and temporal modulation reduction [4]. For moderate syllabic compression (compression ratio = 2) in only a few channels, the average reduction of intelligibility is very small and some listeners may even profit. However, from these data it can not be determined whether this profit is accidental or real.

References

1. ANSI S3.79-199X, Am. Nat. Standard, Methods for the calculation of the Speech Intelligibility Index, Draft V3.1 (1993).
2. Dillon H., Compression? Yes, but for low or high frequencies, for low or high intensities, and with what response times?, *Ear & Hearing* **17**, (1996) pp. 287-307.
3. Drullman R., Festen J.M. and Plomp R., Effect of reducing slow temporal modulations on speech reception, *J. Acoust. Soc. Am.* **95**, (1994) pp. 2670-2680.
4. Drullman R., Festen J.M. and Houtgast T., Effect of temporal modulation reduction on spectral contrasts in speech, *J. Acoust. Soc. Am.* **99**, (1996) pp. 2358-2364.
5. Festen J.M., Temporal resolution and the importance of temporal envelope cues for speech perception, In *Psychoacoustics Speech and Hearing Aids,* ed. by B. Kollmeier (World Scientific, Singapore, 1996) pp. 95-102.
6. Hohmann V. and Kollmeier B., The effect of multichannel dynamic compression on speech intelligibility, *J. Acoust. Soc. Am.* **97**, (1995) pp. 1191-1195.
7. Houtgast T. and Steeneken H.J.M., A review of the MTF concept in room acoustics and its use for estimating speech intelligibility in auditoria, *J. Acoust. Soc. Am.* **77**, (1985) pp.1069-1077.
8. Plomp R., Noise, amplification, and compression: considerations of three main issues in hearing aid design, *Ear & Hearing* **15**, (1994) pp. 2-12
9. Stone M.A. and Moore B.C.J., Syllabic compression: effective compression ratios for signals modulated at different rates, *Brit. J. of Audiol.* **26**, (1992) pp. 351-361.
10. Verschuure J., Maas A.J.J., Strikvoort E., de Jong R.M., Goedegebure A. and Dreschler W.A., Compression and its effect on the speech signal, *Ear & Hearing* **17**, (1996) pp. 162-175.

A NEURAL NETWORK FOR SOUND SOURCE SEPARATION

JÖRN ANEMÜLLER AND TINO GRAMSS (†)
Graduate School in Psychoacoustics and Department of Physics
Carl von Ossietzky University, Oldenburg, Germany
e-mail: ane@uni-oldenburg.de

1 Introduction

The brain possesses the remarkable capability to filter incoming signals of multiple speakers in such a way that the subject's attention can be focused on a single sound source, the other sources being suppressed.

Much effort has been spent in order to mimic this behaviour by machines or clever algorithms which try to reconstruct the separated sources. A key issue in such attempts is the question 'Which is the quantity to optimise', i.e., which function tells us whether the original sources have been separated. An appealing answer to this question might be given by the concept of Independent Component Analysis. Here, the mutual statistical independence of the different source signals is exploited. I.e., one of the source signals does not give us information about any of the other sources.

However, when different mixtures of the source signals are detected, e.g., at the two ears of a human subject, the mixtures are strongly correlated, i.e., *not* independent of each other. Therefore, in source separation one attempts to filter recorded mixtures in such a way that mutually independent signals are obtained which resemble the original source signals.

Since learning rules for this task exist which employ simple, 'neuronal' operations only, it is conceivable that the brain exploits the principle of mutual independence of the sources in order to separate them.

2 Source Separation in the time-domain

In this section, we sketch a simple algorithm for separating an instantaneous mixture of independent signals. For a review refer to, e.g., [1] .

Imagine N original source signals, $\mathbf{s}(t) = (s_1(t), \dots, s_N(t))$, which are emitted from, say, different speakers. These source signals are superimposed by the invertible mixing matrix $\mathbf{A}$, resulting in N recorded mixtures

$$\mathbf{m}(t) = (m_1(t), \dots, m_N(t)), \qquad \mathbf{m}(t) = \mathbf{A}\mathbf{s}(t). \tag{1}$$

The goal is to find a weight-matrix $\mathbf{W}$ such that the reconstructed signals

$$\mathbf{x}(t) = (x_1(t), \ldots, x_N(t)), \qquad \mathbf{x}(t) = \mathbf{W}\mathbf{m}(t)$$

are mutually statistically independent. It can be shown that $\mathbf{x}(t)$ resembles $\mathbf{s}(t)$ up to a constant permutation and rescaling of the sources.

A neural learning rule finds $\mathbf{W}$ by first computing a non-linear function of the reconstructed signals:

$$u_i(t) = \tanh(x_i(t)), \qquad i = 1, \ldots, N. \tag{2}$$

Afterwards, the weight-matrix is changed by an increment

$$\Delta\mathbf{W}^{(t)} \propto \mathbf{W}^{(t)} - \mathbf{u}(t)\mathbf{x}^T(t)\mathbf{W}^{(t)}, \qquad \mathbf{W}^{(t+1)} = \mathbf{W}^{(t)} + \Delta\mathbf{W}^{(t)}. \tag{3}$$

This learning scheme is iterated at each time-step t.

3 Sound Source Separation in the frequency-domain

In the case of acoustic superposition of sound sources we are dealing with echoes and reflections. Hence, the multiplication in eq. (1) has to be replaced by the convolution of the source signals with the room's impulse responses. However, in the frequency domain, we rediscover multiple versions of the instantaneous source separation problem, one for each frequency band $f = f_1, \ldots, f_M$:

$$\hat{\mathbf{m}}(f,T) = \hat{\mathbf{A}}(f)\hat{\mathbf{s}}(f,T).$$

By $\hat{\mathbf{m}}(f,T)$ and $\hat{\mathbf{s}}(f,T)$ we denote short-time spectra obtained from $\mathbf{m}(t)$ and $\mathbf{s}(t)$, respectively, at consecutive times $T = T_1, T_2, \ldots$. Hence, the goal is to solve M source separation problems of the form of eq. (1), resulting in weights-matrices $\hat{\mathbf{W}}(f_1), \ldots, \hat{\mathbf{W}}(f_M)$.

Since the spectral components $\hat{\mathbf{m}}(f,T)$ and $\hat{\mathbf{s}}(f,T)$ are complex valued, the learning algorithm (eqs. 2, 3) has to be modified. From a maximum likelihood approach [2] we derive the learning rule

$$\hat{u}_i(f,T) = \frac{\hat{x}_i(f,T)}{||\hat{x}_i(f,T)||}\tanh(||\hat{x}_i(f,T)||), \qquad i = 1, \ldots, N,$$

$$\Delta\hat{\mathbf{W}}(f) \propto \hat{\mathbf{W}}(f) - \hat{\mathbf{u}}(f,T)\hat{\mathbf{x}}^H(f,T)\hat{\mathbf{W}}(f), \qquad f = f_1, \ldots, f_M.$$

The operator $\mathbf{x}^H$ denotes the complex conjugate transpose of vector $\mathbf{x}$, and $||x||$ denotes the norm of the complex number x.

Without additional precautions, reconstruction of the source signals in the time-domain is hindered by the inherent permutation invariance of source separation algorithms, i.e., we can't tell which network output resembles which source. E.g., at frequency f_1 the reconstructed signal $\hat{x}_1(f_1, T)$ might belong to source s_1 while

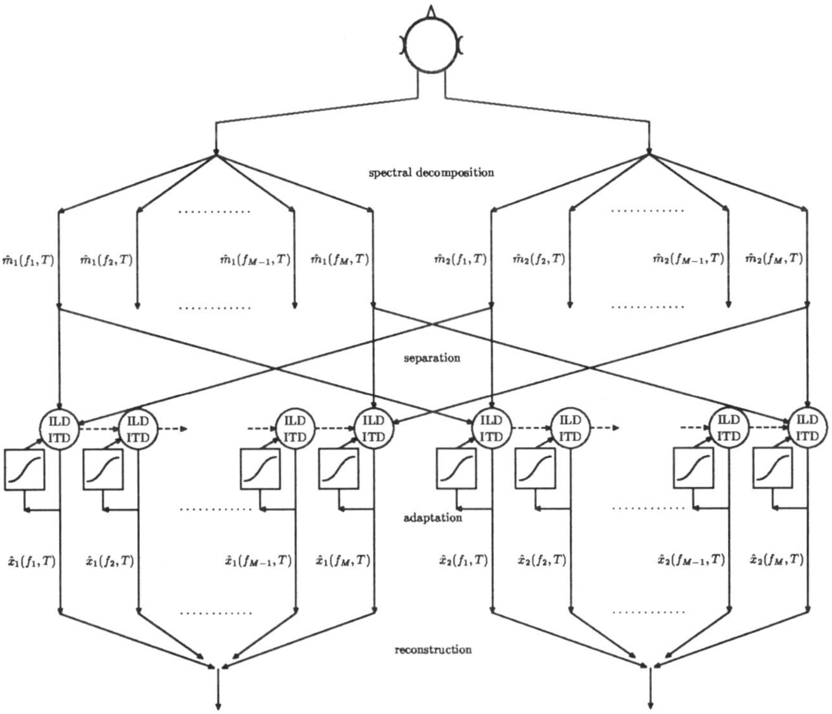

Figure 1: The processing stages of our source separation algorithm.

at frequency f_2 the component $\hat{x}_1(f_2, T)$ belongs to the opposite source, s_2. We solve this problem for the case where the mixing process introduces time- and level-differences only, which is fulfilled in an ideal anechoic chamber. In this situation, the coefficients $\hat{A}_{ij}(f)$ of the mixing matrices at different frequencies can be derived from the propagation-times τ_{ij} and -levels A_{ij} from source j to microphone i via the formula

$$\hat{A}_{ij}(f) = A_{ij} e^{-2\pi i f \tau_{ij}}.$$

The algorithm starts adaptation at low frequencies and propagates time- and level-differences towards higher frequencies, see also fig. 1. Hence, a rough estimate of the sources' parameters is obtained at low frequencies while the information obtained in high frequency components improves on this estimate. For more details refer to [3].

Finally, we resolve the scaling invariance by fixing the diagonal elements of the estimated *mixing* matrices $\hat{\mathbf{W}}^{-1}(f_1), \ldots, \hat{\mathbf{W}}^{-1}(f_M)$ to unity.

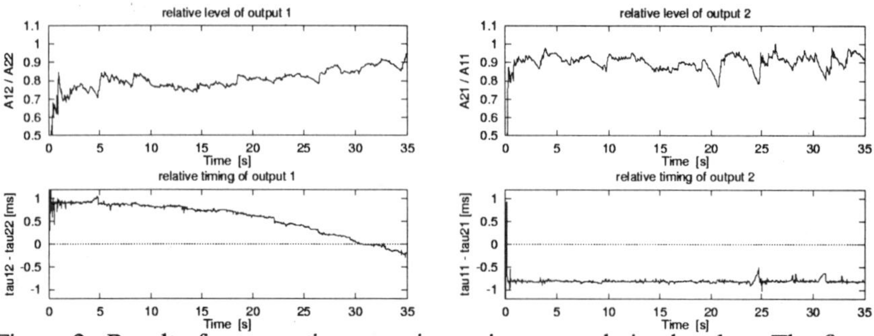

Figure 2: Result of a separation experiment in an anechoic chamber. The figure shows the time-course of the relative timing and level for both reconstructed signals. It is clearly visible that initial convergence of the algorithm takes only about one second of the recorded signals. The algorithm's estimate for the relative timing ('ITD') is accurate and displays the movement of one speaker. The estimates for the relative level ('ILD') show larger fluctuations.

4 Moving sources in an anechoic chamber

Signals from a moving and a standing speaker were recorded in stereo in the anechoic chamber of the University of Oldenburg. The proposed algorithm was used to separate the mixed signals.

The experimental setup was as follows: Two microphones were placed 35 cm apart. The stationary speaker was standing in a distance of 3 m at 60 degrees left of the mid-perpendicular of the microphones. The moving speaker started at a distance of 4 m at 70 degrees to the right. He walked in a straight line parallel to the microphones until he reached a position at about 30 degrees left. The algorithm started off from the (wrong) assumption that no mixing of the signals occurs.

Figure 2 illustrates the outcome of the experiment.

References

1. Jean-Pierre Nadal and Nestor Parga. Redundancy reduction and independent component analysis: Conditions on cumulants and adaptive approaches. *Neural Computation*, 9:1421–1456, 1997.
2. D. J. C. MacKay. Maximum likelihood and covariant algorithms for independent component analysis, draft 3.7. URL: ftp://wol.ra.phy.cam.ac.uk/pub/-mackay/ica.ps.gz, Dec. 1996.
3. Jörn Anemüller and Tino Gramss. On-line blind separation of moving sound sources. In *ICA '99*, Aussois, France, Jan. 1999.

DETECTION OF SPEECH AT LOW SIGNAL-TO-NOISE-RATIO IN A FIXED PCA SUBSPACE

FINN DUBBELBOER, TAMMO HOUTGAST, AND JOOST M. FESTEN

Dept. of Otolaryngology, University Hospital VU, P.O. Box 7057, 1007 MB Amsterdam, The Netherlands

E-mail: F.Dubbelboer@azvu.nl

Speech signals can be represented in the time-frequency domain and visualised by a spectrogram, i.e. the energy distribution in a number of spectral bands for successive speech frames (typically 10-20 ms fragments). This spectral representation is highly redundant. Principal component analysis (PCA) applied to such spectral representations of speech sounds reveals that a large proportion of the variance in speech can be described by only a few underlying components, or new dimensions. These dimensions define a subspace in which, by definition, the spectral variability of speech sounds is optimally represented. Projection of acoustical signals, speech and non-speech sounds, within this subspace shows different patterns. A speech detection scheme is presented which uses these differences to detect speech at low signal-to-noise-ratios, only by its spectral variation.

1 Introduction

Many hearing impaired persons suffer from reduced speech intelligibility in everyday situations. Reduction of background noise by means of digital signal processing may compensate for this problem. In order to reduce the noise it is important to have an algorithm which can distinguish between noise and speech embedded in noise. As both speech and noise have their characteristic patterns in time and frequency, a successful speech detection algorithm should operate in the spectro-temporal domain to make full use of the differences. A widely used spectro-temporal representation of speech is given in a spectrogram, i.e. the energy distribution in a number of spectral bands for successive speech frames (typically 10-20 ms fragments). Each single frame can be plotted as point in a multi-dimensional space with a separate dimension for the energy in each frequency band [3]. The variance in these frequency bands shows a high correlation, indicating that for speech this spectral representation is redundant. In other words, the number of dimensions in which speech is described can be reduced without much loss of information. Principal Component Analysis (PCA) applied to such a spectral representation of speech sounds reveals that a large proportion of the variance can be described by only a few underlying components, or new dimensions. The dimensions obtained from PCA define a subspace in which, by definition, the spectral variation of speech sounds is optimally represented. For many different noise types however, the representation of spectral variation in the subspace for speech is sub-optimal. The effect is that the variability for the representation of noise in this subspace may be very limited. These differences form the basis for an algorithm that detects speech frames in a noisy background.

2 PCA applied to speech signals

Since different phonemes are characterised by different spectral shapes a large set of speech material is used to build a subspace in which the spectral variations of that occur in running speech are optimally represented. A recording of 30 seconds of speech from a single speaker sampled at 15625 Hz is used as data set [2]. From this set power spectra are determined by calculating the FFT (Fast Fourier Transform) for 256-point (16 ms) Hanning-windowed speech segments. The window is repeatedly shifted over 128 points to obtain the next time frame, until the entire set is analysed. For each time frame the energy is summed over 1/3-octave bands and converted to a log scale. The spectral variance and the correlation among these frequency bands is reflected in a variance-covariance matrix from which the eigenvalues and eigenvectors are determined. These eigenvectors (or components) form a new orthogonal framework as linear combinations of the original vectors. All these components together span a new space with the same dimensionality as the original space. For a multi-dimensional data set the algorithms for finding underlying principal components arrange the output in order of decreasing importance, i.e. the first component explains most variance, the second component explain most of the remaining variance, and so on. Former studies in this area [1,4] show that the first four components are speaker-independent and fairly good interpretable. Roughly, the first component is related to the overall level of all spectral bands, the second component reflects the spectral tilt, while the third and fourth represent the first two formants. Higher order components represent speaker-dependent spectral properties while the highest components bear only minor variations. Since the first PCA component is not related to spectral-shape properties, the dimensions two, three and four are selected to span a subspace relevant for speech.

3 Projection of signals into a fixed speech subspace

New signals can be projected into this subspace after power spectra are calculated, as described in the previous paragraph. Projection means simple scalar products of this data set with three selected principal components. Figure 1 shows a projection of a sentence of the same speaker (a) of the same talker as used to define the subspace and a low-frequency stationary noise, recorded in a printing work shop (b). For speech, the time frames are widely spread within this space while the frames of the noise are more condensed, indicating a major difference in spectral variations between these signals.

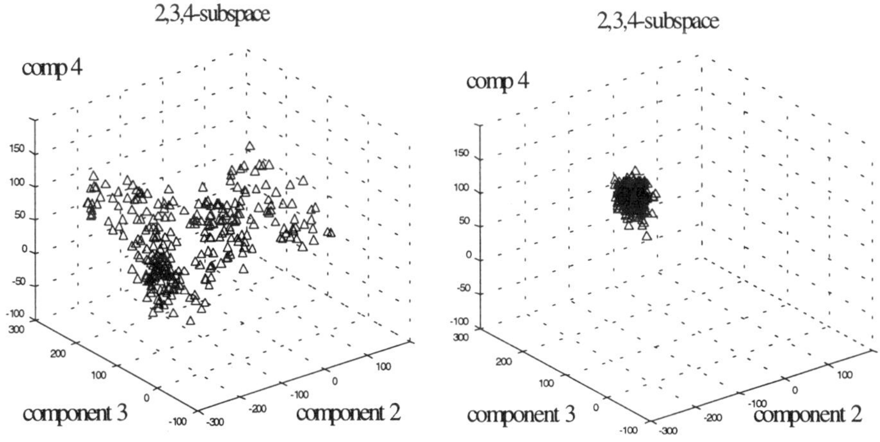

a. b.

Figure 1. (a) Projection of a speech signal, recorded from a female speaker, into a speech-based PCA-subspace (b) Projection of a noise, recorded in a printing work shop, into a speech-based PCA-subspace.

Projection of a mixture of speech and noise into this subspace shows an intermediate form between pure speech and pure noise, depending on the signal-to-noise-ratio. The cubes (a) and (b) in Figure 1 can be regarded as extreme cases for a very high and very low signal-to-noise-ratio, respectively.

4 Detection of speech frames

Since the noise frames form a cluster, a sphere enclosing most of the noise frames has a radius which is small compared to the variation among speech frames. This difference in "spread of frames" (as shown in Figure 1) can be used for detection of speech in noise. As an example, Figure 2 shows noise from running water and washing hands to which, after three seconds, a voice is added at S/N ratio of -6 dB. The first half second of the noise is used to locate the noise's "centre of gravity" and the standard deviation which determines the radius of the imaginary noise sphere. For each new time frame a vector which is larger than this radius is selected as speech while frames which are projected within the sphere are determined as noise frames. The resulting instantaneous detection of speech versus non-speech is in the seconds trace. A windowed version of this trace in which the proportion of speech frames must reach a preselected criterion presents a very reliable speech key.

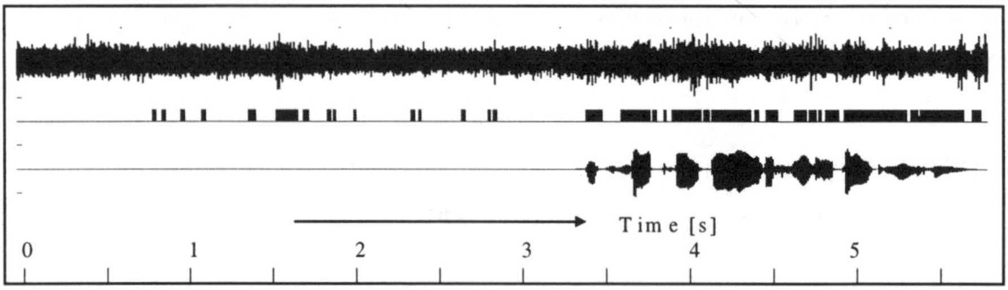

Figure 2. Detection of speech parts in background noise. The upper trace depicts 2.5 seconds of speech in a background of running water (while washing hands) with a S/N ratio of -6 dB. In the middle trace, the time frames selected to contain speech are plotted and at the bottom the uncorrupted speech signal is known for comparison.

5 Discussion

In this study a possible speech-detection algorithm is presented based on characteristic spectral properties of speech. Within a PCA subspace defined by these properties, speech is characterised by a typical pattern of successive speech frames while successive frames of various noise types form clusters within this subspace. This phenomenon can be used to discriminate between noise and speech embedded in noise.

6 Acknowledgements

This work is part of the SPACE-project (DE 3012) initiated by the European Community.

References

1. Bustamante, D. K., and Braida, L.D., Principal-component amplitude compression for the hearing impaired, *J. Acoust. Soc. Am* **82** (1987) pp. 1227-1242
2. Li, K.P., Hughes, G.W., and House, A.S., Correlation characteristics and dimensionality of speech spectra, *J. Acoust. Soc. Am.* **46** (1969) pp. 1019-1025
3. Plomp, R., Pols, L.C.W., and van de Geer, J. P., Dimensional analysis of vowel spectra, *J. Acoust. Soc. Am.* **41** (1967) pp. 707-712
4. Zahorian, S.A., and Rothenberg, M., Principal-components analysis for low redundancy encoding of speech spectra, *J. Acoust. Soc. Am.* **69** (1981) pp. 832–845

STATISTICS OF BINAURAL PARAMETERS AND LOCALIZATION IN NOISE

J. NIX AND V. HOHMANN

AG Medizinische Physik, Universität Oldenburg, Germany

E-mail: jnix@medi.physik.uni-oldenburg.de,vh@medi.physik.uni-oldenburg.de

The statistics of the binaural difference parameters were investigated by analyzing binaural signals from different directions in various sound environments. By using these statistics as a priori probabilities, an a posteriori probability of the location can be calculated using a Bayesian approach. A real-time localization algorithm based on this approach was implemented and tested. First results show a good azimuth and elevation localization accuracy in quiet environments and a robust azimuth estimation at high noise levels up to -5 dB Signal-to-Noise-Ratio (SNR).

1 Objectives

Interaural time and level differences are among the most important cues for sound source localization. Small time and level differences exist also in the median plane [1]. These cues can be used to resolve front/back ambiguities in models of auditory localization [2,3]. However, when using these cues in noisy or reverberant environments, they will show strong stochastic deviations from the expected values on a short time basis which should be taken into account. Psychoacoustical results show that a quite exact estimation of sound source azimuth in noise is possible near the detection threshold [4,5]. Current models of sound source localization do not account for (external) noise, and there is little knowledge about the statistics of binaural parameters under the influence of noise in realistic situations [6].

Therefore, the aim of this work was to measure the statistics of binaural parameters in noise and to design a robust localization algorithm which provides directional information for use in noise suppression strategies (e.g. [7]).

2 Measuring statistics of binaural parameters in noise

Speech sounds from 430 different incident angles were presented in the free field to a human subject equipped with In-the-Ear (ITE) hearing aids. Each signal was of 26 sec duration. Directions included azimuths from 0° to 355° in 5° steps and elevations of -15°,0°,15°,30°,45°. The microphone signals from the ITE devices were recorded on DAT. Various noise environments were recorded using the same setup, including cafeteria noise, traffic noise and a metal workshop. Noise and target speech were mixed digitally and evaluated with a signal processor system. Interaural level and phase differences of the mixed signals were calculated every 8 ms. These parameters were averaged across 43 1/2-ERB-spaced frequency groups of 1/2 ERB bandwidth each, ranging from 0.5 to 22 ERB. Histograms of the binaural parameters for each frequency and each direction were counted, each histogram including approximately 3200 cases.

The statistical properties of these parameters were characterized with the moment coefficients of expectation value, standard deviation, skew and curtosis. For the phase variable, the trigonometric moments were used to calculate analogous quantities [8]. Results of this analysis show that there is a strong dependency of all moment coefficients from direction, frequency and signal-to-noise ratio (SNR). As an example, Figure 1 shows the vector strength of the interaural phase difference (IPD) as a function of azimuth and elevation. The values for skew and curtosis indicate that the parameters are not normally distributed in general.

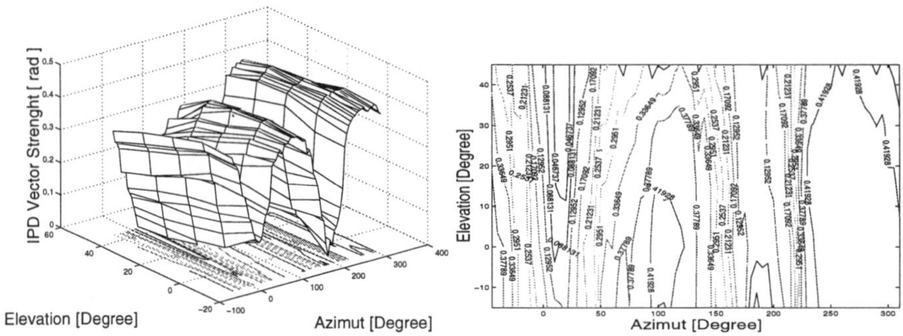

Figure 1. Vector Strength of the interaural phase differences (IPD) at 5.5 ERB as a function of azimuth and elevation. Speech mixed with cafeteria noise at 0 dB SNR.

3 Using the statistics of binaural parameters for 'bayesian localization'

Phase and level differences for a noisy target signal from a known spatial direction calculated from time frames of 8 ms length were treated as a multidimensional random variable. The binaural differences in the frequency groups for one direction are used as a pattern vector $\vec{x}$. The multidimensional probability density function (PDF) of $\vec{x}$ is described by the marginal distributions $f_{l,b}(x_b)$ for each frequency group b and direction l. The measured histograms were used as estimates for this PDF. According to Bayes' formula the a posteriori probability for the presence of one direction l of L possible directions is calculated from the actual pattern vector $\vec{x}$ and from $f_{l,b}(x_b)$. The probabilities are combined across frequency by multiplication assuming independent variables.

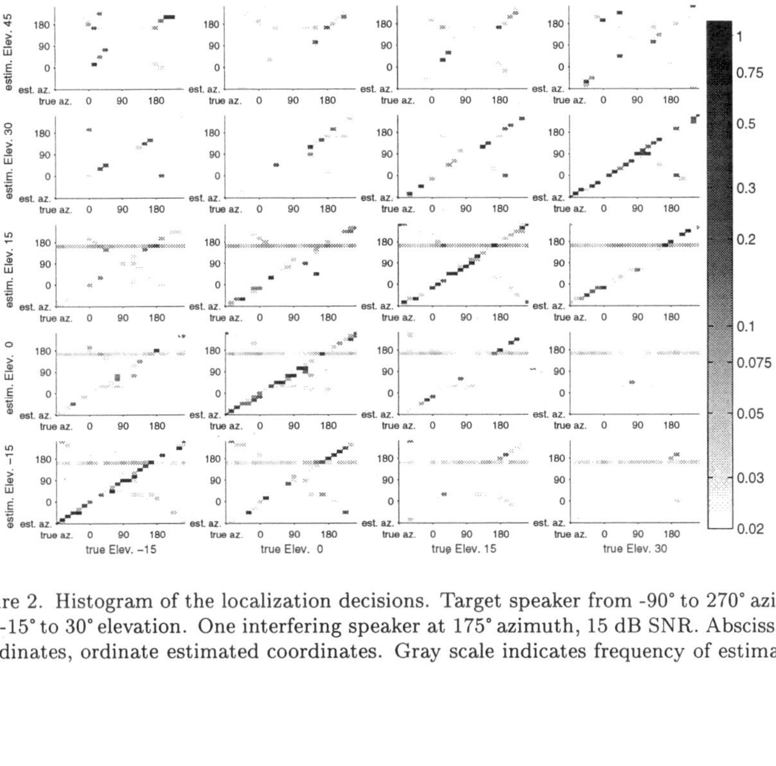

Figure 2. Histogram of the localization decisions. Target speaker from -90° to 270° azimuth and -15° to 30° elevation. One interfering speaker at 175° azimuth, 15 dB SNR. Abscissa real coordinates, ordinate estimated coordinates. Gray scale indicates frequency of estimation.

Based on the Bayesian classification, a localization algorithm was implemented on a digital signal processor system as described in [7]. The implementation operates in real time and calculates every 8 ms a probability map for 430 directions. The map is smoothed using a moving average with 100 ms time constant. The direction of the strongest source is estimated from the maximum in the map. The values of $f_{l,b}(x_b)$ were used as noise-specific references.

4 Tests of localization performance

The implementation was tested in 85 situations with different 'trained' references, different classes of noise and SNRs. In each situation, direction estimates of one target speaker (different speech material than was used to determine the references) from 96 directions were counted for 27 sec.

Generally, the algorithm shows a very good performance in quiet environments. Especially it is able to estimate the front/back directions with very few errors. With a high level of diffuse noise, the estimation of azimuth remains exact. However at SNRs below 5 dB the front/back discrimination degrades rapidly. With one interfering talker it is possible to estimate exactly the azimuth of the interfering talker (see Figure 2). Elevation is estimated well in many cases, but especially for sources in the median plane elevation errors or front/back errors may occur. Performance is best when references are similar to the noise signals and SNRs used during the test.

Acknowledgments

This research was supported by DFG Ho 1697/2-1.

References

1. John C. Middlebrooks, James C. Makous, and Richard M. Green. Directional sensitivity of sound-pressure levels in the human ear. *J. Acoust. Soc. Am.*, 86(1):89–108, jul 1989.
2. Stephan Albani, Jürgen Peissig, and Birger Kollmeier. Model of binaural localization resolving multiple sources and spatial ambiguities. In Birger Kollmeier, editor, *Psychoacoustics, Speech and Hearing Aids*, pages 227–232, Singapore, 1996. World Scientific Publishing Co.
3. R. O. Duda. Elevation dependence of the interaural transfer function. In Robert H. Gilkey and Timothy R. Anderson, editors, *Binaural and Spatial Hearing in Real and Virtual Environments*, pages 49–75. Lawrence Erlbaum Assoc., Mahwah, New Jersey, 1997.
4. T. Jacobsen. Localization in noise. Technical Report 10, The Acoustics Laboratory, Technical University of Denmark, 1976.
5. Michael D. Good and Robert H. Gilkey. Sound localization in noise: The effect of signal-to-noise ratio. *J. Acoust. Soc. Am.*, 99:1108–1117, February 1996.
6. P. M. Zurek. Probability distributions of interaural phase and level differences in binaural detection stimuli. *J. Acoust. Soc. Am.*, 90(4):1927–1932, October 1991.
7. Thomas Wittkop, Stephan Albani, Volker Hohmann, Jürgen Peissig, William S. Woods, and Birger Kollmeier. Speech processing for hearing aids: Noise reduction motivated by models of binaural interaction. *Acustica united with Acta Acustica*, 83(4):684–699, 1997.
8. Nicholas Irving Fisher. *Statistical analysis of circular data*. Cambridge University Press, Cambridge, New York, 1993.

COMBINING MONAURAL NOISE REDUCTION ALGORITHMS AND PERCEPTIVE PREPROCESSING FOR ROBUST SPEECH RECOGNITION

MICHAEL KLEINSCHMIDT, MARK MARZINZIK AND BIRGER KOLLMEIER
AG Medizinische Physik
Carl von Ossietzky Universität Oldenburg
D-26111 Oldenburg
Germany
E-mail: michael@medi.physik.uni-oldenburg.de

In noisy and reverberant environments the performance of automatic speech recognition systems drops below acceptable levels. It has been shown before, that using a psychoacoustical model of the peripheral auditory processing, as introduced by Dau [1], yields much higher recognition rates in noisy background situations than standard pre-processing methods [2]. To further improve the robustness of a speaker-independent digit recognition system, single-microphone noise reduction procedures [3] have been combined with the auditory model feature extractor and continuous Hidden Markov Models as well as Locally-recurrent Neural Networks. The study shows that the recognition rates improve significantly for certain noise conditions, while the performance for clean speech is not negatively effected.

1 Noise reduction

In this study a number of noise reduction algorithms were examined, that represent a class of minimum mean square error (MMSE) short-term spectral amplitude estimators (STSA) proposed by Ephraim and Malah [4,5]. Both, the current SNR and the average SNR of the most recent frames, are estimated and evaluated. These suppression rules allow a significant reduction of noise while avoiding the *musical tones* phenomenon, often encountered when applying, e.g., spectral subtraction. The original MMSE STSA algorithm (**e7**) was applied to noisy speech signals. Additionally, modified versions were utilised, taking into account the uncertainty of signal presence (**e30**) or working on logarithmic spectra (**elog**) respectively [3].

2 Auditory Model

The quantitative model of auditory perception (PEMO) was originally developed to simulate human perception in psychoacoustical experiments [1]. It has been successfully applied to a number of temporal masking and modulation perception experiments. The *'internal representation'* (Fig.1 b, d, f), i.e., the output vector of this model, also serves well as a basis for transmission quality and robust speech recognition tasks [2]. Following the initial pre-emphasis, the input signal is split up into 19 frequency bands by an auditory gammatone filterbank. After a further half-way rectification and low-pass filtering, five non-linear elements, so-called *'adaptive compression loops'* perform a near logarithmic compression to steady-state signals, while fast fluctuations are transformed rather linearly. This stage has been shown to be essential for robust speech recognition. The signal is filtered by a 4 Hz modulation lowpass, which leads to an overall modulation transfer function similar to the modulation spectrum of spoken language. Finally, 10 ms averages are calculated.

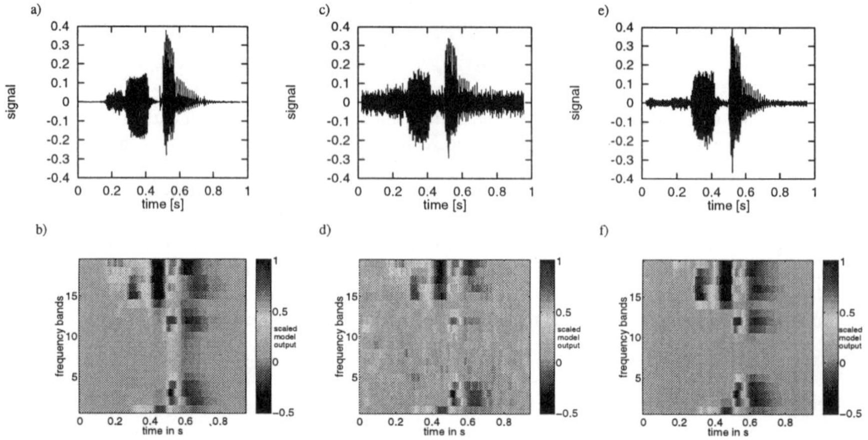

Figure 1. *a) Time signal and b) internal representation of the German word 'sieben' uttered by a male speaker. c) and d) with CCITT noise at 5 dB SNR. e) and f) after processing with the* ***e7*** *algorithm.*

3 Experiment

In this study a speaker independent digit recognition experiment was carried out on clean, noisy and enhanced speech signals. The Zifkom database (*Deutsche Telekom*), consisting of 200 utterances of each German digit by 100 male and 100 female speakers, was split equally into training and test set. The signals were mixed with CCITT noise at signal-to-noise ratios (SNR) of –10, -5, 0, 5, 10, and 15 dB, and then enhanced by the three noise reduction algorithms described above. **Lo**cally-**r**ecurrent **N**eural **N**etworks (LRNN; *developed and implemented at IAP, University of Frankfurt, see [2]*) and continuous **H**idden **M**arkov **M**odels (HMM) were used as classifiers and trained on unprocessed clean speech signals. The recognition performance was evaluated for PEMO pre-processing and compared to the widely used **mel-f**requency **c**epstral **c**oefficients (MFCC).

Table 1. Recognition rates [in %] of different recognition systems applied on clean test data.

algorithm	MFCC / HMM	MFCC / LRNN	PEMO / HMM	PEMO /LRNN
without	98.7	98.0	97.6	98.0
e30	87.7	95.9	97.3	97.5
e7	97.7	97.7	97.4	97.6
elog	94.0	96.6	97.3	97.6

4 Results

- All three Ephraim-Malah algorithms lead to significantly increased recognition rates for most signal-to-noise ratios (Tab. 2).
- For clean speech, the noise reduction does *not* deteriorate the performance of PEMO-based recognition systems (Tab. 1).
- Generally, speech enhancement PEMO-based systems show a higher overall performance than MFCC pre-processing.
- The combination of **e7**, PEMO and LRNN is most successful and may be used as a highly robust digit recognition system with tolerable error rates at SNR as low as 5 or even 0 dB.

Table 2. Recognition rates [in %] of different recognition systems for noisy speech with and without speech enhancement.

	MFCC / HMM : CCITT noise [dB SNR]					
algorithm	-10	-5	0	5	10	15
without	7.2	10.9	9.8	11.5	34.0	66.5
e30	11.4	23.4	49.1	69.3	76.5	83.2
e7	9.6	10.9	39.7	77.0	94.7	96.7
elog	11.0	21.3	55.9	77.7	86.5	90.4

	MFCC / LRNN : CCITT noise [dB SNR]					
algorithm	-10	-5	0	5	10	15
without	9.5	10.0	10.0	10.9	16.5	29.2
e30	9.5	23.3	47.1	75.3	89.1	92.2
e7	9.8	15.4	28.9	54.3	82.2	93.7
elog	9.5	23.3	47.2	74.0	90.6	94.8

	PEMO / HMM : CCITT noise [dB SNR]					
algorithm	-10	-5	0	5	10	15
without	13.6	13.7	10.3	12.1	25.5	65.7
e30	12.7	17.0	49.8	76.1	82.1	88.0
e7	12.9	14.8	26.3	61.7	90.1	93.2
elog	14.5	18.7	45.1	80.3	87.4	92.7

	PEMO / LRNN : CCITT noise [dB SNR]					
algorithm	-10	-5	0	5	10	15
without	11.6	12.8	17.2	41.1	79.0	93.3
e30	14.0	40.4	59.3	76.2	86.5	92.3
e7	14.0	40.1	76.0	88.7	94.0	95.7
Elog	15.1	43.5	66.2	82.5	91.8	95.5

The ranking of the Ephraim-Malah algorithms varies for different SNR, noises and recognition systems. For conditions with already high recognition rates the **e7** seems to be most effective. At lower performance levels **elog** and **e30** are superior. This agrees with the stochastical derivation and the findings in [3] that **e7** gives the smallest amount of noise reduction, while causing relatively low distortions on the speech signal itself. It can

be concluded, that on one hand reducing as much noise as possible is a good strategy at low performance levels, but on the other hand a better 'fine-tuning' is needed when the recognition rates are initially high.
MFCC-based recognition systems are less robust, when the test material is evaluated under conditions different to the training corpus. This can be seen in the results for noisy speech without noise suppression algorithms (see Tab. 2) and also for processed clean speech (see Tab. 1). At low SNR the combination of Ephraim-Malah and PEMO/LRNN system is most robust. This is probably attributed to the inherent robustness of PEMO and LRNN, which reduces the negative effects of residual noise and signal distortions in these cases.

5 Conclusion

The combination of the original Ephraim-Malah MMSE short-term spectral amplitude estimator (e7) and an auditory-based feature extraction (PEMO) with a Locally-recurrent Neural Network classifier was found to be most robust in all investigated noise situations. At the same time the performance in the absence of noise is not significantly affected. This system is therefore a promising candidate for further automatic speech recognition studies and for practical applications, e.g., noise-robust speech recognition in 'difficult' acoustical environments.

This work was supported by DFG (Deutsche Forschungsgemeinschaft).

References

1. Dau, T. , Püschel, D., and Kohlrausch, A. (1996). A quantitative model of the 'effective' signal processing in the auditory system: I. Model structure. *J. Acoust. Soc. Am.* **99(6)**, 3615-3622 .
2. Tchorz, J. , Kasper, K. , Reininger, H., and Kollmeier, B. (1997). On the Interplay between auditory-based features and locally recurrent neural networks for robust speech recognition in noise. *Eurospeech '97, p. 2075-2078, ESCA, Patras, Greece, 1997.*
3. Marzinzik, M. and Kollmeier, B. (1998). Development and evaluation of single-microphone noise reduction algorithms for digital hearing aids. *Presented at this meeting.*
4. Ephraim, Y. and Malah, D. (1984). Speech enhancement using a minimum mean-square error short-time spectral amplitude estimator. *IEEE Trans. Acoust., Speech, Signal Processing* **32**, 1109-1121.
5. Ephraim, Y. and Malah, D. (1985). Speech enhancement using a minimum mean-square error log-spectral amplitude estimator. *IEEE Trans. Acoust., Speech, Signal Processing* **33**, 443-445.

MINIMUM BANDWIDTH REQUIRED FOR INTELLIGIBILITY AS A TOOL FOR STUDYING SPEECH-PROCESSING IMPAIRMENT

INGRID M. NOORDHOEK, TAMMO HOUTGAST AND JOOST M. FESTEN

Dept. of Otolaryngology, University Hospital VU, P.O. Box 7057, 1007 MB Amsterdam, The Netherlands

E-mail: jm.festen@azvu.nl

In an adaptive listening test, the bandwidth of speech in complementary notched noise was varied. The bandwidth (center frequency 1 kHz) required for 50% speech intelligibility is called Speech Reception *Bandwidth* Threshold (SR*B*T). The SR*B*T was measured for 10 normal-hearing and 30 hearing-impaired listeners. The average SR*B*T of the normal-hearing listeners is 1.4 octave. The performance of seven hearing-impaired listeners is considered normal, whereas 23 hearing-impaired listeners have a wider-than-normal SR*B*T. The SR*B*T of a hearing-impaired listener may be wider than normal, due to inaudibility of a part of the speech band, or to an impairment in the processing of speech. The Speech Intelligibility Index (SII) is used to separate these effects. The SII may be regarded as the proportion of the total speech information that is available to the listener. Each individual SR*B*T is converted to an SII value. For the normal-hearing listeners, the SII is about 0.3. For 21 hearing-impaired listeners, the SII is higher. This points to a speech-processing impairment in the 1-kHz frequency region. The deviation of an individual SII value from 0.3 can be used to "quantify" the degree of processing impairment.

1 Introduction

Listeners with a sensorineural hearing impairment often need a higher speech-to-noise ratio than normal-hearing listeners to understand speech, even if the speech is presented well above the hearing threshold. It has been suggested that these difficulties are due to suprathreshold impairments, such as reduced temporal or spectral resolution, and abnormal loudness growth. However, clear correlations between speech processing and psychoacoustic abilities have not been found. The reason may be that the performance on various psychoacoustic tests is usually measured at only one or two frequencies, while for understanding speech a broad frequency range is important.

An adaptive test was developed that measures the bandwidth of speech with a center frequency of 1 kHz required for 50% intelligibility (Speech Reception *Bandwidth* Threshold, or SR*B*T). This test is meant as a tool for studying the speech processing in a limited frequency region around 1 kHz. The SR*B*T is a good candidate for correlation with several auditory functions at 1 kHz.

In the present study, the SR*B*T of 10 normal-hearing listeners and 30 hearing-impaired listeners is measured. The aim is to investigate whether a wider-than-normal SR*B*T is caused by inaudibility of part of the speech signal, or by impaired speech processing.

2 Method

For each listener, the SR*B*T was measured twice, using two lists of 13 meaningful Dutch sentences. The bandwidth of the speech was changed according to an adaptive up-down procedure, with a step size corresponding to multiplication or division of the bandwidth (in Hz) by 1.37. The speech was bandpass filtered around 1 kHz. Complementary bandstop noise was added to the speech (Fig. 1), to ensure that the speech was only audible within the desired frequency band. To keep the signal above hearing threshold at all frequencies, speech and noise spectra were presented halfway hearing threshold and uncomfortable loudness level (UCL).

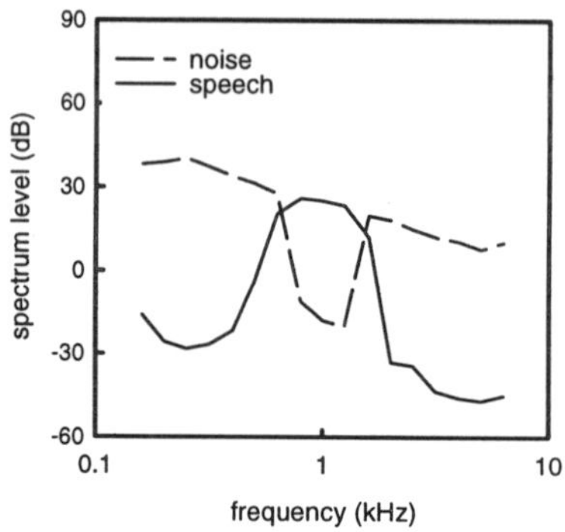

Figure 1. Spectral configuration of the stimulus.

3 Speech Intelligibility Index

The Speech Intelligibility Index is a physical measure between 0 and 1 that may be interpreted as the proportion of the total speech information available to the listener [1]. The SII is defined by:

$$\mathrm{SII} = \sum \mathrm{I_i A_i}$$

where $\mathbf{I_i}$ is the importance of frequency band **i** to intelligibility, and $\mathbf{A_i}$ is the band-audibility function, i.e., the proportion of a 30-dB dynamic range of the speech within band **i** that is not masked by noise or presented below the hearing threshold.

With the speech material of this experiment, normal-hearing listeners usually reach 50% intelligibility at a speech-to-noise ratio of about -5 dB. At this S/N ratio, 10 dB of the original 30-dB dynamic range of the speech still contributes to intelligibility in each frequency band. This results in an SII of 0.33. By definition, the SR*B*T also corresponds to 50% intelligibility. Therefore, at the SR*B*T of the normal-hearing listeners an SII of 0.33 is expected.

4 Results and discussion

Figure 2 shows the SII corresponding to each SR*B*T for the normal-hearing listeners (open circles) and hearing-impaired listeners (closed circles). The average SR*B*T of the normal-hearing listeners is 1.4 octave and their average SII is 0.31, close to the expected value of 0.33. The one-tailed 95-percent confidence interval for the data of the normal-hearing listeners is chosen as the boundary between normal and "impaired" for both SR*B*T and SII (lines in Fig. 2).

The SR*B*T of seven hearing-impaired listeners can be considered normal, whereas 23 listeners have a wider-than-normal SR*B*T. Two hearing-impaired listeners had a wider-than-normal SR*B*T, but a normal SII. Thus, their impaired SR*B*T can be explained by the inaudibility of a part of the speech signal. For these two listeners, the difference between the threshold and UCL was too small to contain the full relevant dynamic range of the speech.

The SII of 9 hearing-impaired listeners is normal. They can reach 50% intelligibility with the same amount of speech information as the normal-hearing listeners. The remaining 21 hearing-impaired listeners with a higher-than-normal SII seem to demonstrate impaired speech processing, to a degree indicated by the individual SII values.

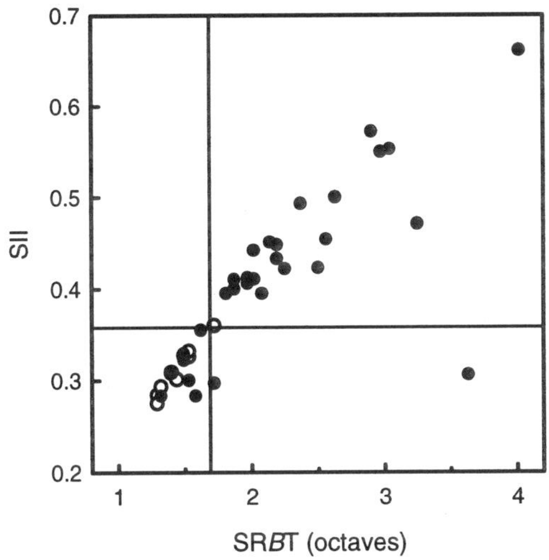

Figure 2. Speech Intelligibility Index versus SR*B*T for normal-hearing listeners (open circles) and hearing-impaired listeners (filled circles). Lines indicate boundaries of normal and "impaired" for both SII and SR*B*T.

Pavlovic *et al.* [2] proposed to use a "speech desensitization factor" that decreases as a function of hearing loss to account for impaired speech processing. Figure 3 shows the individual SII values as a function of the hearing loss at 1 kHz. Our data do not show a one-to-one relation between SII and hearing loss. Therefore, a hearing-loss dependent desensitization factor does not seem appropriate.

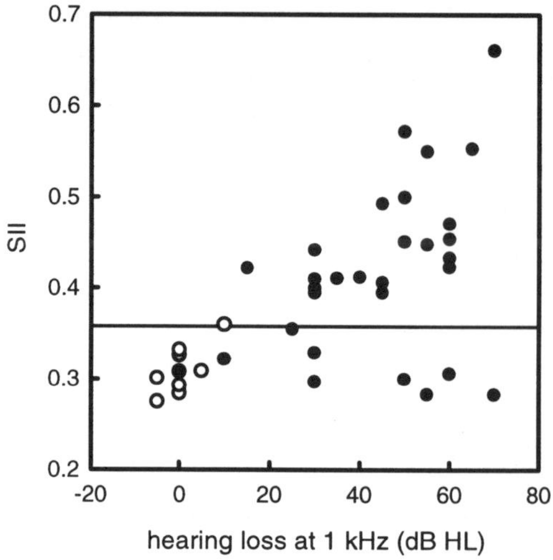

Figure 3. Speech Intelligibility Index versus hearing loss at 1 kHz for normal-hearing listeners (open circles) and hearing-impaired listeners (filled circles). The line indicates the boundary between a normal and higher-than-normal SII.

5 Acknowledgments

This research was supported by the Foundation "Heinsius-Houbolt Fonds," The Netherlands.

References

1. ANSI S3.5, American national standard methods for calculation of the speech intelligibility index (American National Standards Institute, New York, 1997).
2. Pavlovic, C. V., An articulation based procedure for predicting the speech recognition performance of hearing-impaired individuals. *J. Acoust. Soc. Am.* **80** (1986) pp. 50–57.

USING AMPLITUDE MODULATION INFORMATION FOR SOUND CLASSIFICATION

JÜRGEN TCHORZ AND BIRGER KOLLMEIER
AG Medizinische Physik, Universität Oldenburg, 26111 Oldenburg, Germany
E-mail: tch@medi.physik.uni-oldenburg.de

1 Introduction

Digital processing in modern hearing aids allows the implementation of a wide range of sound processing schemes. Monaural and binaural noise suppression can enhance speech quality and speech intelligibility in adverse acoustical environments. However, the performance of these algorithms is strongly dependent on a proper and reliable classification of the acoustical situation. Noise suppression techniques like spectral subtraction, for example, are strongly dependent on a fast and proper estimate of the present noise level. Thus, an important task of the classification algorithm is to decide whether speech or noise is present. If both speech and noise are present, an estimate of the signal-to-noise ratio is desired. Furthermore, the classification of the monitoring algorithm should work on short time scales, because most noise suppression algorithms need an exact detection of speech pauses for good performance. The classification algorithm which is presented in this paper bases on so-called amplitude modulation spectrograms (AMS). Its basic idea is that both spectral and temporal information of the signal is used to attain a separation between "acoustical objects" within the signal. The AMS approach is motivated by neurophysiological experiments on periodicity coding in the auditory cortex of mammals, where neurons tuned to different center frequencies were found to be organized almost orthogonal to neurons which are tuned to different modulation frequencies [1]. Kollmeier and Koch [2] implemented a binaural noise suppression scheme which bases on AMS sound representation. They could demonstrate a benefit in terms of speech intelligibility, in comparison to unprocessed speech.

2 Generating AMS patterns

In a first processing step, the input signal is long-term level adjusted, i.e., changes in the overall level are compensated for, whereas short-term level differences (e.g., those between successive phonemes) are maintained to serve as

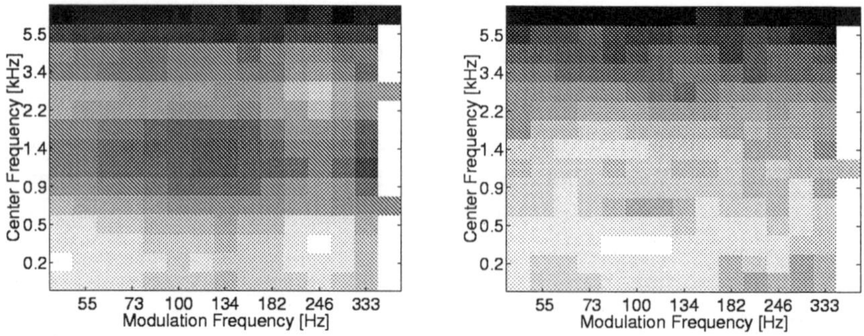

Figure 1. AMS pattern generated from voiced speech (left) and speech simulating noise (right). Bright and dark shading indicates high and low amplitude, respectively. Each AMS pattern represents a 32-ms portion of the input signal.

additional cues for classification. This level adjustment is realized by dividing the input signal by its 2 Hz-low pass filtered rms function. The level-adjusted signal is then subdivided into overlapping segments of 4.0 ms duration with a progression of 0.25 ms for each new segment, and transformed into a complex spectrum with a FFT. The resulting 64 complex samples are considered as a function of time, i.e., as bandpass-filtered complex time signal. Their respective envelopes are extracted by squaring. This envelope signal is again segmented into overlapping segments of 128 samples (32ms) with an overlap of 64 samples. A further FFT is computed and supplies a modulation spectrum in each frequency channel. By an appropriate summation of neighboring FFT bins, both axes are scaled logarithmically with a resolution of 15 channels for center frequency (100-7300 Hz) and 15 channels for modulation frequency (50-400 Hz). In a last processing step, the amplitude range is log-compressed. Examples for AMS patterns can be seen in Fig. 1. The left AMS pattern was generated from a voiced speech portion. The formant structure is represented, as well as the periodicity at the fundamental frequency (i.e., approx. 250 Hz) in each center frequency band. The AMS pattern on the right was generated from speech simulating noise. The typical spectral tilt can be seen, but no structure across modulation frequencies.

3 Neural Network Training and Classification Experiments

Speech/noise classification of AMS patterns is performed with a standard feed-forward neural network. It consists of an input layer with 225 (15x15) neurons,

Table 1. Classification results in pure speech/pure noise conditions

	database	# of AMS patterns	% correct
training data:			
speech	PHONDAT	30948	98.4
noise	various	30948	99.5
test data:			
speech	PHONDAT	10383	95.7
	ZIFKOM	9829	93.6
	TIMIT	1573	91.7
noise	NOISEX	10383	96.8
	ALIFE	2507	92.7

a hidden layer with 20 neurons, and an output layer with 1 output neuron. The target activity of the output neuron for training is set to 0.95 or 0.05 for AMS patterns generated from speech or noise, respectively. For training, 495 s of speech from 28 different talkers and 495 s of noise from 14 different noise sources were used, yielding 61896 AMS training patterns in total. After training, AMS from "unknown" signal segments are classified with an output neuron activity threshold of 0.5: If the network responds to an AMS pattern with an output neuron activity above 0.5, the segment is classified as "speech", otherwise as "noise". Classification results on different databases are shown in Tab. 1. For speech, the best classification results were gained when the test data originated from the same database as the training material (PHONDAT). Degraded performance for TIMIT-data can be explained by different long-term spectra of these databases due to differences in the recording conditions.

4 SNR estimation in mixed situations

In situations where both speech and noise are present, segment classification as either "speech" or "noise" is not appropriate. Instead, an estimation of the present signal-to-noise ratio (SNR) is desired. To achieve this, the network is trained with AMS patterns generated from mixtures of speech and noise. The target activity then depends on the "local" SNR of the according AMS analysis frame. The SNR range from 25 to -10 dB is linearly transformed to target activities between 0.95 or 0.05. SNRs above 25 dB and below -10 dB are transformed to 0.95 and 0.05, respectively. The training AMS patterns were generated from a mixture of the training material described in Sec. 3. After training, the output neuron activity of the network when presenting "un-

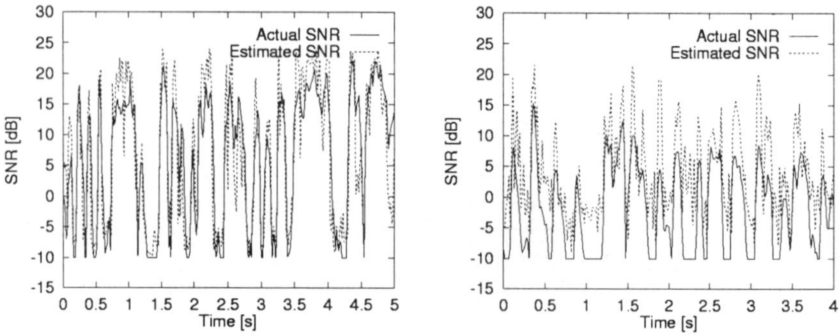

Figure 2. Estimation of SNR in car noise (left) and cafeteria noise (right).

known" AMS patterns serves as estimate for the present SNR. In Fig. 2, two examples of SNR estimations are illustrated. The solid lines show the actual SNR, the dotted lines show the estimate of the SNR (after re-transforming output neuron activity to SNR). On the left, the input signal was a mixture of speech and car noise. In that condition, the algorithm can predict the present SNR with satisfactory accuracy. On the right, the input signal was a mixture of speech and cafeteria noise. Here, the estimated SNR is higher than the actual SNR most of the time. This might be due to the "speech-like" characteristic of cafeteria noise.

5 Conclusion

It was demonstrated that the combination of spectral and temporal information of acoustic signals can be exploited for automatic classification of the acoustical situation. The precision of SNR estimation in situations where both speech and noise are present is dependent on the background noise. The performance might be improved by extending the amount of training data and modifying the SNR - target activity transformation function.

References

1. G. Langner, M. Sams, P. Heil, and H. Schulze, J. Comp. Physiol. A **181**, 665-676 (1997).
2. B. Kollmeier and R. Koch, J. Acoust. Soc. Am. **95**, 1593-1602 (1994).

DEVELOPMENT AND EVALUATION OF SINGLE-MICROPHONE NOISE REDUCTION ALGORITHMS FOR DIGITAL HEARING AIDS

MARK MARZINZIK* AND BIRGER KOLLMEIER

AG Medizinische Physik
Carl von Ossietzky Universität Oldenburg
D-26111 Oldenburg, Germany
* *E-mail: mark@medi.physik.uni-oldenburg.de*

In this study single-microphone noise reduction procedures were investigated for use in digital hearing aids. One widely reported artifact of most noise suppression systems, the musical noise phenomenon, can partly be overcome by the Ephraim-Malah noise suppression algorithms [1,2]. Based on these algorithms, three different versions have been implemented together with a procedure for automatically updating the noise-spectrum estimate. To evaluate the algorithms, different tests have been performed with six normal-hearing and six hearing-impaired subjects. With 'standard' measurement methods no increase in speech intelligibility was found compared to the unprocessed signal. However, benefits with respect to reductions in listener fatigue and in the mental effort needed to listen to speech in noise over longer periods of time were found in this study by use of a newly developed ease-of-listening test. Subsequent paired comparison tests also revealed a clear preference of the hearing-impaired subjects for the noise-reduced signals in situations with rather stationary noise. In the case of strongly fluctuating noise at low SNR, however, the subjects preferred the unprocessed signal due to speech distortions caused by the noise reduction algorithms.

1 Algorithms

Almost all noise suppression algorithms considered so far in the literature are reported to produce artifacts in the residual noise that exhibits a very unnatural disturbing quality, often called "musical tones". Ephraim and Malahs' minimum mean-square error (MMSE) short-time spectral amplitude (STSA) estimator [1,2] is supposed to obtain a significant noise reduction while eliminating the musical tones phenomenon. This makes it a good candidate for hearing aid applications.

For further evaluation, the MMSE STSA estimator (E7) was implemented in the framework of a signal-processing workstation. Also, Ephraim and Malahs' modified estimator under uncertainty of signal presence (E30), and the MMSE log-spectra estimator (EL) were implemented. For application of these algorithms in hearing aids, a noise spectrum estimate has to be updated during speech pauses. Hence, a reliable speech pause detection algorithm with low computational demands was combined with these algorithms in order to construct a self-adaptive noise reduction scheme for stationary parts of the background noise.

2 Evaluation

A test battery was designed to assess the supposed benefits of the noise suppression algorithms when used in noisy environments. Noise samples recorded from a drilling machine were employed that yield a rather stationary noise, and a cafeteria noise with babble in the background which is more fluctuating. Six hearing-impaired subjects with moderate hearing losses as well as six normal-hearing subjects participated in the measurements. To compensate the hearing loss of the hearing-impaired subjects, two graphic equalizers were used together with a stereo amplifier. The required gain in each frequency band was determined from the audiogram for each subject using the one-half gain rule.
The tests described in the following sections are performed for the unprocessed signal, and the signal processed with the three different Ephraim-Malah algorithms (E7, EL, and E30). Presentation of the signals was via headphones (Sennheiser HDA200 for the sentence test, HD25 for the other tests).

2.1 Speech Intelligibility

Speech reception thresholds (SRT) in drill noise and in cafeteria babble were measured with the Göttingen sentence test using at least three test lists (each consisting of 10 sentences) at three different signal-to-noise ratios (SNR) for each condition. The noise level was adjusted in a preliminary test run for each subject so that the overall loudness impression was at the top border of the comfortable loudness range.

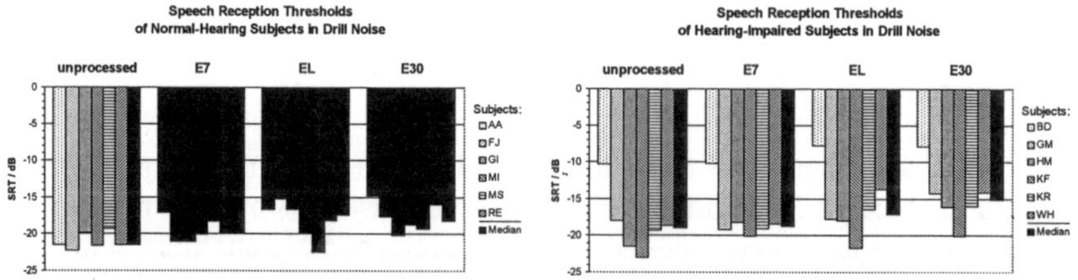

Figure 1. Speech reception thresholds using a sentence test with drilling machine noise for normal-hearing subjects (left panel) and hearing-impaired subjects (right panel).

The sentence test measurements do not show any improvements in speech intelligibility with the noise suppression algorithms (Fig. 1). This can partly be due to the extreme low SNRs at which the speech intelligibility measurements were performed. SNRs as low as -18 dB overload the performance of the noise suppression algorithms. Speech distortions are the consequence, so that additional degradations of the speech counterbalance positive effects of the noise suppression. Note, however, that the Ephraim-Malah algorithms do not result in significantly worse SRTs for normal-hearing listeners; this holds even for hearing-impaired subjects with respect to the algorithm E7. Similar results are obtained with the cafeteria noise: No significant differences were found for normal-hearing subjects. But the SRTs of the hearing-impaired subjects are in median 1.8 dB worse in cafeteria noise with the Ephraim-Malah algorithm compared to the unprocessed situation.

2.2 *Ease of Listening*

One potential benefit that can be expected from single-microphone noise suppression algorithms is to lower the listening effort at SNR values well above the SRT. To assess this feature, an ease-of-listening test was developed as follows:

Step 0: The subject has to listen to a radio newscast of 2½ minutes duration, mixed with background noise (drilling machine at -5 dB SNR). The task is to listen carefully to the news and to repeat afterwards as many news as can be remembered. The repetition of the news will be recorded with a dictaphone. Actually, the content of the subject's report of the newscast will not be evaluated. This procedure only serves to force the subject to listen carefully.

Step 1: The subject has to listen again to the same newscast, but including the possibility to switch between different algorithms. The subject is instructed to try all 4 programs and to judge the effort required to understand the news according to the given scale.

	Score
Complete relaxation possible; no effort required	(5)
Attention necessary; no appreciable effort required	(4)
Moderate effort required	(3)
Considerable effort required	(2)
No meaning understood with any feasible effort	(1)

Step 2: The subject has to listen to a further and yet unknown newscast. All 4 programs can be used. The task is to repeat again as many news as can be remembered. After the repetition of the news, the subject has to judge the listening effort again.

Step 3: Same as step 2, but now a female newscaster is speaking and the background is cafeteria noise at 0 dB SNR.

The results of the ease-of-listening test show clear improvements for the hearing-impaired subjects with the Ephraim-Malah algorithms E7 and EL compared to the algorithm without noise suppression when used in situations with the rather stationary background noise (Figure 2). In the fluctuating cafeteria noise no significant effects were found.

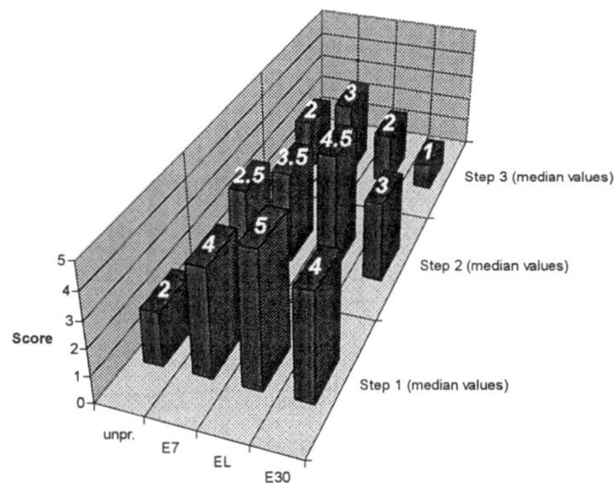

Figure 2. Results of the ease-of-listening test with the hearing-impaired subjects in drilling machine noise.

2.3 Paired Comparison

A complete paired comparison test was carried out to better assess the perceived differences between the algorithms. In each trial, two programs were compared with regard to three different criteria using the same speech signal (approx. 4 sec.) throughout the test with the same noise signals as before at SNRs of -5 and +5 dB. As a result, the hearing-impaired subjects did not perceive any distortions in the speech caused by the processing of the noise reduction algorithms (except for cafeteria noise at -5 dB SNR), while the normal-hearing subjects did notice distortions. The hearing-impaired subjects judged the naturalness of the speech in drill noise to be better with the Ephraim-Malah algorithms than unprocessed; the normal-hearing subjects stated the opposite. The overall judgement of the Ephraim-Malah algorithms was significantly better than unprocessed (except for cafeteria at -5 dB). The effect of the noise reduction is very clear in the drill noise (both SNRs) as well as in cafeteria noise at +5 dB. E30 gives the biggest amount of noise reduction, followed by EL and E7. However, the algorithms with the biggest amount of noise reduction show also the highest speech distortions in unfavorable noise situations (cafeteria at -5 dB SNR). In such situations, E7 preserves the naturalness of the speech best.

3 Conclusion

The proposed single-microphone noise reduction algorithms are able to reduce the mental effort needed to listen to speech in noise over longer periods of time. Moreover, the sound quality improves significantly in environments with rather stationary noise. Improvements with respect to "objective" speech intelligibility as measured with standard methods were not found in this study. The relatively low complexity of the algorithms allows an application in digital hearing aids in the near future.

4 Acknowledgements

This study was supported in parts by a research contract with AudioLogic, Danavox, and ReSound.

References

1. Ephraim, Y. and Malah, D. (1984). Speech enhancement using a minimum mean-square error short-time spectral amplitude estimator. *IEEE Trans. Acoust., Speech, Signal Processing* **32**, 1109-1121.
2. Ephraim, Y. and Malah, D. (1985). Speech enhancement using a minimum mean-square error log-spectral amplitude estimator. *IEEE Trans. Acoust., Speech, Signal Processing* **33**, 443-445.

AUTHOR INDEX

NAME INDEX